Lecture Notes in Applied and Computational Mechanics

Volume 96

This series aims to report new developments in applied and computational mechanics - quickly, informally and at a high level. This includes the fields of fluid, solid and structural mechanics, dynamics and control, and related disciplines. The applied methods can be of analytical, numerical and computational nature. The series scope includes monographs, professional books, selected contributions from specialized conferences or workshops, edited volumes, as well as outstanding advanced textbooks.

Indexed by EI-Compendex, SCOPUS, Zentralblatt Math, Ulrich's, Current Mathematical Publications, Mathematical Reviews and MetaPress.

More information about this series at http://www.springer.com/series/4623

Michael Kaliske · Markus Oeser · Lutz Eckstein ·
Sabine Leischner · Wolfram Ressel ·
Frohmut Wellner
Editors

Coupled System Pavement—Tire—Vehicle

A Holistic Computational Approach

Springer

Editors
Michael Kaliske
Institute for Structural Analysis
TU Dresden
Dresden, Germany

Markus Oeser
Institute of Highway Engineering
RWTH Aachen University
Aachen, Germany

Lutz Eckstein
Institute for Automotive Engineering
RWTH Aachen University
Aachen, Germany

Sabine Leischner
Institute of Urban and Pavement Engineering
TU Dresden
Dresden, Germany

Wolfram Ressel
Institute for Road and Transport Science
University of Stuttgart
Stuttgart, Germany

Frohmut Wellner
Institute of Urban and Pavement Engineering
TU Dresden
Dresden, Germany

ISSN 1613-7736 ISSN 1860-0816 (electronic)
Lecture Notes in Applied and Computational Mechanics
ISBN 978-3-030-75485-3 ISBN 978-3-030-75486-0 (eBook)
https://doi.org/10.1007/978-3-030-75486-0

This Springer imprint is published by the registered company Springer Nature Switzerland AG
The registered company address is: Gewerbestrasse 11, 6330 Cham, Switzerland

Preface

Road infrastructure is essential for the establishment and maintenance of competitive and successful industrialized societies. Further, infrastructural investments represent a huge economic value. While the development of new vehicles and intelligent transportation concepts in Germany is primarily driven by industry, there have not been significant innovations in the field of pavement structures in recent decades. This deficit originates from two sources. First, this research lacks research funds that—in contrast to commercial automotive products—are supplied predominantly by public resources. Second, relatively strict and inflexible regulations hinder creativity, the transfer of knowledge, and the innovative capacity of German industry, engineers, and scientists. These circumstances contribute to the fact that so far, the approaches of progressive engineering sciences in the construction and maintenance of pavement infrastructures have not or have rarely been used. Consequently, current solutions are often inadequate and lack durability.

To overcome this problem and to prepare road infrastructure for future requirements, a paradigm shift towards dimensioning, structural realization, and the maintenance of pavements is needed. Research Unit FOR 2089, funded by the German Research Foundation (DFG), aimed to develop the scientific base for this shift. The main goal of Research Unit FOR 2089 is to provide a coupled thermo-mechanical model for a holistic physical analysis of the pavement-tire-vehicle system. Based on this model, pavement structures and materials can be optimized so that new demands become compatible with the main goal—durability of the structures and the materials.

The development of the scientific base for these new and qualitatively improved modeling approaches requires a holistic procedure through the coupling of theoretical numerical and experimental approaches as well as an interdisciplinary and closely linked handling of the coupled pavement-tire-vehicle system. This interdisciplinary research provided a deeper understanding of the physics of the full system through complex, coupled simulation approaches and progress in terms of improved, and therefore, more durable and sustainable structures.

The inter-and multi-disciplinary research required to approach the challenging topics to be addressed by Research Unit FOR 2089 has been carried out by five closely linked sub-projects carried out at the Institute for Structural Analysis (TU Dresden), the Institute of Highway Engineering (RWTH Aachen), the Institute for

Road and Transport Science (University of Stuttgart), the Institute of Urban and Pavement Engineering (TU Dresden), and the Institute for Automotive Engineering (RWTH Aachen).

All reported contributions in this book are outcomes of Research Unit FOR 2089. The financial support of the German Research Foundation is gratefully acknowledged.

Dresden, Germany
Aachen, Germany
April 2021

Michael Kaliske
Markus Oeser

Contents

Multi-physical and Multi-scale Theoretical-Numerical Modeling of Tire-Pavement Interaction

Michael Kaliske, Ronny Behnke, Felix Hartung, and Ines Wollny

Abstract In this chapter, the tire-pavement system as one subsystem of the complex vehicle-tire-pavement system is investigated in detail. As basic framework, the finite element method (FEM) is used for both, tire and pavement simulation, to obtain a detailed representation of the dynamic system, where the special case of steady state motion of the rolling tire is considered. The finite element (FE) discretization further enables to study the tire-pavement interface in terms of transmitted stresses and friction characteristics for different tire and surface properties. For the modeling of this complex subsystem, new FE based analysis methods have been derived using the Arbitrary Lagrangian-Eulerian (ALE) framework for tire and pavement. With the help of the ALE framework, the relative motion of tire and pavement is captured in a computationally efficient way. Friction in the tire-pavement interface is numerically represented by a homogenization approach of the friction interface over several length scales. With the help of a time homogenization technique, spatially detailed long-term predictions regarding rutting of the pavement become feasible by considering different time scales of the thermo-mechanical investigation.

Keywords Tire · Pavement · Interaction · Friction · Simulation · Prediction

1 Introduction

As part of our infrastructure, the road network (Fig. 1) fulfills several important functions to guarantee our today's road-bound mobility. During the last decades, new fundamental developments of the automobile population took place and, at the moment, a further transformation from fossil-fuel-powered vehicles to electrically driven vehicles is expected. Regarding the pavement structure, less innovations in

Funded by the German Research Foundation (DFG) under grant KA 1163/30.

M. Kaliske (✉) · R. Behnke · F. Hartung · I. Wollny
Institute for Structural Analysis, Technische Universität Dresden, Dresden, Germany
e-mail: michael.kaliske@tu-dresden.de

M. Kaliske et al. (eds.), *Coupled System Pavement—Tire—Vehicle*,
Lecture Notes in Applied and Computational Mechanics 96,
https://doi.org/10.1007/978-3-030-75486-0_1

Fig. 1 Road network in Germany, motorway A4 (Dresden)

terms of new materials or construction principles can be observed. To stimulate fundamental developments of new pavement structures and to increase their durability/performance [1, 26, 29, 39, 48], new numerical methods have been proposed for a holistic analysis of the vehicle-tire-pavement system [20] to understand the underlying basic interaction principles within a larger system approach.

The research presented in this chapter was conducted within a subproject of the Research Unit FOR 2089 "Durable Pavement Constructions for Future Traffic Loads: Coupled System Pavement-Tire-Vehicle" funded by the German Research Foundation (DFG). In this subproject and the present chapter, the tire-pavement subsystem is analyzed in more detail, see Fig. 2. Here, the tire-pavement interaction plays an important role for the handling and safety of the vehicles [16, 18] but also for the correct assessment of the mechanical loading of the pavement [22, 38].

The objective is to propose a numerically efficient continuum mechanical macroscopic and thermo-mechanical finite element (FE) formulation of the coupled tire-pavement model based on a stationary Arbitrary Lagrangian-Eulerian (ALE) formulation for both, tire and pavement. The consistently coupled models consist of an inelastic thermo-mechanical ALE FE tire model, a homogenized friction model considering different length scales of the friction surface and an inelastic thermo-mechanical ALE FE pavement model. This global FE approach enables to study the deformation of tire and pavement at steady state rolling contact. Macroscopic material parameters for the asphalt materials are identified based on experimental tests presented in chapter "Experimental Methods for the Mechanical Characterization of Asphalt Concrete at Different Length Scales: Bitumen, Mastic, Mortar and Asphalt Mixture".

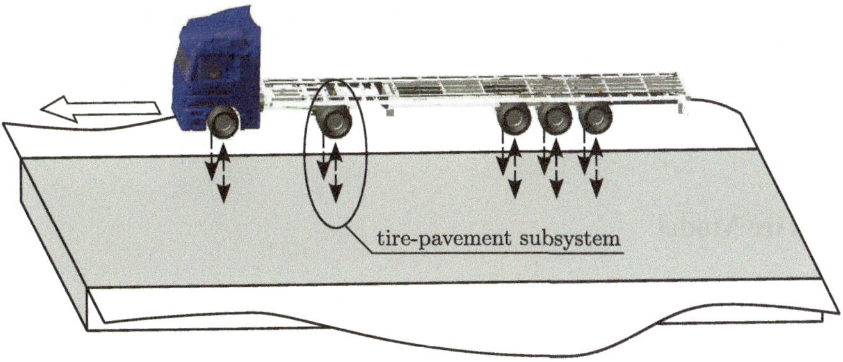

Fig. 2 Vehicle-tire-pavement system with tire-pavement subsystem (tire, tire-pavement interface with friction characteristics and pavement)

The thermo-mechanical contact formulation uses the numerical framework of a multi-scale friction analysis. In this framework, a general macroscopic friction law considering the frictional effects of hysteresis and adhesion of the contacting partners (rubber and pavement surface) at different length scales is obtained by numerical homogenization. A comparison of this friction model to friction tests is included in chapter "Numerical Friction Models Compared to Experiments on Real and Artificial Surfaces".

Special attention is paid in the present chapter to the numerical modeling and assessment of long-term processes of the pavement subjected to rolling traffic load and changing temperature conditions (climate). The numerically efficient treatment of the long-term processes influencing the durability of the pavement structure (i.e. rut formation) is accomplished by a temporal multi-scale formulation of the tire-pavement system. With the help of a time homogenization technique, repeated mechanical impact on the pavement (passing of tires) in the short term as well as climate effects (varying temperature fields due to day-night change, seasonal change) in the long term are computed to assess the consequences for tire and pavement.

A coupling of the macroscopic pavement model to a microscopic asphalt model is shown in chapter "Multi-scale Computational Approaches for Asphalt Pavements under Rolling Tire Load". The models presented in this chapter are, further, included in chapter "Simulation Chain: From the Material Behavior to the Thermo-mechanical Long-term Response of Asphalt Pavements and the Alteration of Functional Properties (Surface Drainage)" to obtain an overall coupled simulation approach.

Outline. In Sect. 2, the FE discretized tire model and the thermo-mechanical framework for its analysis at steady state rolling are introduced. In Sect. 3, friction in the tire-pavement interface is assessed by a developed numerical framework of multi-scale friction analysis. The loading of the inelastic and deformable pavement structure by rolling tires is then investigated in Sect. 4. Here, the ALE FE pavement model for short-term loading is introduced for thermo-mechanical analysis. In Sect. 5, the afore-discussed submodels are combined to study tire-pavement interaction phenomena. In Sect. 6, a computationally efficient method for the long-term analysis of the pavement structure is presented and discussed. A conclusion and outlook of the chapter is given in Sect. 7.

2 Tire Model

In Fig. 3, different approaches to represent the tire within a numerical simulation are illustrated. While simplified rheological models with a reduced set of degrees of freedom (DOF) are mainly employed in analytical vehicle simulations, more detailed information on the tire response can be obtained by the belt-spring tire model used in multi-body simulations of vehicles driving on a flat or uneven surface (see chapter "Experimental and Simulative Methods for the Analysis of Vehicle-Tire-Pavement Interaction") or via an FE discretized tire model. However, in the latter cases, the number of DOF and the computational effort increase. Hence, detailed information regarding the tire contact patch or the tire structure itself might be expected by an FE model of the tire, but normally, the computational effort of an FE discretized model for a dynamic tire simulation is too high.

2.1 Thermo-mechanical FE Tire Model

In order to obtain a detailed FE representation of the tire at low computational cost, an inelastic ALE FE approach has been developed and applied to the tire models used in this study. The ALE FE approach allows to reduce the DOF of the tire

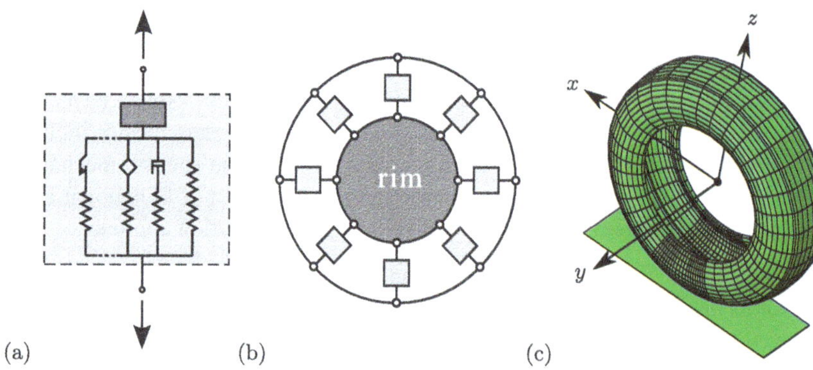

Fig. 3 Different approaches for the numerical representation of the tire within the tire-pavement system: **a** 1-mass model with reduced rheology, **b** belt-spring models with simplified rubber ring, tread stiffness and rigid rim, **c** full 3D FE model of the tire, see [10]

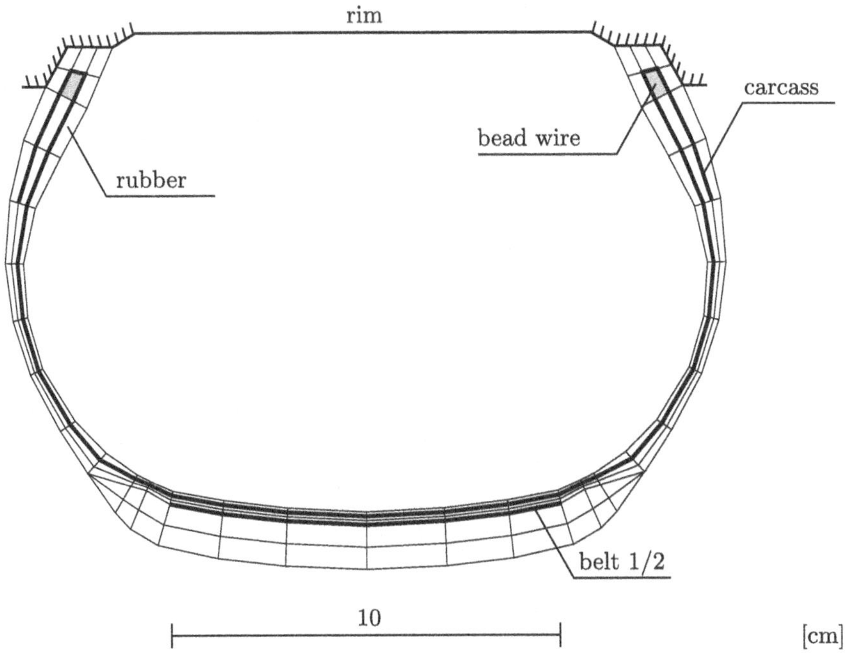

Fig. 4 FE discretized cross-section of the 175 SR 14 passenger car tire (PCT) used as benchmark tire, see [6] (carcass layer with polyethylene terephthalate (PET) cords)

since only the contact region of the tire requires a finer discretization by FE and the motion of the tire (rolling) is represented by flow of the material through the fixed FE mesh of the tire structure rather than using the tire as a moving body/load in a transient framework [11]. In [5], the procedure is described for the incorporation of inelastic effects stemming from dissipative rubber compounds. In the following, a brief overview of the methodology is provided.

In Fig. 4, the cross-section of the tire model used for benchmark tests in the further analyses is shown. Additional information on the discretization in circumferential direction and the components/material characteristics of the simple passenger car tire (PCT) is available in [6]. For the rubber compounds and cords, nonlinear material models have been developed and used to obtain a thermo-mechanical description of the tire structure at large strains. For dimensioning-relevant tire-pavement configurations, different FE discretized truck tire models are employed.

For its thermo-mechanical analysis at steady state rolling, a sequentially coupled, modular analysis scheme has been implemented, see Fig. 5. The analysis consists of a mechanical module and a thermal module. In the mechanical module, the energy dissipation stemming from the inelastic (i.e. viscoelastic) rubber compounds is computed from the current steady state motion of the rolling tire at fixed cross-sectional temperature profile. The cross-sectional temperature profile of the tire is then computed and updated via the current information on the dissipated energy of cross-

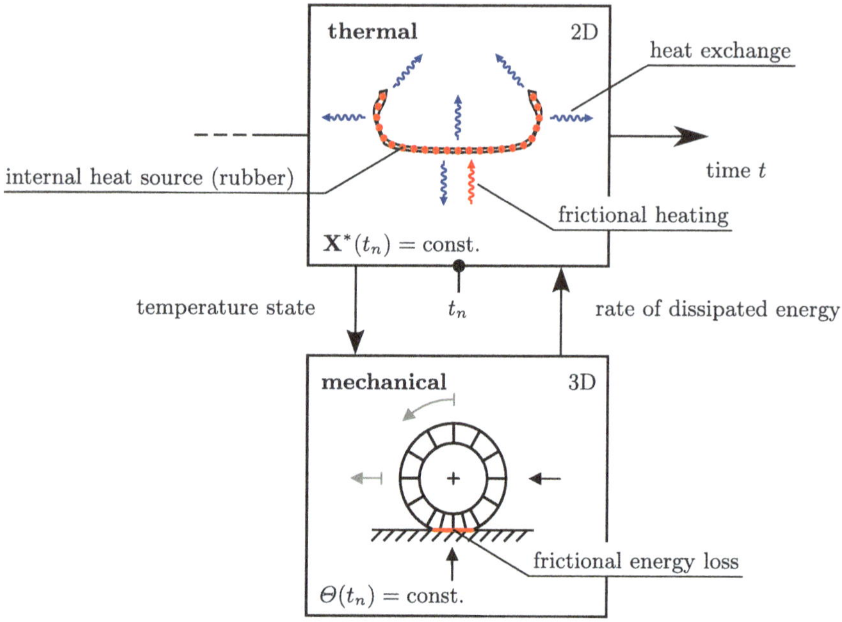

Fig. 5 Sequentially coupled thermo-mechanical simulation approach for the thermo-mechanical investigation of steady state rolling inelastic tires, see [5]

sectional points of the tire. The method is described in detail in [5], where different tires have been analyzed at different rolling conditions. Special attention has been paid to the correct implementation of thermal boundary conditions as a function of the temperature of the environment (air, road) [35]. Infrared surface temperature measurements of the rotating tire have been carried out on a drum test rig to validate the developed simulation strategy, see [5].

From the dissipated energy at rolling, the rolling resistance of the tire is computed by the sequentially coupled simulation approach as a function of the elapsing time and the current temperature state of the tire. Due to the incorporation of inelastic effects within the mechanical simulation of the tire, the rolling resistance can be obtained as reaction force or moment as a direct outcome of the tire simulation or as an integrated quantity from an energetic approach. These different procedures are described in more detail in [5].

3 Friction

Friction (defined as force resisting the relative motion of solid surfaces, fluid layers and material elements sliding against each other) is a complex phenomenon and has significant importance in daily life. This challenging field of research is also associated with tire industry, because tire-pavement interaction affects every driving maneuver. Friction in general consists of different contributions: hysteresis, adhesion as well as viscous friction and interlocking effects. Hysteresis friction due to internal dissipation of the viscoelastic material and adhesion friction in consequence of molecular bonding to the surface represent the main friction contributions. Adhesion as well as hysteresis friction strongly depend on e.g. contact pressure, sliding velocity and temperature conditions. In Sect. 3.1, an adhesion model is introduced which can be coupled to a developed multi-scale approach for hysteresis friction (see Sect. 3.2).

3.1 Adhesion Friction

Adhesion in contrast to cohesion mainly describes the molecular bonds between two different surfaces. The debonding process can also be called decohesion between two different materials [46]. In the FEM, adhesion friction can e.g. be modeled by a nonlinear traction-separation-law with the adhesional stress vector

$$\boldsymbol{\sigma} = (1 - D) \cdot K \cdot \boldsymbol{\delta} \tag{1}$$

that is a function of the relative separation vector $\boldsymbol{\delta}$ between two contact points in normal and two tangential directions, the initial adhesional stiffness K and the damage function D. The evolution of damage is comparable to the devolution of the intensity of adhesion proposed in [32]. Different analytical functions to describe the evolution of the damage value, e.g. a bilinear formulation, can be used. A nonlinear approach

$$D = \frac{G}{G_{\text{tot}}} = \frac{1}{G_{\text{tot}}} \int_0^{\delta_{\text{max}}} \hat{\sigma} \, \mathrm{d}\delta \tag{2}$$

with only two unknown parameters (K, G_{tot}) is chosen to decrease the numerical effort of parameter identification. As soon as the damage value is zero, no stresses can be transferred between the contact points (total debonding). In Eq. (2), G and G_{tot} represent the current and total fracture energy (corresponds to the area underneath the transferred stress, see Fig. 6), whereas $\delta_{\text{max}} = \max(\|\boldsymbol{\delta}\|)$ is the largest separation and $\hat{\sigma} = \|\boldsymbol{\sigma}\|$ stands for the absolute adhesion stress.

The adhesion model distinguishes between normal (without contact) and tangential adhesion (with contact). Therefore, only one damage value is required to characterize transmitted stresses. The damage value described in Eq. (2) is reset to zero (healing) as soon as the fracture energy is reached in normal direction. After the points come into contact again (bonding), damage can evolve repeatedly due to further tangential or normal separation. For some scenarios (e.g. remaining stress transmission after $D = 1$), it is suitable to ensure minimal friction after tangential debonding. Therefore, a classical Coulomb law (represented by $\mu_{\text{adh},0}$) is added to the adhesion model via hyperbolic tangent regularization, see [46].

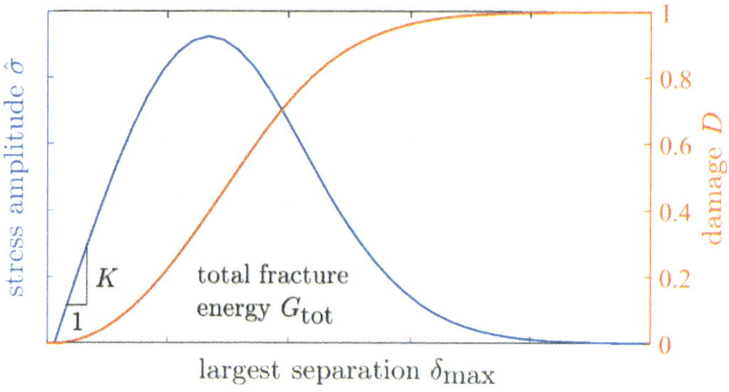

Fig. 6 Traction-separation-law of adhesion model

Fig. 7 Patch test for adhesion model

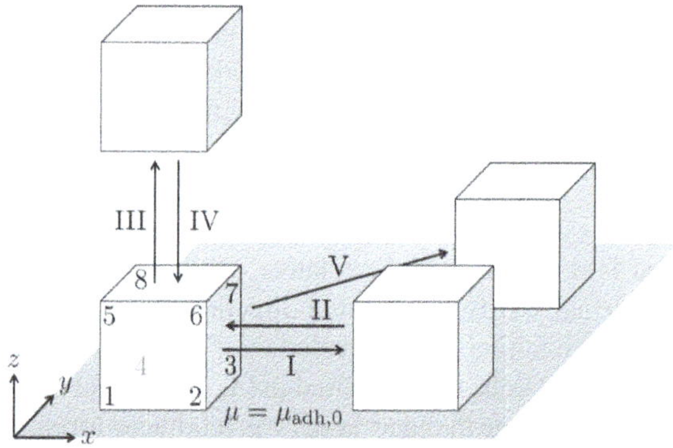

A patch test to represent the functionality of the adhesion model is shown in Fig. 7 that is comparable to [47]. There, a cube with edge length of 1 mm is pressed by 1 N/mm^2 on a rigid surface. Then, the upper nodes (5–8) are moved 0.1 mm along the x-axis (Step I). In the second step, the same nodes are moved back to the original position. Then the upper nodes are lifted up 0.1 mm along the z-axis (Step III). Within Step IV, the cube is moved back to its initial position. Finally, all upper block nodes are shifted diagonally (0.1 mm in x-direction and 0.1 mm in y-direction). The duration of each step is one second. The bottom nodes (1–4) are coupled in z-direction to avoid tipping which could occur at high shear forces. The initial adhesion stiffness K is set to 100 N/mm^3 and the total fracture energy G_{tot} is 0.01 N/mm. The additional Coulomb friction coefficient to characterize friction between the cube and the rigid surface is $\mu_{adh,0} = 0.5$.

The damage value D of Node 1 (identical to Nodes 2–4 due to coupling in z-direction) as well as the reaction forces of the rigid body surface are shown in Fig. 8. During the first step, the adhesion stress (only in first tangential direction) is observed until the total fracture energy is reached. As soon as D is equal to zero, only shear stresses due to Coulomb friction are transmitted. The shear stress changes the sign

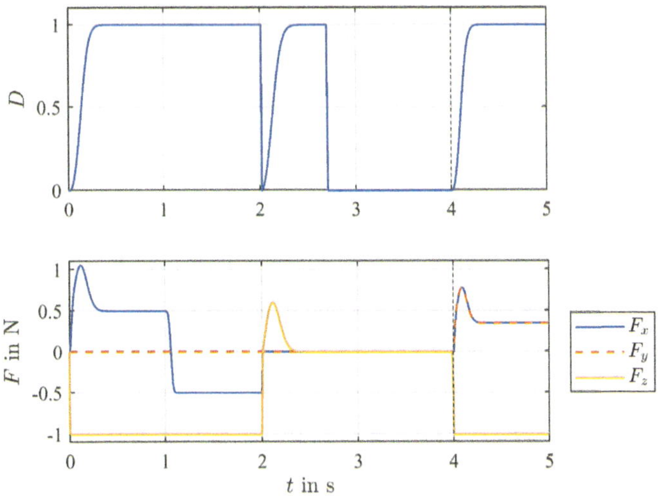

Fig. 8 Damage values and reaction forces during the patch test

while moving back in Step II. In the third step, the adhesion stress in normal direction is activated. Since the bottom nodes of the cube are in contact with the counter surface at the end of Step IV again, adhesion stresses as well as Coulomb friction in both tangential directions appear in the final Step V. Note that the peaks of the adhesion stresses in Step V are lower than in Step I and III, because there is separation in more than one direction. Moreover, the total fracture energy G_{tot} is reached earlier.

Further developments, e.g. usage of different initial adhesion stiffness parameters for normal and tangential adhesion, would be conceivable, but also include increased complexity regarding parameter identification.

3.2 Hysteresis Friction

The so-called hysteresis effect is a consequence of the internal dissipation of the viscoelastic material of e.g. rubber. Hence, the substrate structure is characterized by many asperities and it is important to understand the physical background of hysteresis friction on different length scales. The analytical approaches for multi-scale friction of Persson (see among others [31]) and [23] are common and show suitable results compared to friction measurements. Numerical models like the multi-scale approaches [14, 17, 36, 47], which base for instance on the FEM, give the opportunity to consider additional multi-physical phenomena. In the following part, a scale identification method and homogenization for hysteresis friction are introduced to build up a multi-scale friction approach which is validated numerically.

Scale Identification. Due to the different asperities of a rough surface, e.g. asphalt pavement, it is necessary to consider the entire frequency spectrum or length scales. The height difference correlation function (HDCF)

$$C_{\text{HDC}}(\zeta) = \left\langle \left(z\,(x+\zeta) - z\,(x)\right)^2 \right\rangle, \tag{3}$$

which compares the height z of two points with the distance ζ, is one method to characterize a multi-scale surface texture (compare [14, 47]). The brackets $\langle\ldots\rangle$ in Eq. (3) denote the mean value of the expression associated with. By applying e.g. sine waves to describe the surface on each length scale analytically, an approximation of the HDCF

$$\begin{aligned}\tilde{C}_{\text{HDC}}(\zeta) &= \sum_{i=1}^{n} \frac{B_i}{\pi} \int_{0}^{\pi/B_i} \left[A_i \sin\left(B_i\, x + B_i\, \zeta\right) - A_i \sin\left(B_i\, x\right)\right]^2 \mathrm{d}x \\ &= \sum_{i=1}^{n} 2\, A_i^2\, \sin^2\left(\frac{B_i\, \zeta}{2}\right)\end{aligned} \tag{4}$$

can be used to identify the sine wave parameters A_i and B_i within a fitting algorithm explained in [17]. Other surface characterization methods like the power spectral density function proposed in [36] or bandpass filters introduced in [37] are also applied in multi-scale rubber friction models.

Friction Homogenization and Scale-Dependent Friction Law. Friction features on a specific length scale can be homogenized to generate friction characteristics for a next coarser length scale. For this purpose, FE simulations are performed on the block level. During every simulation, a rubber block is pressed on a periodic rigid surface that is generated by the scale identification algorithm. The block length is identical to the current wave length λ_i. Then, the block is sliding over the rough rigid surface with a constant velocity. Periodic boundary conditions at the leading and trailing block edge are applied. The top nodes of the block are coupled in z-direction to ensure uniform vertical displacements. Temperature evolution is neglected within the FE simulation. The ratio of the total horizontal and vertical reaction forces forms the friction coefficient $\mu_{\text{hyst}}(t)$ as a function of time. A time homogenization algorithm of the friction coefficient

$$\mu_{\text{hyst,hom}} = \frac{1}{t_{\text{tot}} - t_{\text{st}}} \int_{t_{\text{st}}}^{t_{\text{tot}}} \mu_{\text{hyst}}(t)\, \mathrm{d}t \tag{5}$$

is introduced to take only the steady state part, starting at $t = t_{\text{st}}$, into account. Via an abort criterion

$$Q \leq \left| 1 - \frac{\bar{\mu}_{\text{hyst},k}}{\bar{\mu}_{\text{hyst},k-1}} \right| \tag{6}$$

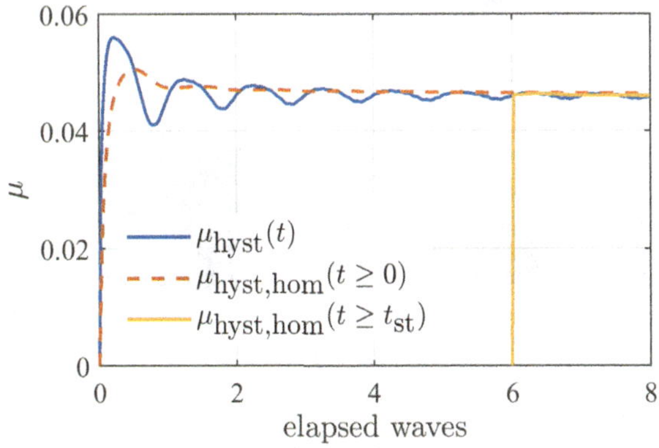

Fig. 9 Time homogenization of the friction coefficient

with

$$\bar{\mu}_{\text{hyst,k}} = \frac{v}{\lambda} \int_{(k-1)\cdot\frac{\lambda}{v}}^{k\cdot\frac{\lambda}{v}} \mu_{\text{hyst}}(t)\,\mathrm{d}t\,, \tag{7}$$

the steady state time $t_{\text{st}} = k \cdot {}^{\lambda}/_{v}$ is defined. The required height and discretization of the rubber block as well as the time in which the imposed pressure is applied influence the friction coefficient significantly and need to be identified within a parameter study [12]. In Eq. (7), parameter λ represents the current wave length and k stands for the number of elapsed waves. In Fig. 9, t_{st} is reached after six periods. A friction law for the next upper scale is created by piecewise cubic spline interpolation

$$\mu_{\text{hyst,spl}}(p, v) = \sum_{i=1}^{n_p} \sum_{j=1}^{n_v} c_{i,j} \cdot \left(p - \xi_p\right)^{n_p - i} \cdot \left(v - \xi_v\right)^{n_v - j} \tag{8}$$

with $n_p = n_v = 4$ (cubic spline generation), the breakpoints ξ_p and ξ_v (load and velocity conditions in each block simulation) and the spline coefficients $c_{i,j}$. Figure 10 shows the homogenized friction coefficients (red dots) at breakpoints ξ_p and ξ_v as well as the spline evaluation (friction map) for a micro- and mesoscale exemplarily (compare [36]).

It has to be ensured that the friction map consists of an adequate range of pressure and velocity breakpoints so that no friction coefficients outside the fitted map are used during the block simulations on the next coarser length scale. The application of artificial neural networks (ANN) to interpolate between breakpoints like in [34] is found to be working alternatively.

Multi-scale Hysteresis Friction. The combination of the length scale decomposition and the introduced time homogenization leads to a multi-scale friction procedure to compute macroscopic friction features. Figure 10 gives an overview of all required steps of the multi-scale simulations.

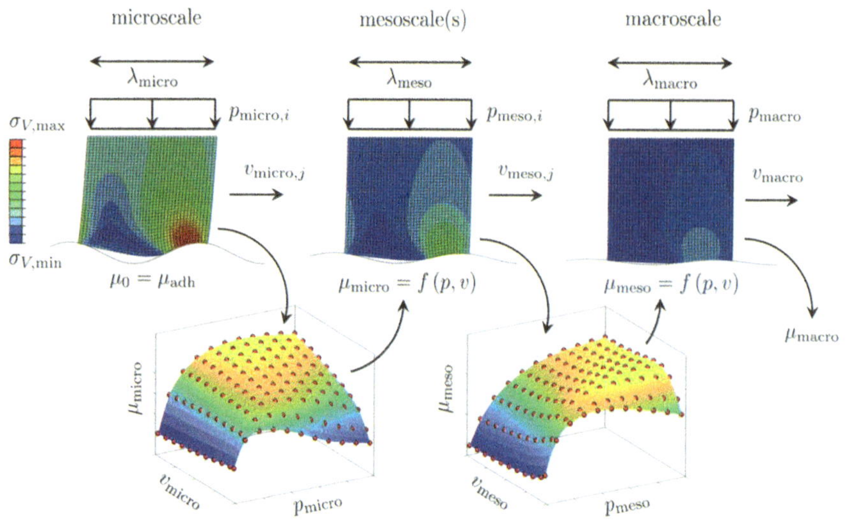

Fig. 10 Schematic outline of the multi-scale friction approach

The multi-scale procedure starts on the microscopic scale, where multiple FE simulations are executed in parallel to compute the interaction between the rubber block and road surface at different loads p_{micro} and sliding velocities v_{micro}. The ranges of p_{micro} and v_{micro} are the output of a preliminary study and may need to be adjusted iteratively. If adhesion friction is considered, the adhesion model described in Sect. 3.1 is applied on the microscale. For every microscale block simulation, the introduced time homogenization adds a breakpoint (see red dots within the diagrams in Fig. 10) into the friction map which is used as a pressure- and velocity-dependent friction law for the mesoscale via spline interpolation. On the mesoscale(s) (depending on the scale identification results), the friction law for the macroscale is formed by performing a sufficient number of block simulations at scale-dependent loads p_{meso} and velocities v_{meso}. Finally, the macroscopic friction coefficient μ_{macro} is gained at the requested boundary conditions p_{macro} and v_{macro} applying the friction map of the coarsest mesoscale as friction law.

Numerical Validation of Multi-scale Approach. The multi-scale approach for hysteresis friction is numerically validated by a simple 2D academic example using two different scales represented by sine waves

$$z(x) = \sum_{i=1}^{2} A_i \sin\left(\frac{2\pi}{\lambda_i} x\right), \tag{9}$$

which is based on [36]. On each scale, a rubber block with the length λ_i is sliding over the corresponding sine wave. The material properties of the rubber block as well as the mesh sizes are mainly taken from [36]. The macroscopic load and slip

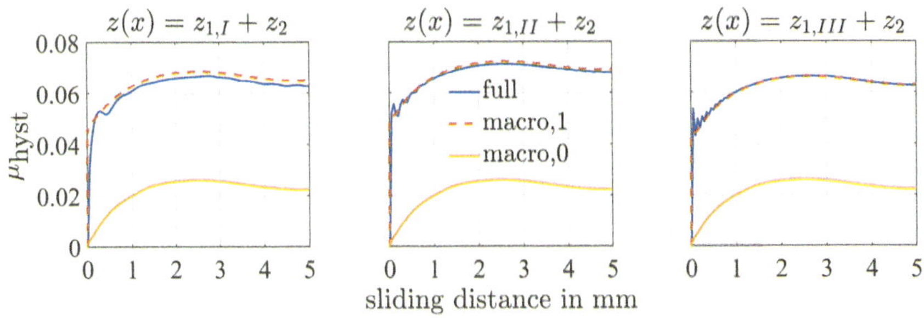

Fig. 11 Numerical validation of the multi-scale approach using two scales

velocity are 1 N/mm² and 500 mm/s, respectively. The adhesion model is not applied in this example.

The macroscopic wave is described by $z_2(x) = 0.075\,\text{mm} \cdot \sin\left(\frac{2\pi}{5\,\text{mm}}x\right)$ with a wavelength of 5 mm. In contrast to [36], three different microscopic scales $z_{1,I}$, $z_{1,II}$ and $z_{1,III}$ with wavelengths of 0.5 mm, 0.2 mm and 0.1 mm are investigated to identify the representative microscale for the predefined macroscale. The ratio between amplitude and wavelength A_i/λ_i of each microscopic wave is defined by 0.02. Three diagrams in Fig. 11 display the macroscopic friction coefficients of the reference models (full) with $z_i(x) = z_{1,i}(x) + z_2(x)$, $i = I, II, III$ (from left to right) in comparison to the resulting friction coefficients of the macroscopic scale with ($\mu_{\text{hyst,macro,1}}$) and without ($\mu_{\text{hyst,macro,0}}$) microscopic friction.

The multi-scale approach slightly overestimates the coefficient of friction of the reference model using a wavelength factor between micro- and macroscale of 10. In this example, a representative microscale must be at least larger than 25. Similar findings emerge if the macroscopic load and slip velocity are changed to 0.5 N/mm², 2 N/mm², 100 mm/s and 1000 mm/s. Whereby, the error of the friction coefficient between reference and multi-scale model increases at lower wave length factors if higher load or sliding velocity are applied.

The numerical validation example proves that the quality of the homogenization method depends on the distance between adjacent scales. This fact should be considered within the scale identification to ensure the validity of the multi-scale approach.

4 Pavement Model for Short-Term Loading by Rolling Tires

The FE pavement model, which is presented in this section, is the third important submodel (considering also tire and friction model) that is required to achieve a realistic coupled tire-pavement-interaction description. Thereby, the pavement model has to capture the layered pavement structure, the material properties and the bonding behavior between the single pavement layers realistically. The layered 3D structure of pavements is modeled by 3D finite elements. The material properties are, therein, described by constitutive formulations, whereby this section focuses on the inelastic behavior of asphalt. Nevertheless, the overall procedure can be adopted to other material formulations (e.g. concrete) as well by implementing the corresponding constitutive formulations. To account for the fact that the single pavement layers are not bonded rigidly to each other, interface elements are included additionally at the boundaries between the single layers [42]. The bonding behavior between the single pavement layers is, then, represented by a viscoelastic, temperature- and (normal) pressure-dependent traction-separation law, which acts as a constitutive formulation for the interface element.

4.1 Constitutive Material Formulation for the Short-Term Behavior of Asphalt

Asphalt Material Model. Asphalt is composed of aggregates, bituminous binder, air voids and additives and, thus, is a heterogeneous material. However, for macroscopic computations, which are done on the scale of the whole pavement structure, asphalt is treated as continuum and its macroscopic material behavior is represented by constitutive formulations. Due to its composition and inner structure, temperature-dependent elastic, viscous and plastic deformation components are observed in asphalt material. To be consistent with the tire model and to account for large deformations that occur in the asphalt in case of ruts, the finite strain constitutive formulation of [50] is applied here to the short-term behavior of asphalt. The formulation bases on the multiplicative split of the deformation gradient

$$\mathbf{F} = \mathbf{F}_{\text{vol}}\,\mathbf{F}_{\text{iso}} = \left(J^{1/3}\mathbf{1}\right)\left(\mathbf{F}^{\text{e}}_{\text{iso}}\,\mathbf{F}^{\text{i}}_{\text{iso}}\right) \tag{10}$$

into a volumetric (index vol) and an isochoric part (index iso). Thereby, the isochoric part consists of an elastic part (index e) and an inelastic part (index i). The Jacobian J represents the volume change ratio related to the reference configuration. The derivative of the strain energy density function Ψ with respect to the left Cauchy-Green tensor $\mathbf{b}$ yields the Kirchhoff stress tensor

$$\boldsymbol{\tau} = J\,\boldsymbol{\sigma} = \boldsymbol{\tau}_{\text{vol}} + \boldsymbol{\tau}_{\text{iso}}\,, \tag{11}$$

which consists of a volumetric and an isochoric part. $\boldsymbol{\sigma}$ is the Cauchy stress tensor, see [19]. The rheology of both parts of the applied asphalt material formulation is illustrated in Fig. 12. Thereby, the volumetric deformation contribution of asphalt (compacted) is assumed to be elastic and is modeled by a spring with bulk modulus κ. The isochoric contribution is modeled by five branches in parallel. The first one is a Neo-Hookean spring with stiffness $C_{10,1}$ and the second branch is a vis-

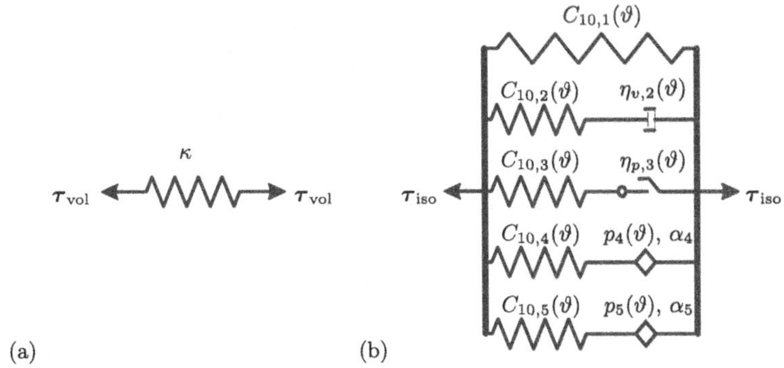

Fig. 12 Rheology of the asphalt material model: **a** volumetric part, **b** isochoric part, see [50]

coelastic Maxwell element with a Neo-Hookean spring $C_{10,2}$ and a dashpot with $\eta_{v,2}$ in series. Rate-independent elastoplastic behavior is represented by the third branch with a Neo-Hookean spring with $C_{10,3}$ and an endochronic frictional element with the plastic parameter $\eta_{p,3}$ in series. The forth and fifth branch capture rate-dependent viscoelastic behavior by fractional Maxwell elements that consist each of a Neo-Hookean spring $C_{10,4}$ and $C_{10,5}$ as well as a fractional element with parameters p_4, α_4 and p_5, α_5, respectively. While p_4 and p_5 are comparable to the viscosity, α_4 and α_5 prescribe the order of the time derivatives.

To capture the temperature-dependent asphalt material response, all parameters depicted in Fig. 12 (except for α_4 and α_5) are monotonous functions of the temperature ϑ. Details of the applied constitutive formulation can be found in [43, 44, 50].

Identification of Material Parameters. The identification of parameters for asphalt mixtures can be done in different ways. First, parameters can be identified based on results of experimental tests of asphalt specimens. Another option is to investigate the behavior of the single asphalt components, especially of the bituminous binder, and to predict the material behavior of asphalt mixtures based on its inner structure and the properties of the constituents. This step can be done e.g. based on microscopic or mesoscopic models of asphalt mixes as demonstrated in chapter "Multi-scale Computational Approaches for Asphalt Pavements under Rolling Tire Load". In this section, the parameters are identified in two steps based on results of repeated load triaxial tests (RLTT) of asphalt specimens that are described in detail in chapter "Experimental Methods for the Mechanical Characterization of Asphalt Concrete at Different Length Scales: Bitumen, Mastic, Mortar and Asphalt Mixture" and in [44].

In the first step of the identification procedure (see [43, 44] for details), parameter sets for the discrete temperature values of the experimental tests are obtained. Therefore, the strain obtained by use of the material model and the strain measured in the experiments for the same loading are compared to each other. The difference between both is minimized by an optimization procedure based on a so-called particle swarm optimization (PSO). By distinguishing between particle and swarm behavior, the PSO approach avoids sticking in single local minima. In the second step of the identification procedure, monotonously increasing or decreasing functions are adjusted to obtain continuous temperature-dependent parameter functions based on the material parameters at the discrete temperatures from the first step.

4.2 Cohesive Zone Model for Bonding Layers

The bonding behavior between the different asphalt layers influences the overall structural behavior of the pavement essentially. Experimental cyclic tests of the layer bond behavior between two asphalt layers caused by a bituminous emulsion showed that the bonding behavior is not rigid. In contrary, it depends on the loading frequency, the temperature as well as on the present normal pressure [40]. To capture all these dependencies, a viscoelastic cohesive zone model (see [51]) is enhanced to describe the constitutive behavior between bonding traction vector $\mathbf{T}$ and the separation vector $\boldsymbol{\Delta}$ of the interface element. For all normal (tension) and tangential separations (shear) in the interface layer, the bonding traction $\mathbf{T} = \mathbf{T}^{\mathrm{e}} + \mathbf{T}^{\mathrm{v}}$ consists of an elastic and a viscous contribution.

In case that the normal pressure acts on the interface, a contact algorithm that increases the normal contact stiffness is applied in order to minimize the penetration of the two bonded layers [51]. A normal pressure-dependent shear stiffness is obtained in [43] by the implementation of an additional shear traction in the contact algorithm. Further, the corresponding layer bond material parameters are identified based on experimental test results in [43].

4.3 Mechanical ALE FE Pavement Model

Different possibilities are available to model the pavement loaded by a rolling tire. In a classical Lagrangian formulation with respect to the coordinate system $\mathbf{e}_i^{\mathrm{L}}$, which is fixed in space, the tire load is stepwise shifted over the pavement in many time steps, which is numerically expensive and time consuming. An alternative is the application of an ALE formulation. Thereby, a moving reference coordinate system $\mathbf{e}_i^{\mathrm{ALE}}$ is introduced that moves together with the tire through the space, see Fig. 13. In case of steady state rolling tire and pavement, which is homogeneous in longitudinal direction, the deformation state of the pavement becomes steady state as well related to the moved reference frame. This enables time-independent and numerically efficient computations. However, introducing this moving reference frame leads to the fact that the material is no more fixed to the mesh but flows along streamlines through it, which has to be considered in case of inelastic material formulations.

ALE Kinematics. In addition to the initial $\mathcal{B}$ and the current configuration $\Phi(\mathcal{B})$, a reference configuration $\chi(\mathcal{B})$ that includes all rigid body motions is introduced in the ALE kinematics, see Fig. 14. The mapping from initial to current configuration reads then

$$\mathbf{x} = \Phi(\mathbf{X}, t) = \hat{\Phi}(\chi(\mathbf{X}, t)) \, . \tag{12}$$

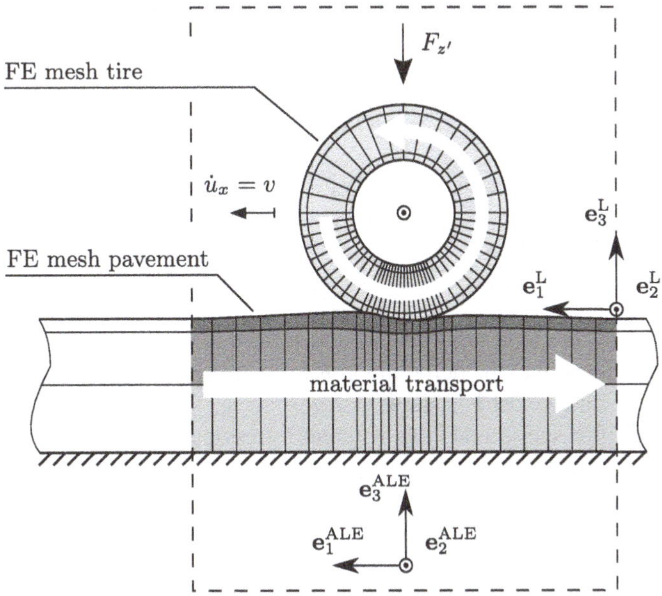

Fig. 13 ALE approach for tire and pavement: FE discretized structures and coordinate systems

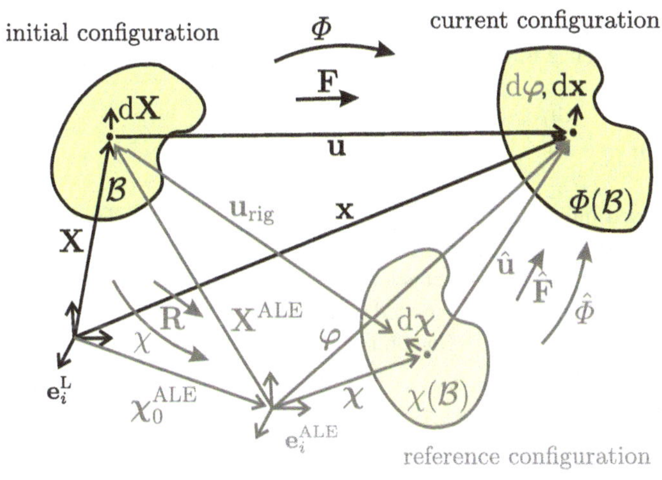

Fig. 14 ALE kinematics

Further, a moving coordinate system $\mathbf{e}_i^{\mathrm{ALE}}$ is used, which describes the reference configuration by the coordinates $\boldsymbol{\chi}$ and the current configuration by coordinates $\boldsymbol{\varphi}$. Thereby, the time-dependent position of the moving coordinate system $\mathbf{e}_i^{\mathrm{ALE}}$ with respect to the fixed Lagrangian coordinate system $\mathbf{e}_i^{\mathrm{L}}$ is given by the vector $\boldsymbol{\chi}_0^{\mathrm{ALE}}(t)$.

Based on the two different introduced coordinate systems, the current position of a point P is given by

$$\mathbf{x} = \mathbf{X} + \mathbf{u} = \boldsymbol{\chi}_0^{\mathrm{ALE}} + \boldsymbol{\varphi} = \boldsymbol{\chi}_0^{\mathrm{ALE}} + \boldsymbol{\chi} + \hat{\mathbf{u}} = \mathbf{X} + \mathbf{u}_{\mathrm{rig}} + \hat{\mathbf{u}} \,. \tag{13}$$

Then, using $\mathbf{x} = \boldsymbol{\chi}_0^{\mathrm{ALE}} + \boldsymbol{\varphi}$ as well as $\partial \boldsymbol{\chi}_0^{\mathrm{ALE}} / \partial \mathbf{X} = 0$, the deformation gradient

$$\mathbf{F} = \frac{\partial \mathbf{x}}{\partial \mathbf{X}} = \mathrm{GRAD}\, \mathbf{x} = \frac{\partial \boldsymbol{\varphi}}{\partial \mathbf{X}} = \frac{\partial \boldsymbol{\varphi}}{\partial \boldsymbol{\chi}} \frac{\partial \boldsymbol{\chi}}{\partial \mathbf{X}} = \hat{\mathbf{F}}\, \mathbf{R} \tag{14}$$

can be split into the rigid body rotations $\mathbf{R}$ from initial to reference configuration with $\det \mathbf{R} = 1$ and into the deformation $\hat{\mathbf{F}}$ from reference to current configuration with the Jacobian $J = \det \mathbf{F} = \det \hat{\mathbf{F}} = \hat{J} > 0$.

Further, the first and second Piola-Kirchhoff stress tensors are defined with respect to the reference frame as

$$\hat{\mathbf{P}} = \hat{J}\, \boldsymbol{\sigma}\, \hat{\mathbf{F}}^{-\mathrm{T}} \quad \text{and} \quad \hat{\mathbf{S}} = \hat{\mathbf{F}}^{-1}\, \hat{\mathbf{P}} \tag{15}$$

in addition to the standard continuum mechanical stress measures of the Cauchy stress tensor $\boldsymbol{\sigma}$, the first Piola-Kirchhoff stress tensor $\mathbf{P} = J\, \boldsymbol{\sigma}\, \mathbf{F}^{-\mathrm{T}}$ and the second Piola-Kirchhoff stress tensor $\mathbf{S} = \mathbf{F}^{-1}\, \mathbf{P}$ (compare e.g. [19]), respectively.

One key issue of ALE kinematics is the material time derivative. Due to the introduced reference frame, the material time derivative of a scalar value f

$$\dot{f} = \left. \frac{\partial f}{\partial t} \right|_{\mathbf{X}} = \left. \frac{\partial f}{\partial t} \right|_{\boldsymbol{\chi}} + \mathrm{Grad}\, f \cdot \mathbf{w} \quad \text{with} \quad \mathrm{Grad}\, f = \frac{\partial f}{\partial \boldsymbol{\chi}}, \quad \mathbf{w} = \left. \frac{\partial \boldsymbol{\chi}}{\partial t} \right|_{\mathbf{X}} \tag{16}$$

is decomposed into a relative and a convective part, see [13]. Thereby, $\mathbf{w}$ is called guiding velocity and corresponds to the velocity with that the material flows through the reference frame. For pavements loaded by steady state rolling tires, the reference frame and coordinate system is moved with the translational tire velocity through the space $\boldsymbol{\chi}_0^{\mathrm{ALE}} = \mathbf{v}_{\mathrm{tire}} \cdot t$. Further, in contrast to the tire, the pavement performs no rigid body motion. Thus, with $\mathbf{u}_{\mathrm{rig,pav}} = 0$ and Eqs. (13) and (16), the guiding velocity of the pavement

$$\mathbf{w}_{\mathrm{pav}} = \left. \frac{\partial \boldsymbol{\chi}}{\partial t} \right|_{\mathbf{X}} = \left. \frac{\partial \left(\mathbf{X} + \mathbf{u}_{\mathrm{rig}} - \boldsymbol{\chi}_0^{\mathrm{ALE}} \right)}{\partial t} \right|_{\mathbf{X}} = - \left. \frac{\partial \boldsymbol{\chi}_0^{\mathrm{ALE}}}{\partial t} \right|_{\mathbf{X}} = -\mathbf{v}_{\mathrm{tire}} \tag{17}$$

is known a priori and is equal to minus one times the translational tire velocity, see [43].

ALE FE Pavement Model. Base for the mechanical FE equation is the balance of momentum in the reference frame given in the weak form

$$\int_{\chi(\mathcal{B})} \hat{\rho}\,\dot{\mathbf{v}} \cdot \boldsymbol{\eta}\,\mathrm{d}\hat{v} + \int_{\chi(\mathcal{B})} \hat{\mathbf{P}} : \mathrm{Grad}\,\boldsymbol{\eta}\,\mathrm{d}\hat{v} = \int_{\chi(\mathcal{B})} \hat{\rho}\,\mathbf{b} \cdot \boldsymbol{\eta}\,\mathrm{d}\hat{v} + \int_{\partial\chi(\mathcal{B})} \hat{\mathbf{T}} \cdot \boldsymbol{\eta}\,\mathrm{d}\hat{a}\,, \tag{18}$$

see [27], whereby $\hat{\rho}$ is the density with respect to the reference frame, $\dot{\mathbf{v}} = \ddot{\mathbf{x}}$ is the material time derivative of the material velocity, $\boldsymbol{\eta}$ are the so-called test functions, $\mathbf{b}$ represents the volume loads and $\hat{\mathbf{T}}$ is the surface traction due to prescribed loads on the body surface as well as due to interfacial traction. It is worth noting that for the steady state case, the inertia term

$$\begin{aligned}\int_{\chi(\mathcal{B})} \hat{\rho}\,\dot{\mathbf{v}} \cdot \boldsymbol{\eta}\,\mathrm{d}\hat{v} = &\int_{\partial\chi(\mathcal{B})} \hat{\rho}\boldsymbol{\eta} \cdot (\mathrm{Grad}\,\boldsymbol{\varphi} \cdot \mathbf{w})\,\mathbf{w} \cdot \hat{\mathbf{n}}\,\mathrm{d}\hat{a} \\ &- \int_{\chi(\mathcal{B})} \hat{\rho}\,(\mathrm{Grad}\,\boldsymbol{\varphi} \cdot \mathbf{w}) \cdot (\mathrm{Grad}\,\boldsymbol{\eta} \cdot \mathbf{w})\;\mathrm{d}\hat{v}\end{aligned} \tag{19}$$

becomes independent of time, as shown in [27].

The standard steps of linearization $\boldsymbol{\varphi}_{(i+1)} = \boldsymbol{\varphi}_{(i)} + \Delta\varphi$ and discretization lead finally to the FE equation for steady state motion

$$\left(\mathbf{K}_{(i)} - \mathbf{W}\right) \Delta\tilde{\varphi} = \hat{\mathbf{f}}_{\mathrm{ext\,(i+1)}} - \hat{\mathbf{f}}_{\sigma\,(i)} - \hat{\mathbf{f}}_{\mathrm{i}\,(i)}\,, \tag{20}$$

whereby the inertia of the material is considered and represented by the time-independent inertia matrix $\mathbf{W}$ and the inertia forces $\hat{\mathbf{f}}_{\mathrm{i}}$, compare [27, 43] for further details. Since the ALE FE equation is independent of time in the steady state case, no time consuming time step algorithm is required for the solution, which is one big advantage of the ALE formulation.

Treatment of Inelastic Materials. Inelastic material formulations typically involve evolution equations of the internal variables $\boldsymbol{\alpha}$, whereby the material time derivative of the internal variables

$$\dot{\boldsymbol{\alpha}} = \left.\frac{\partial \boldsymbol{\alpha}}{\partial t}\right|_{\mathbf{X}} = f(\mathbf{F}, \boldsymbol{\alpha}) \tag{21}$$

depends on the current deformation as well as on the internal variables themselves. In Lagrangian computations, where the material is fixed to the FE mesh, the evolution equation can be solved by numerical time integration

$$\boldsymbol{\alpha}\,(P, t_{n+1}) = f\,(\mathbf{F}\,(P, t_{n+1})\,, \boldsymbol{\alpha}\,(P, t_n)\,, \Delta t)\,. \tag{22}$$

In FE implementations, the evolution equation is solved for each integration point. Thus, the material history $\boldsymbol{\alpha}\,(P, t_n)$ of an integration point P is obtained from the internal variables of the same integration point from the previous time step t_n in the Lagrangian frame.

In ALE formulations, the material is no more fixed to the FE mesh but flows, instead, through the reference frame. The evolution equation of the internal variables reads

$$\dot{\boldsymbol{\alpha}} = \left.\frac{\partial \boldsymbol{\alpha}}{\partial t}\right|_{\mathbf{X}} = \left.\frac{\partial \boldsymbol{\alpha}}{\partial t}\right|_{\chi} + \operatorname{Grad} \boldsymbol{\alpha} \cdot \mathbf{w} = f(\mathbf{F}, \boldsymbol{\alpha}) \tag{23}$$

in the ALE frame. Typical solution strategies to solve this evolution equation are unsplit techniques (see e.g. [3]) and operator split techniques, where the solution is split into a Lagrangian and an Eulerian step (see e.g. [9, 49]). For the application to inelastic pavements loaded by steady state rolling tires, the split approach of [49] is adopted in [45] and, further, an approximated unsplit approach is proposed in [43, 45]. The latter one is computationally more efficient as shown in [45]. A validation of the approximated unsplit inelastic ALE approach by comparison to a transient Lagrangian computation is further included in [44]. In the special case of pavements loaded by steady state rolling tires, a pavement material particle takes the time $\Delta t = |\Delta \chi|/|\mathbf{w}|$ to flow with the guiding velocity $\mathbf{w}$ a distance of $\Delta \chi$ along the material streamline through the reference configuration. Then, for the steady state case, where the relative part of the evolution equation given in Eq. (23) vanishes, the material time derivative of the internal variables can be approximated in the unsplit strategy [43, 45] by

$$\dot{\boldsymbol{\alpha}}(P) = \frac{\partial \boldsymbol{\alpha}}{\partial \chi} \cdot \mathbf{w} = f(\mathbf{F}(P), \boldsymbol{\alpha}(P)) \approx \frac{\Delta \boldsymbol{\alpha}(P)}{|\Delta \chi|} \cdot |\mathbf{w}| = \frac{\Delta \boldsymbol{\alpha}(P)}{|\Delta t|} . \tag{24}$$

This formulation allows a Lagrangian like numerical time integration for each integration point k

$$\boldsymbol{\alpha}^k = f\left(\mathbf{F}^k, \boldsymbol{\alpha}^{k-1}, \Delta t^k\right) . \tag{25}$$

Thereby, the history of the material particle is now taken from the integration point $k-1$ that the particle passed previously and the time that the particle took to pass from integration point $k-1$ to integration point k is obtained from the distance between both points as $\Delta t^k = |\Delta \chi^k|/|\mathbf{w}|$, see Fig. 15. Prerequisite for this method is a regular FE mesh, where the integration points are lying chain-like on the material streamlines in the reference configuration.

4.4 Transient, Thermal FE Pavement Model

The pavement temperature state mainly depends on the climatic conditions. In contrast to the tire, dissipation due to friction and inelastic material behavior in the pavement has a minor effect on the temperature of the pavement and, therefore, is neglected. For the short term of one single tire overrun, it is, thus, assumed that the

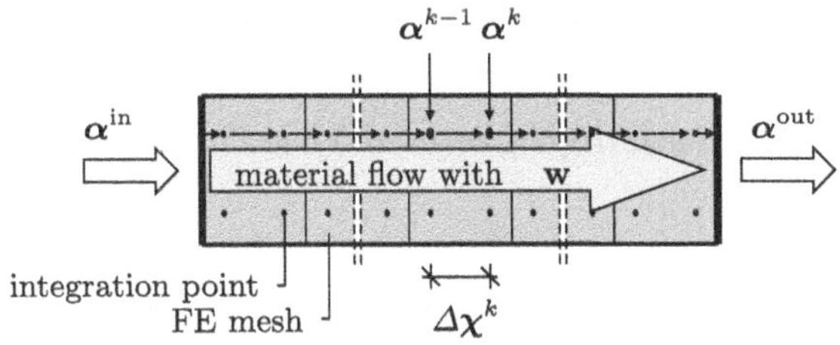

Fig. 15 Transport of the material history [43]

temperature field in the pavement structure is stationary in longitudinal direction. To save computational cost, the pavement temperature field is computed in a thermal 2D FE cross-section model [43]. Base for the thermal computation is the heat balance equation (see e.g. [33]), which leads after linearization and discretization to the FE form

$$\mathbf{C}\,\dot{\tilde{\boldsymbol{\Theta}}} + \mathbf{W}\,\tilde{\boldsymbol{\Theta}} = \tilde{\mathbf{Q}}. \tag{26}$$

Thereby, $\mathbf{C}$ is the heat capacity matrix, $\tilde{\boldsymbol{\Theta}}$ is the nodal temperature vector, $\mathbf{W}$ is the heat conductivity matrix and $\tilde{\mathbf{Q}}$ is the vector of the nodal heat flux. To solve this transient equation, an implicit Euler backward time step algorithm is applied. As time-dependent boundary conditions, temperature (Dirichlet boundary condition) and heat flux (Neumann condition) can be prescribed. The latter one also enables the prescription of convection boundary conditions $q_{\text{c}} = h_{\text{c}} \cdot (\vartheta - \vartheta_{\infty})$ and radiation boundary conditions $q_{\text{r}} = h_{\text{r}}(\vartheta) \cdot (\vartheta - \vartheta_{\text{r}})$ (see e.g. [33]).

4.5 *Thermo-mechanical Pavement Model*

The coupled thermo-mechanical pavement model consists of the thermal module and the mechanical ALE module [43]. In the thermal module, the time-dependent temperature field of the pavement cross-section is calculated first. Then, the mechanical ALE module can be applied at any prescribed time t^{ALE}. Therefore, it reads the corresponding nodal temperatures $\tilde{\boldsymbol{\Theta}}(t^{\text{ALE}})$ of the thermal cross-section model and transmits them to the equivalent nodes of the mechanical ALE model. Then, the mechanical ALE computation is able to consider the temperature-dependent material properties within the FE bulk and cohesive zone elements.

5 Tire-Pavement Interaction

5.1 Sequential Coupled Tire-Pavement Model

The submodels of the tire, the friction and of the pavement, which are introduced in the previous sections, are now coupled to the tire-pavement interaction model. The coupling is sequentially realized via a program interface as illustrated in Fig. 16 and described in [20, 42]. The coupled tire-pavement computation starts with the tire and contact simulation that is conducted assuming an undeformed and rigid contact surface. The resulting nodal contact forces are then forwarded by the program interface to the pavement simulation and, there, applied as external loads in the ALE pavement computation. When the pavement simulation finishes, the deformation of the pavement surface is given back to the tire and contact simulation such that the next tire and contact computation is conducted on the rigid contact surface that has now a deformed shape. The updated nodal contact forces are the input for the second run of the pavement computation. This sequential procedure is continued until the exchanged contact forces and pavement deformations do not (significantly) change any more. To avoid slow and oscillating convergence of the sequentially coupled tire-pavement model in case of soft pavement structures, a stabilization procedure is proposed in [42].

The coupling procedure for the tire-pavement model requires, due the exchange of nodal forces and deformations, compatible FE meshes of the contact surface (in the tire and contact simulation) and of the pavement surface (in the pavement simulation). Additionally, the contact algorithm requires an FE contact surface that consists of linear surface elements. For pavement meshes which are composed of linear 8-node 3D elements, this is easily achieved by prescribing the contact surface with linear 4-node 2D elements corresponding to the meshing of the pavement surface. Since linear 8-node 3D elements are known to show locking effects, a contact interface is introduced in [44] that enables the application of quadratic 20-node 3D elements for the pavement computation.

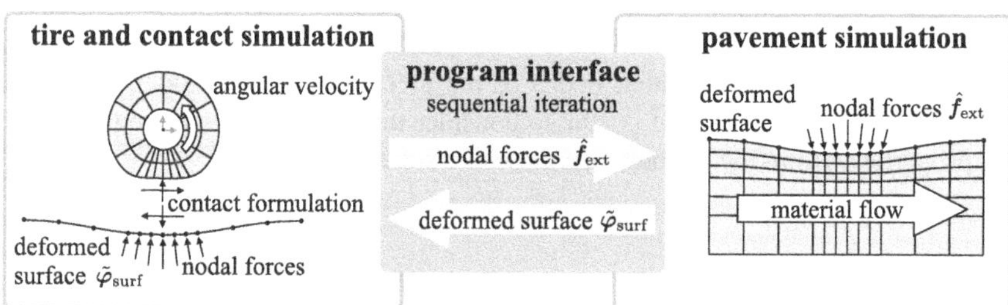

Fig. 16 Sequentially coupled tire-pavement model

5.2 Numerical Examples

Thermo-mechanical Asphalt Pavement Computation. This example studies an asphalt pavement structure under the load of a rolling truck tire. The FE model of the layered asphalt pavement as well as the utilized truck tire are shown in Fig. 17. The pavement mesh uses isoparametric 20-node elements for the bulk material and 16-node elements for the interface layers. The asphalt layers are described by the temperature-dependent asphalt material model and the material parameters given in [43]. The unbound base layer is assumed to be elastic (Young's modulus $E = 150\,000\ \text{kN/m}^2$ and Poisson's ratio $\nu = 0.35$). The subbase and subgrade are assumed to be negligible in this study. Thus, as boundary condition, the displacements at the bottom of the unbound base layer are fixed. The interface layers 1 and 2 between the asphalt layers are represented by the temperature-dependent cohesive zone model (see [43]). Interface layer 3 between the asphalt base layer and the unbound base layer is modeled as elastic with a low shear stiffness. The truck tire, which is a trailer tire of type 385/65 R22.5, is assumed to be hyperelastic and is loaded by 4.5 tons ($\hat{=}\,44.145$ kN). The tire model includes all relevant structural components (rubber parts, steel cords, textile reinforcements) by brick and rebar finite elements, respectively. In this example, which concentrates on the structural behavior of the pavement, the influence of the deformed contact surface on the tire simulation is assumed to be negligible (one-way coupling). Thus, the load on the pavement is taken from the tire simulation on the undeformed contact surface.

The structural behavior of the asphalt pavement is examined at different temperature states that arise from climatic boundary conditions. The time-dependent

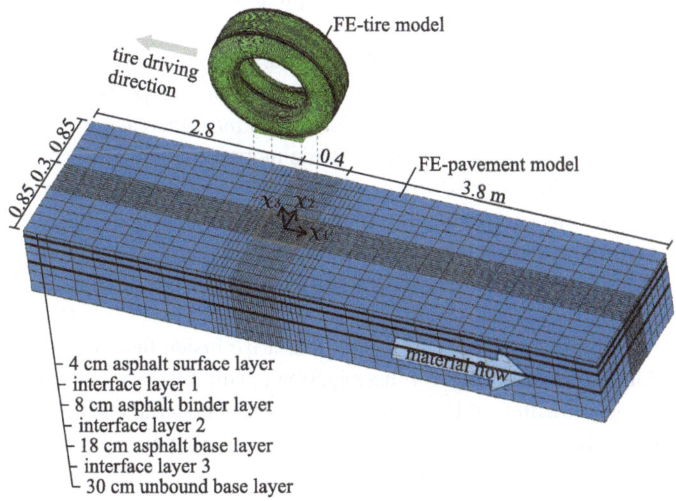

Fig. 17 Coupled FE model of tire and pavement [41]

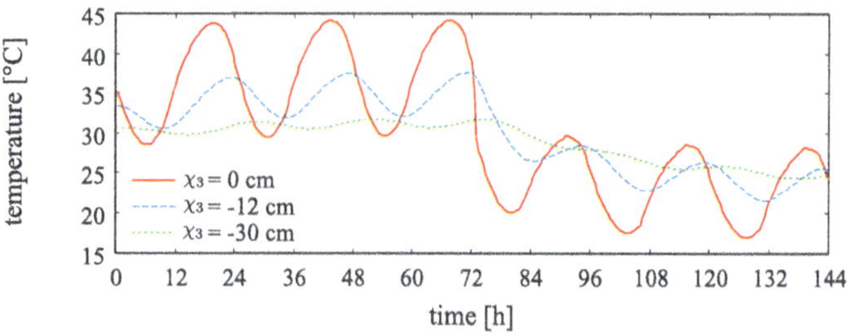

Fig. 18 Pavement temperatures versus time

temperatures of the pavement cross-section are obtained from a transient thermal simulation. The thermal boundary conditions represent a series of three summer days followed by cooling due to a rain shower and three colder days, compare [43] for the corresponding boundary conditions. The resulting pavement temperatures at several depths of the pavement cross-section versus time are given in Fig. 18. The rain shower at $t = 72$ h causes a significant reduction of the pavement temperatures. Especially, the pavement surface that transmits the heat to the environment cools down fast. Thus, the pavement temperatures inside the pavement structure are partially higher than those at the pavement surface.

To study the temperature-dependent structural behavior of the pavement, the temperature distributions at $t_1 = 67$ h and $t_2 = 80$ h are exemplary chosen for the mechanical ALE computation of the pavement at tire load. The results for a driving velocity of 5 km/h are illustrated in Fig. 19. According to the expectations, the pavement behaves softer at time t_1, when the pavement temperatures are higher. This is the result of the softer material behavior of asphalt as well as the softer behavior of the layer bond at higher temperatures.

The vertical displacements of the pavement surface along the middle of the driving lane for the temperature states at t_1, t_2 and a constant temperature state of 15 °C are given in Fig. 20 for the velocities of 5 and 80 km/h. Increasing pavement temperatures cause increasing displacements while increasing velocities cause decreasing displacements. The latter effect is caused by the viscoelastic material behavior. Higher velocity means shorter loading time for the pavement and, thus, less viscous deformation. Furthermore, a non-symmetric displacement distribution with respect to the tire axle (at $\chi_1 = 0$) is visible, which is caused by viscoelastic and plastic behavior.

The investigation of stresses and strains that arise inside the pavement structure yield important information, e.g. for the selection of suitable pavement materials and for the definition of loading and boundary conditions in material tests. The horizontal and the vertical stresses in the asphalt surface at a constant pavement temperature of 15 °C are illustrated in Fig. 21 for both velocities.

At first appearance, stresses inside the pavement seem to be almost independent of driving velocities. However, regarding the corresponding cycle duration of one load due to the rolling tire, which can be calculated from the length of the load

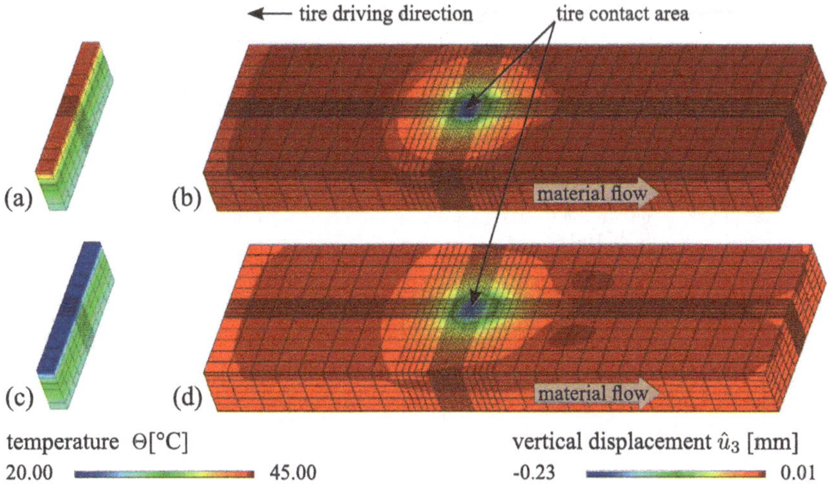

Fig. 19 Results of the FE pavement computations: **a** temperature at t_1, **b** vertical displacements at t_1, **c** temperature at t_2, **d** vertical displacements at t_2

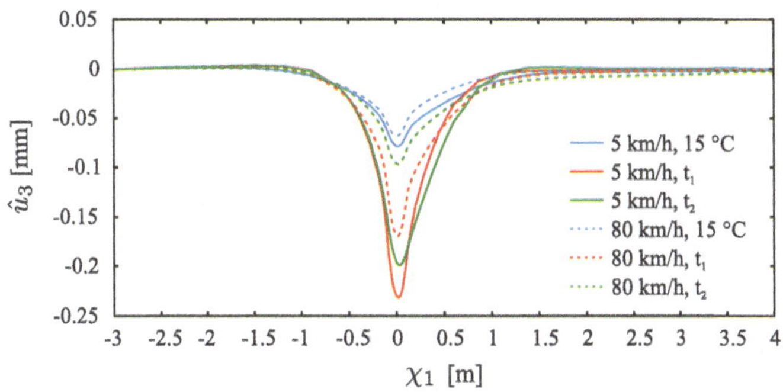

Fig. 20 Vertical displacement of the pavement surface along the driving lane at different temperatures and velocities

impulse divided by the translational tire velocity, enormous differences are observed. Furthermore, it is interesting to note that corresponding loading frequencies of the horizontal and the vertical stresses are different due to the structural behavior of the pavement. To our knowledge, this effect is not yet considered in standard asphalt material tests.

This example demonstrates the applicability of the proposed models to enable realistic numerical investigations of layered pavement structures under rolling tire load at various thermal conditions. The results and knowledge about the temperature-dependent structural behavior of layered pavements obtained from such computations is one key issue for future design of durable pavements.

Further computations and results (e.g. regarding the influence of the applied finite elements (3D isoparametric 8-nodes or 20-nodes), the influence of the pavement

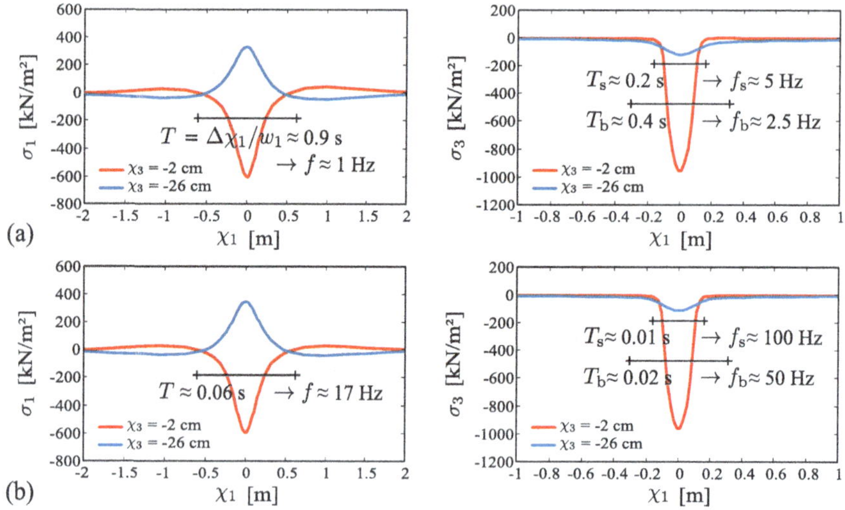

Fig. 21 Horizontal stress σ_1 and vertical stress σ_3 in the surface and base layer along the middle of the driving lane: **a** at 5 km/h translational tire velocity, **b** at 80 km/h translational tire velocity

deformation on the tire model, the influence of the tire properties (inflation pressure) on the pavement and a further study on the temperature-dependent structural behavior of layered pavements) can be found in [41, 44].

Coupled Tire-Soft Subsoil Computation. In the previous example, the deformation of the pavement surface at one single tire overrun is small and, thus, neglected for the tire and contact simulation (one-way coupling of tire model to pavement model). This second example, compare [43], demonstrates now that the coupled tire-pavement computation can capture large pavement deformations as well (two-way coupling of tire and pavement model). Therefore, an elastoplastic soft subsoil segment (0.1 m high, 0.7 m wide and 2 m long) is loaded by a rolling truck tire with 4.4 tons ($\hat{=}$ 43.164 kN) tire load and 5 km/h translational driving velocity. The material behavior of the subsoil is represented by an elastic volumetric contribution with $\kappa = 1000\,\text{kN/m}^2$ and an elastoplastic isochoric part that consists of a Neo-Hookean spring with $C_{10} = 10000\,\text{kN/m}^2$ and an endochronic frictional element $\eta_p = 5000\,\text{kN s/m}^2$ in series (see branch 3 in Fig. 12b). Conducting the coupled tire-soft subsoil computation without stabilization leads to a slow and oscillating convergence. The convergence of the staggered tire-soft subsoil coupling is significantly improved by applying a stabilization scheme [43]. The resulting vertical displacements of the soft subsoil and of the tire are illustrated in Fig. 22. The tire overrun leads to a visible rut that remains in the subsoil on the right hand side of the contact area, see Fig. 22a. In contrast to the first example, the deformed shape of the contact surface affects the rolling tire significantly [41, 43].

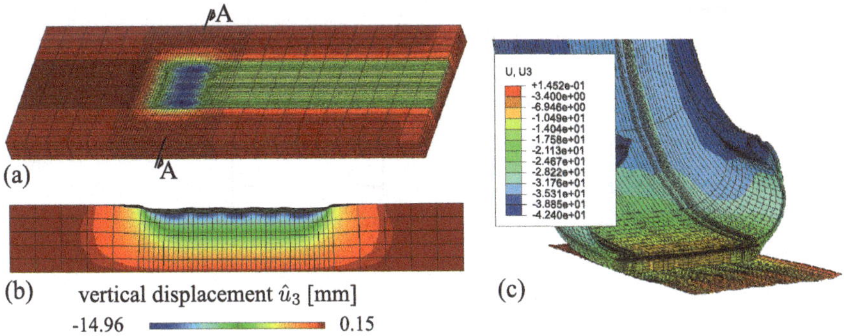

Fig. 22 Vertical displacements: **a** soft subsoil under tire load, **b** soft subsoil in cut A-A, **c** tire on the deformed contact surface

6 Long-Term Pavement Simulation

Repeated loading by tires and varying climate conditions will lead in the long term to an alteration of the pavement structure with several influencing factors [21]. In the previous sections, a detailed FE based tire-pavement model for the short-term loading (single overrun) is developed, see Fig. 13. In this section, the objective is to propose a computationally efficient numerical method [8] to reduce the computation time for long-term predictions using the still detailed FE based tire-pavement model.

6.1 Material Formulation to Capture the Long-Term Behavior of Asphalt

To represent the complex material behavior of asphalt including its temperature dependency [25], different continuum mechanically based material models [4, 50] have been developed and used within the afore-introduced numerical FE framework. Mainly viscoelastic and elastoplastic material features can be captured by the proposed models. While the short-term material response to mechanical tire loads is characterized by small strains, large strains and deviations from the initial geometry occur in the long term due to repetitive loading (accumulation of inelastic deformations). In consequence, the material models employed are formulated in terms of large strains. For a representative but still numerically efficient modeling of the material behavior, a continuum mechanical approach is used on the macroscale of an FE discretized pavement structure instead of other possible approaches (e.g. spatial homogenization, ANN etc.). For a general overview of different strategies, the reader is referred e.g. to [24].

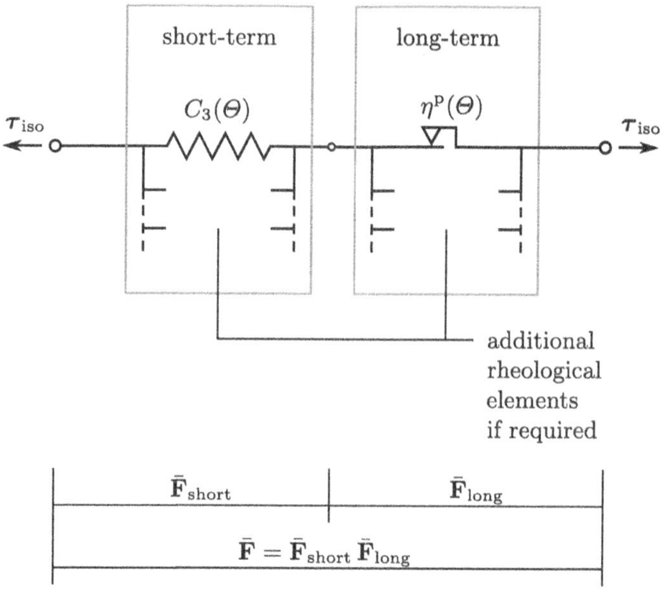

Fig. 23 1D representation of the rheology (isochoric part) used for the continuum mechanical description of the pavement materials (short-term and long-term approach)

In the framework of large strain theory, the deformation gradient $\mathbf{F}$ (with hydrostatic and deviatoric contributions) is multiplicatively decomposed into an isochoric (volume-preserving) part $\bar{\mathbf{F}}$ and a volumetric part $\mathbf{F}_{\mathrm{vol}}$,

$$\mathbf{F} = \bar{\mathbf{F}}\,\mathbf{F}_{\mathrm{vol}}\,, \quad \bar{\mathbf{F}} = J^{-\frac{1}{3}}\,\mathbf{F}\,, \quad \mathbf{F}_{\mathrm{vol}} = J^{\frac{1}{3}}\,\mathbf{1}\,. \tag{27}$$

The isochoric part $\bar{\mathbf{F}}$ is further decomposed into so-called short-term and long-term contributions,

$$\bar{\mathbf{F}} = \bar{\mathbf{F}}_{\mathrm{short}}\,\bar{\mathbf{F}}_{\mathrm{long}}\,, \tag{28}$$

see Fig. 23. The volumetric deformations are assumed as purely elastic for the sake of simplicity. Note that for the modeling of compaction (i.e. inelastic volume change), a multiplicative decomposition of the infinitesimal volume change

$$J = \det \mathbf{F} \tag{29}$$

into elastic and inelastic parts can be accomplished as well.

A unit reference volume of the material at the absolute temperature Θ is considered. For the rheology depicted in Fig. 23, the isothermal volumetric-isochoric Helmholtz free energy function yields

$$\Psi = U(J, \Theta) + \bar{\Psi}(\bar{\mathbf{F}}_{\mathrm{short}}, \bar{\mathbf{F}}_{\mathrm{long}}, \Theta)\,, \tag{30}$$

where $U(J, \Theta)$ stands for the free energy function of the volumetric part and $\bar{\Psi}(\bar{\mathbf{F}}_{\text{short}}, \bar{\mathbf{F}}_{\text{long}}, \Theta)$ denotes the free energy function of the isochoric part with the quantities $\bar{\mathbf{C}} = \bar{\mathbf{F}}^{\text{T}}\,\bar{\mathbf{F}}$ and $\bar{\mathbf{C}}_{\text{short}} = \bar{\mathbf{F}}^{\text{T}}_{\text{short}}\,\bar{\mathbf{F}}_{\text{short}}$ (unimodular part of the right Cauchy-Green tensor and the short-term right Cauchy-Green tensor). As constitutive relations, the functions

$$U(J, \Theta) = \kappa(\Theta)\,(J - \ln(J) - 1) \tag{31}$$

and

$$\bar{\Psi}(\bar{\mathbf{C}}_{\text{short}}, \Theta) = C_3(\Theta)\left(\bar{I}_1^{\text{short}} - 3\right) \tag{32}$$

are used in the following benchmark example (see Sect. 6.3). $\kappa(\Theta)$ and $C_3(\Theta)$ are temperature-dependent material parameters of the absolute temperature Θ and $\bar{I}_1^{\text{short}}$ is the first invariant of $\bar{\mathbf{C}}_{\text{short}}$. More details of the material model are provided in [50]. In general, the model parameters are unknown and have been identified from material tests carried out on compacted asphalt specimens. If details of the asphalt mixture are already known (e.g. aggregate size distribution, binder content, void content etc.), these information can be directly used to simplify the model parameter identification as demonstrated in [4].

The deformation part $\bar{\mathbf{F}}_{\text{long}}$ represents inelastic deformations of the material. Depending on the material model considered, the evolution of the inelastic deformation has been modeled by different approaches (endochronic plasticity and fractional derivatives [50] or nonlinear creep [4]). The evolution law of the inelastic device (see Fig. 23) is governed by the plasticity or viscosity parameter $\eta^{\text{p}}(\Theta)$ depending on the material model considered. A detailed explanation of the underlying theories is provided in [4, 28]. The temperature-dependent material behavior is modeled by sets of model parameters, which have been identified for different discrete testing temperatures. A closed-form expression over a certain temperature range is obtained by linear interpolation of the model parameters according to the current temperature at the material point considered. Furthermore, the thermal conductivity k and the volumetric heat capacity c_v are used for the thermal pavement analysis.

As already introduced for the tire, the relative motion of the pavement with respect to the steady state rolling tire is also tracked via an inelastic ALE FE formulation for the pavement. In this context, an initial configuration with the Lagrangian frame $\mathbf{e}_i^{\text{L}}$, a moving reference configuration fixed to the tire axle with coordinate system $\mathbf{e}_i^{\text{ALE}}$ as well as the current configuration are considered. The translation of the pavement under the steady state rolling tire is modeled by a material flow through the fixed FE mesh of a representative part of the pavement around the tire (near field). In this case, the inelastic material behavior of the asphalt is evaluated along straight streamlines, which are formed by consecutive integration points of the regular FE mesh of the pavement as described in Sect. 4, see [43, 45].

6.2 Temporal Homogenization Procedure

The pavement is idealized to have an infinite extension in longitudinal (driving) direction. This partially unbounded domain is investigated in a simplified manner by focusing on the near field consisting of the tire in contact with a representative part of the pavement (near field). The near field has to capture significant pavement deformations induced by the rolling tire and constitutes in longitudinal direction the characteristic distance L of consecutive tire loads. The infinite longitudinal extension is taken into account by an inflow and an outflow boundary and the transport of the pavement material (translation of the pavement part) through the FE mesh of the pavement during the steady state rolling of the tire. The evolution of the inelastic material (in terms of its material history) is computed based on a 2D reference cross-section as illustrated in Fig. 24. The FE discretization of the reference cross-section coincides with the FE discretization of the 3D near field of the pavement. Based on the 2D reference cross-section, the displacement boundary value problem (DBVP) and the temperature boundary value problem (TBVP) are solved assuming plane strain conditions (due to the infinite longitudinal extension of the pavement) and a longitudinally constant temperature state (only temperature variations within the pavement cross-section), respectively. As temperature boundary conditions, the absolute surface temperature Θ_s of the pavement and the absolute temperature of the ground Θ_g (corresponding temperatures T_s and T_g are measured in degree Celsius with $\Theta_0 = 273.15\,\mathrm{K}$) are prescribed, see Fig. 24.

For the long-term prediction of the structural pavement behavior, a computationally efficient strategy is required if the whole service life of the pavement is considered. Therefore, time homogenization, see e.g. [7, 15], with multiple time scales has been employed and formulated for the multi-field problem (solution fields: displacement and temperature) [8]. The different time scales of the thermo-mechanical problem are identified from the mechanical loading (short-term loading by rolling tires) and the thermal boundary conditions (day-night temperature change, seasonal temperature change). As illustrated in Fig. 25, the micro-time scale captures the mechanical loading by the tire in terms of contact forces f_c and shows a characteristic time period in the millisecond range. During the passing of the rolling tire, the temperature profile of the pavement cross-section is regarded as constant and no thermo-mechanical coupling effects are addressed, i.e. instantaneous thermo-elastic and thermo-elastoplastic coupling is neglected. On the meso-time scale, the mechanical loading is zero (load pause between consecutive tire passings) and the daily temperature variation is captured. Hence, a characteristic time of one day is considered on the meso-time scale. The macro-time scale further takes into account the annual temperature variations during one year (seasonal changes) with a characteristic time period of one year. Furthermore, additional homogenization is applied on the global-time scale to capture e.g. a general trend of increasing average annual temperature (climate change), see Fig. 25.

The time-homogenized structural response of the pavement is evaluated on the reference cross-section of the pavement by solving the DBVP at the different time

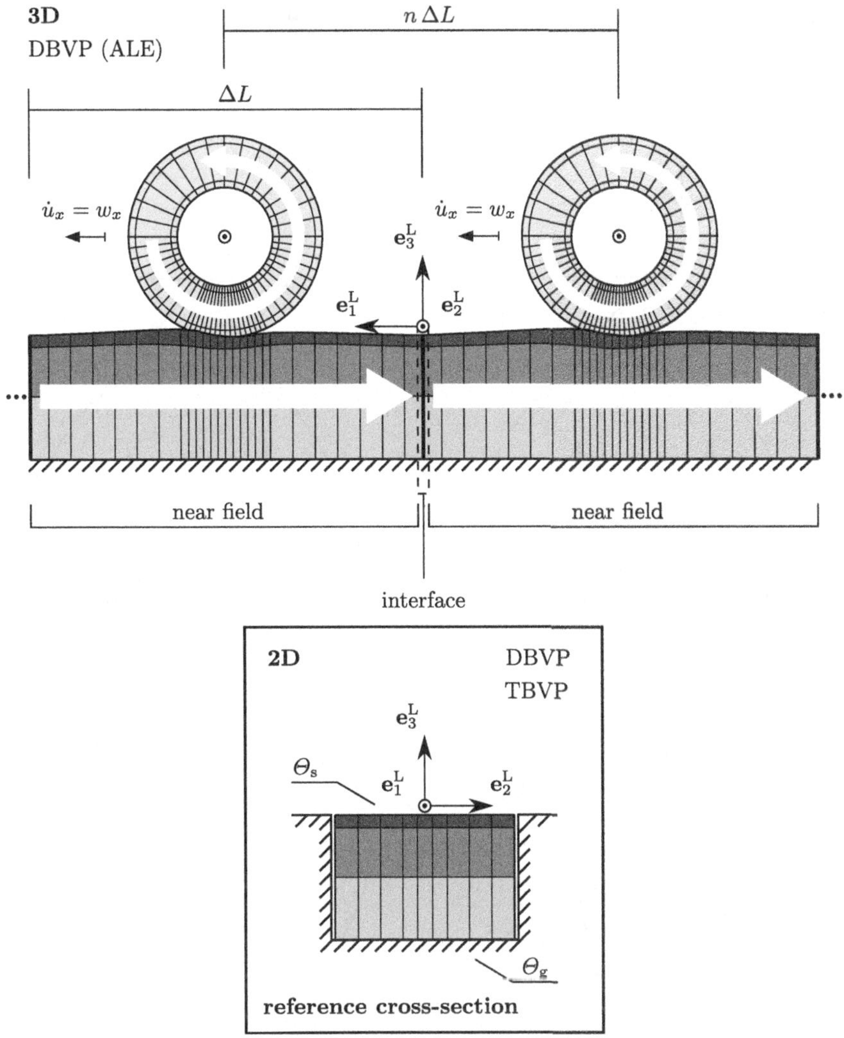

Fig. 24 Reference cross-section (plane strain condition) for time-homogenized evaluation of the inelastic material state for several tire passings via displacement boundary value problem (DBVP ALE) and temperature boundary value problem (TBVP)

scales with a set of homogenized internal variables of the material model, which are grouped in the vector $\mathbf{y}$. On each time scale, their homogenized rates

$$\langle \dot{\mathbf{y}}_0 \rangle_{\text{meso}} = \frac{1}{\tau_{\text{micro}}} \int_0^{\tau_{\text{micro}}} \dot{\mathbf{y}}(\tau_{\text{micro}})\, \mathrm{d}\tau_{\text{micro}}\,, \tag{33}$$

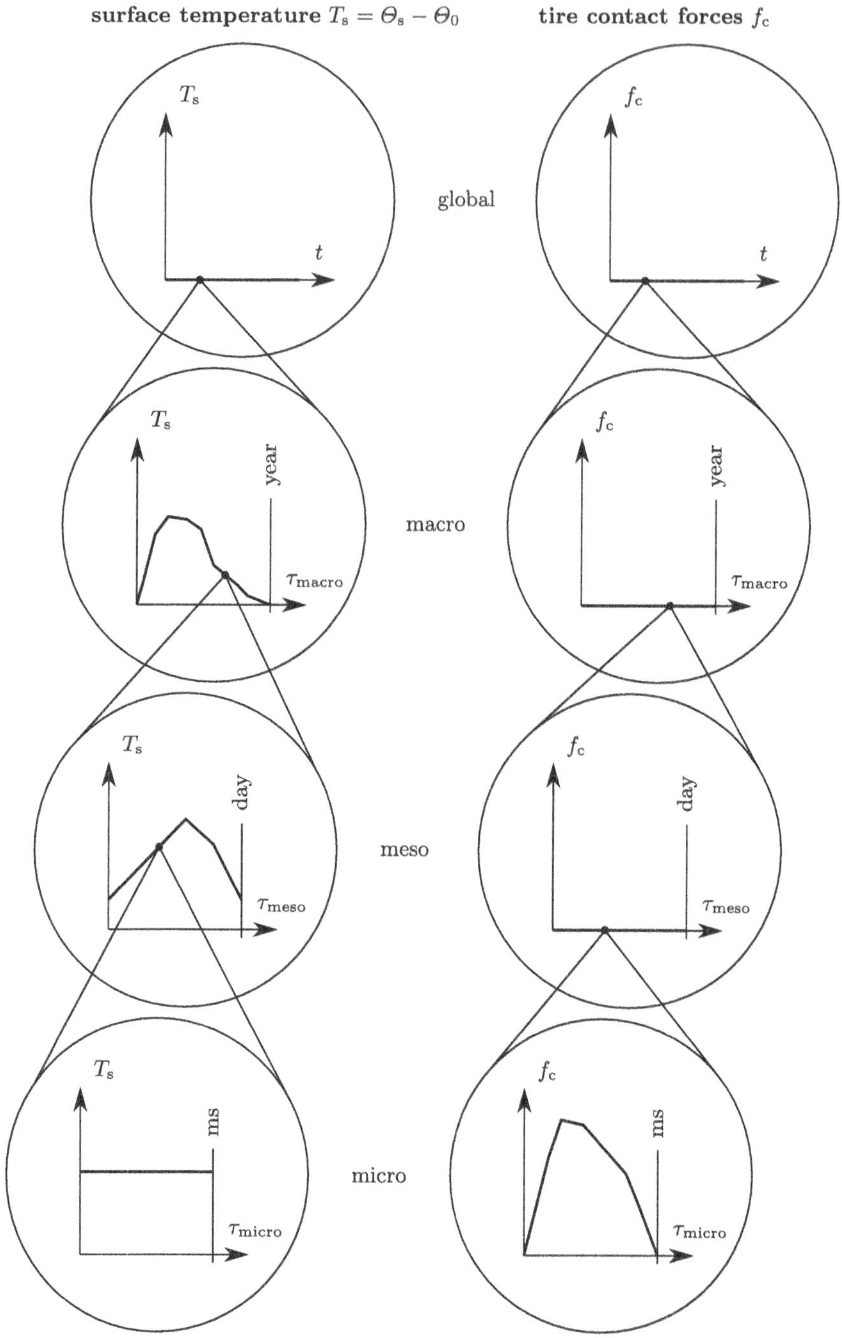

Fig. 25 Different time scales of the thermo-mechanical tire-pavement system

$$\langle \dot{\mathbf{y}}_0 \rangle_{\text{macro}} = \frac{1}{\tau_{\text{meso}}} \int_0^{\tau_{\text{meso}}} \langle \dot{\mathbf{y}}(\tau_{\text{meso}}) \rangle_{\text{meso}} \, \mathrm{d}\tau_{\text{meso}} \,, \tag{34}$$

$$\langle \dot{\mathbf{y}}_0 \rangle_{\text{global}} = \frac{1}{\tau_{\text{macro}}} \int_0^{\tau_{\text{macro}}} \langle \dot{\mathbf{y}}(\tau_{\text{macro}}) \rangle_{\text{macro}} \, \mathrm{d}\tau_{\text{macro}} \tag{35}$$

are computed based on the inflow (subscript 0) and outflow (subscript 1) information for each characteristic time interval τ_i. More details and the computational algorithm are provided in [8].

6.3 Thermo-mechanical Long-Term Simulation of the Pavement Structure Under Repeated Rolling Tire Load

For the considered simple but illustrative example, the benchmark PCT running on a pavement part (thin asphalt layer on deformable subgrade layer), as sketched in Fig. 26, is analyzed. The geometry and material components of the PCT are documented in [6]. The tire is subjected to a vertical load of $F_z = 3300\,\text{N}$ and a steady state translational velocity of 80 km/h in x-direction is considered. The thermo-mechanical material behavior of the asphalt and subgrade layer is described by the material model depicted in Fig. 23 in combination with the temperature-dependent model parameters given in Tables 1 and 2.

The input data for the boundary condition of the surface temperature T_s of the pavement is depicted in Fig. 27. The generated data shall represent the daily and annual temperature variations for a location with continental climate in Germany. Note that location-specific data can be used at this stage if appropriate data is available, e.g. from climate measurement data. For the temperature of the ground, a constant temperature of $T_g = 8°\text{C}$ is assumed, see Fig. 24.

In Fig. 28, the computed long-term pavement response for this illustrative benchmark example is shown. Here, the vertical displacement in cross-sectional direction of the pavement is depicted as a function of the service life. In Fig. 29, the evolution of the tire contact forces is provided for the same time span. It can be clearly seen that due to the alteration of the initially flat pavement surface, the contact forces are also subjected to an alteration since the contact patch of the tire and the pavement surface evolve with elapsing time.

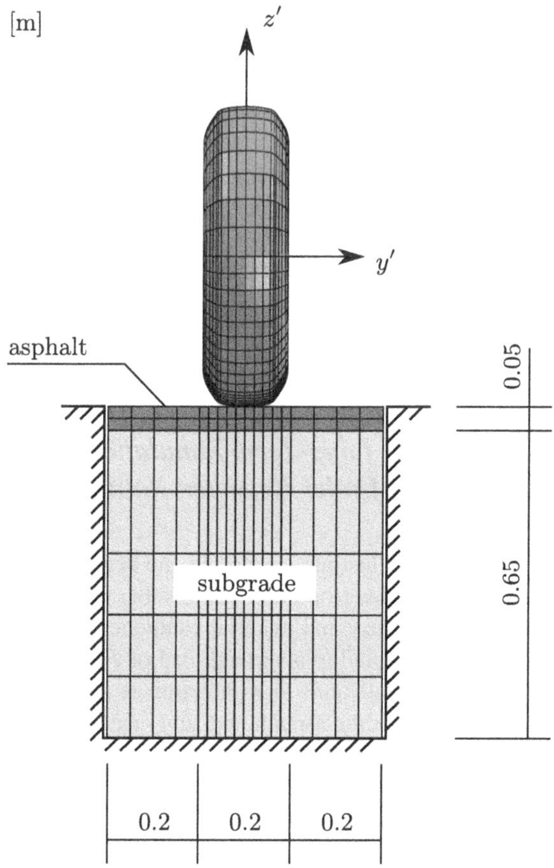

Fig. 26 Benchmark PCT on deformable pavement: FE model with initial geometry of the near field (length in longitudinal direction $L = 10\,\text{m}$)

Table 1 Asphalt: model parameters, see Fig. 23 and [30] as well as [43]

	0 °C	20 °C	40 °C
κ $[\text{N/m}^2]$	1261.0E+06	891.4E+06	881.1E+06
C_3 $[\text{N/m}^2]$	268.2E+06	127.8E+06	60.6E+06
$\eta^{\text{p}}\ \hat{\tau}$ $[\text{N/m}^2]$	175.0E+10	45.0E+10	5.0E+10
ρ $[\text{kg/m}^3]$	2450.0	2450.0	2450.0
k [W/(m K)]	1.3	1.3	1.3
c_{v} $[\text{J/(m}^3\text{K)}]$	2.3765E+06	2.3765E+06	2.3765E+06

Table 2 Subgrade: model parameters, see Fig. 23 and [30] as well as [43]

	Temperature-independent
κ $[\mathrm{N/m^2}]$	8.3E+06
C_3 $[\mathrm{N/m^2}]$	1.9E+06
$\eta^{\mathrm{p}}\,\hat{\tau}$ $[\mathrm{N/m^2}]$	5.0E+09
ρ $[\mathrm{kg/m^3}]$	2000.0
k [W/(m K)]	2.3
c_{v} $[\mathrm{J/(m^3K)}]$	1.9400E+06

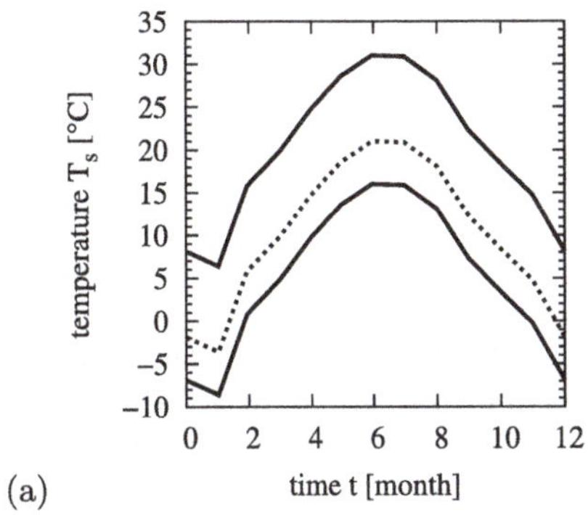

Fig. 27 Surface temperature as thermal boundary condition for the pavement analysis: **a** annual temperature variation, **b** daily temperature variation

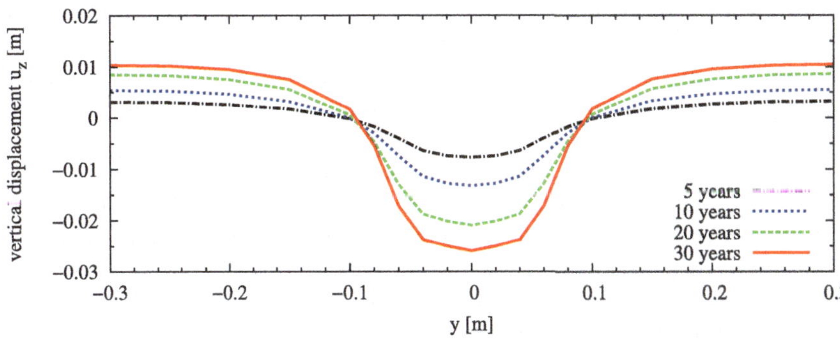

Fig. 28 Evolution of the pavement surface geometry in cross-sectional direction of the pavement as a function of the pavement's service life

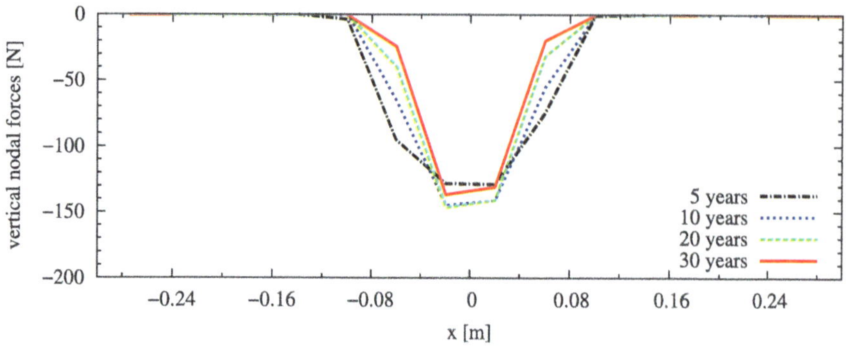

Fig. 29 Evolution of the vertical tire contact forces in longitudinal pavement direction ($y = 0$) as a function of the service life (due to changes of the pavement surface geometry)

7 Conclusions and Outlook

In this chapter, numerical key techniques have been presented to enhance the general understanding of the complex interactions of the vehicle-tire-pavement system, especially focusing on the tire-pavement subsystem. An improvement in numerical modeling of tire and pavement structures could be achieved by developing an inelastic ALE FE approach for tire and pavement, its detailed representation by FE taking into account the tire-pavement interface with frictional contact and a new homogenized friction approach considering different length scales of the pavement surface. With the help of this detailed model, short-term phenomena can be analyzed for various tire loads and steady state driving maneuvers. Regarding the long-term behavior of the pavement structure, a time homogenization method has been derived by considering multiple time scales of the dynamic problem (short-term tire loading, day-night temperature change, seasonal temperature change).

In chapter "Simulation Chain: From the Material Behavior to the Thermomechanical Long-term Response of Asphalt Pavements and the Alteration of Functional Properties (Surface Drainage)", influencing parameters for an improved durability of relevant pavement structures are provided by a study of the global vehicle-tire-pavement system, e.g. also focusing on the evolution of functional properties during the service life of the pavement [2].

References

1. Abed, A., Thom, N., Neves, L.: Probabilistic prediction of asphalt pavement performance. Road Mater. Pave. Design **20**, S247–S264 (2019)
2. Alber, S., Schuck, B., Ressel, W., Behnke, R., Canon Falla, G., Kaliske, M., Leischner, S., Wellner, F.: Modeling of surface drainage during the service life of asphalt pavements showing long-term rutting: a modular hydro-mechanical approach. Adv. Mater. Sci. Eng. **2020**, 8793652 (2020)

3. Bayoumi, H.N., Gadala, M.S.: A complete finite element treatment for the fully coupled implicid ALE formulation. Comput. Mech. **33**, 435–452 (2004)
4. Behnke, R., Canon Falla, G., Leischner, S., Händel, T., Wellner, F., Kaliske, M.: A continuum mechanical model for asphalt based on the particle size distribution: numerical formulation for large deformations and experimental validation. Mech. Mater. **153**, 103703 (2021)
5. Behnke, R., Kaliske, M.: Thermo-mechanically coupled investigation of steady state rolling tires by numerical simulation and experiment. Int. J. Non-Linear Mech. **68**, 101–131 (2015)
6. Behnke, R., Kaliske, M.: Finite element based analysis of reinforcing cords in rolling tires: influence of mechanical and thermal cord properties on tire response. Tire Sci. Technol. **46**, 294–327 (2018)
7. Behnke, R., Kaliske, M.: Square block foundation resting on an unbounded soil layer: long-term prediction of vertical displacement using a time homogenization technique for dynamic loading. Soil Dyn. Earthquake Eng. **115**, 448–471 (2018)
8. Behnke, R., Wollny, I., Hartung, F., Kaliske, M.: Thermo-mechanical finite element prediction of the structural long-term response of asphalt pavements subjected to periodic traffic load: tire-pavement interaction and rutting. Comput. Struct. **218**, 9–31 (2019)
9. Benson, D.: Computational methods in Lagrangian and Eulerian hydrocodes. Comput. Methods Appl. Mech. Eng. **99**, 235–394 (1992)
10. Berger, T., Behnke, R., Kaliske, M.: Viscoelastic linear and nonlinear analysis of steady state rolling rubber wheels: a comparison. Rubber Chem. Technol. **89**, 499–525 (2016)
11. Chabot, A., Chupin, O., Deloffre, L., Duhamel, D.: ViscoRoute 2.0 A. Tool for the simulation of moving load effects on asphalt pavement. Road Mater. Pave. Design **11**, 227–250 (2010)
12. De Lorenzis, L., Wriggers, P.: Computational homogenization of rubber friction on rough rigid surfaces. Comput. Mater. Sci. **77**, 264–280 (2013)
13. Donea, J., Huerta, A., Ponthot, J., Rodriguez-Ferran, A.: Arbitrary Lagrangian-Eulerian methods. In: Stein, E., de Borst, R., Hughes, T. (eds.) Encyclopedia of Computational Mechanics (Vol. 1: Fundamentals), pp. 414–437. Wiley, Chichester, UK (2004)
14. Falk, K., Lang, R., Kaliske, M.: Multiscale simulation to determine rubber friction on asphalt surfaces. Tire Sci. Technol. **44**, 226–247 (2016)
15. Guennouni, T.: Sur une méthode de calcul de structures soumises à des chargements cycliques: L'homogénéisation en temps. Modélisation Mathématique et Analyse Numérique **22**, 417–455 (1988)
16. Guo, M., Zhou, X.: Tire-pavement contact stress characteristics and critical slip ratio at multiple working conditions. Adv. Mater. Sci. Eng. **2019**, 5178516 (2019)
17. Hartung, F., Kienle, R., Götz, T., Winkler, T., Ressel, W., Eckstein, L., Kaliske, M.: Numerical determination of hysteresis friction on different length scales and comparison to experiments. Tribol. Int. **127**, 165–176 (2018)
18. Hernandez, J., Al-Qadi, I.: Tire-pavement interaction modelling: hyperelastic tire and elastic pavement. Road Mater. Pave. Design **18**, 1067–1083 (2017)
19. Holzapfel, G.A.: Nonlinear Solid Mechanics: A Continuum Approach for Engineering. Wiley, Chichester, UK (2000)
20. Kaliske, M., Wollny, I., Behnke, R., Zopf, C.: Holistic analysis of the coupled vehicle-tire-pavement system for the design of durable pavements. Tire Sci. Technol. **2015**, 86–116 (43)
21. Kerali, H., Lawrance, A., Awad, K.: Data analysis procedures for long-term pavement performance prediction. Transport. Res. Record **1524**, 152–159 (1996)
22. Kim, S.M., Darabi, M., Little, D., Abu Al-Rub, R.: Effect of the realistic tire contact pressure on the rutting performance of asphaltic concrete pavements. KSCE J. Civ. Eng. **22**, 2138–2146 (2018)
23. Le Gal, A., Klüppel, M.: Investigation and modelling of rubber stationary friction on rough surfaces. J. Phys. Conden. Matt. **20**, 015007 (2007)
24. Lian-sheng, G., Han-cheng, D., Jia-qi, C.: Research on predicting the rutting of asphalt pavement based on a simplified Burgers creep model. Math. Prob. Eng. **2017**, 3459704 (2017)
25. Liu, Y., Su, P., Li, M., You, Z., Zhao, M.: Review on evolution and evaluation of asphalt pavement structures and materials. J. Traf. Transport. Eng. **7**, 573–599 (2020)

26. Makendran, C., Murugasan, R., Velmurugan, S.: Performance prediction modelling for flexible pavement on low volume roads using multiple linear regression analysis. J. Appl. Math. **2015**, 192485 (2015)
27. Nackenhorst, U.: The ALE-formulation of bodies in rolling contact—theoretical foundations and finite element approach. Comput. Methods Appl. Mech. Eng. **193**, 4299–4322 (2004)
28. Netzker, C., Dal, H., Kaliske, M.: An endochronic plasticity formulation for filled rubber. Int. J. Solids Struct. **47**, 2371–2379 (2010)
29. Norouzi, A., Kim, D., Kim, Y.: Numerical evaluation of pavement design parameters for the fatigue cracking and rutting performance of asphalt pavements. Mater. Struct. **49**, 3619–3634 (2016)
30. Oeser, M.: Numerische Simulation des nichtlinearen Verhaltens flexibler mehrschichtiger Verkehrswegebefestigungen. Ph.D. thesis, Technische Universität Dresden (2004)
31. Persson, B.: Contact mechanics for randomly rough surfaces. Surf. Sci. Rep. **61**, 201–227 (2006)
32. Raous, M., Cangémi, L., Cocu, M.: A consistent model coupling adhesion, friction, and unilateral contact. Comput. Methods Appl. Mech. Eng. **177**, 383–399 (1999)
33. Reddy, J., Gartling, D.: The Finite Element Method in Heat Transfer and Fluid Dynamics. CRC Press, Boca Raton (2000)
34. Serafinska, A., Hassoun, N., Kaliske, M.: Numerical optimization of wear performance—utilizing a metamodel based friction law. Comput. Struct. **165**, 10–23 (2016)
35. Srirangam, S., Anupam, K., Scarpas, A., Kasbergen, C.: Development of a thermomechanical tyre-pavement interaction model. Int. J. Pave. Eng. **16**, 721–729 (2015)
36. Wagner, P., Wriggers, P., Klapproth, C., Prange, C., Wies, B.: Multiscale FEM approach for hysteresis friction of rubber on rough surfaces. Comput. Methods Appl. Mech. Eng. **296**, 150–168 (2015)
37. Wagner, P., Wriggers, P., Veltmaat, L., Clasen, H., Prange, C., Wies, B.: Numerical multiscale modelling and experimental validation of low speed rubber friction on rough road surfaces including hysteretic and adhesive effects. Tribol. Int. **111**, 243–253 (2017)
38. Wang, H., Al-Qadi, I., Stanciulescu, I.: Simulation of tyre-pavement interaction for predicting contact stresses at static and various rolling conditions. Int. J. Pave. Eng. **13**, 310–321 (2012)
39. Wang, Z., Guo, N., Wang, S., Xu, Y.: Prediction of highway asphalt pavement performance based on Markov chain and artificial neural network approach. J. Supercomput. **77**, 1354–1376 (2021)
40. Wellner, F., Hristov, B.: Numerically supported experimental determination of the behavior of the interlayer bond. Transport. Res. Record J. Transport. Res. Board **2506**, 116–125 (2015)
41. Wollny, I.: ALE formulation of inelastic, temperature-dependent and fluid-infiltrated layered pavement structures at loading by steady state rolling tires. Ph.D. thesis, Institute for Structural Analysis, TU Dresden (2018)
42. Wollny, I., Behnke, R., Villaret, K., Kaliske, M.: Numerical modelling of tire-pavement interaction phenomena: coupled structural investigations. Road Mater. Pave. Design **17**, 563–578 (2016)
43. Wollny, I., Hartung, F., Kaliske, M.: Numerical modeling of inelastic structures at loading of steady state rolling—thermo-mechanical asphalt pavement computation. Comput. Mech. **57**, 867–886 (2016)
44. Wollny, I., Hartung, F., Kaliske, M., Canon Falla, G., Wellner, F.: Numerical investigation of inelastic and temperature dependent layered asphalt pavements at loading by rolling tyres. Int. J. Pave. Eng. **22**, 97–117 (2021)
45. Wollny, I., Kaliske, M.: Numerical simulation of pavement structures with inelastic material behaviour under rolling tyres based on an arbitrary Lagrangian Eulerian (ALE) formulation. Road Mater. Pave. Design **14**, 71–89 (2013)
46. Wriggers, P.: Computational Contact Mechanics. Springer, Berlin (2006)
47. Wriggers, P., Reinelt, J.: Multi-scale approach for frictional contact of elastomers on rough rigid surfaces. Comput. Methods Appl. Mech. Eng. **198**, 1996–2008 (2009)

48. Yong-hong, Y., Yuan-hao, J., Xuan-cang, W.: Pavement performance prediction methods and maintenance cost based on the structure load. Procedia Eng. **137**, 41–48 (2016)
49. Ziefle, M., Nackenhorst, U.: Numerical techniques for rolling rubber wheels: treatment of inelastic material properties and frictional contact. Comput. Mech. **42**, 337–356 (2008)
50. Zopf, C., Garcia, M., Kaliske, M.: A continuum mechanical approach to model asphalt. Int. J. Pave. Eng. **16**, 105–124 (2015)
51. Zreid, I., Fleischhauer, R., Kaliske, M.: A thermomechanically coupled viscoelastic cohesive zone model at large deformation. Int. J. Solids Struct. **50**, 4279–4291 (2013)

Numerical Simulation of Asphalt Compaction and Asphalt Performance

Pengfei Liu, Chonghui Wang, Frédéric Otto, Jing Hu, Milad Moharekpour, Dawei Wang, and Markus Oeser

Abstract Asphalt pavement compaction is important, and it can determine the service quality as well as durability of pavement. In recent years, numerical methods have been extensively used to simulate and study the construction process of asphalt pavement and mechanical properties of asphalt mixtures. In the following sections, the compaction process, considering the interaction between the materials and the equipment, is simulated, and the influence of different compaction methods on the mechanical performance of asphalt mixtures is investigated. To achieve this goal, a pre-compaction model is developed using the Discrete Element Method (DEM), and the models of both materials and the paving machine are generated separately. After the pre-compaction simulation, the theory of bounding surface plasticity is combined with the theory of Finite Element Method (FEM) as well as with a kinematic model of a roller drum to simulate the asphalt mixture behavior during a roller pass. In order to ensure consistency both in the laboratory compaction and in-situ compaction, the Aachen compactor has been developed. The effect of different compaction methods (Field, Aachen and Marshall Compactions) on the asphalt specimens is compared and evaluated using the microscale FEM.

Keywords Numerical simulation · Asphalt compaction · Asphalt performance · Discrete element method · Finite element method

Funded by the German Research Foundation (DFG) under grant OE 514/1.

P. Liu (✉) · C. Wang · F. Otto · M. Moharekpour · D. Wang · M. Oeser
Institute of Highway Engineering, RWTH Aachen, Aachen, Germany
e-mail: liu@isac.rwth-aachen.de

J. Hu
School of Transportation, Southeast University, Nanjing, P.R. China

D. Wang
School of Transportation Science and Engineering, Harbin Institute of Technology, Harbin, P.R. China

M. Kaliske et al. (eds.), *Coupled System Pavement—Tire—Vehicle*,
Lecture Notes in Applied and Computational Mechanics 96,
https://doi.org/10.1007/978-3-030-75486-0_2

1 Introduction

To provide the basis for the future design of durable pavements through a holistic physical analysis of the pavement—tire—vehicle system, the focus of this chapter is on developing a basic understanding of the machine-material interaction behavior during the asphalt compaction process and the mechanical performance of asphalt mixtures. In particular, the processes involved in the in-situ paver and roller compaction of asphalt will be systematically researched by numerical simulations at meso- and macro-scales.

Asphalt pavement compaction is one of the most vital phases during road construction, which can determine the service quality as well as the durability of the pavement. During the process of pavement construction, the compaction process consists of two main stages. In the first stage, the compaction is achieved by the paver. It can be called preliminary compaction or paving compaction. In the following stage, the final compaction can be achieved using different combinations of rollers (static, vibrating, pneumatic, etc.) [12]. Therefore, the required density of pavement compaction can be attributed to the initial density compacted by the paver and the final density behind the rollers [59]. According to the construction guidance from different countries, the evaluation of the quality of pavement compaction normally relies on the required density and is always assessed after the final compaction finished by rollers [2, 4, 39]. Xu et al. have proven that adequate preliminary compaction achieved by the paver can guarantee the overall quality of compaction and, furthermore, it can be very helpful for rolling compaction (e.g. reduce the passes of roller compaction) [68].

One of the core parts of a paver that comes into use during the pre-compaction stage is the screed, which works as a vibrating compactor while paving the material using its own weight. Generally, a screed of a paver consists of a tamper and the screed plates, which accommodate the compacting systems to provide high density, smooth surfaces, and durable results [6, 7, 22]. During the pre-compaction, the paving angle is very important; the optimized angle of paving compaction indicates the optimized position of the tamper. Theoretically, the operation of the screed vibration and the paving speed are also determined by the type of material, the mixture gradation, and the paving temperature, as these factors have a direct influence on the quality of paving compaction [62, 64].

Although it is possible to achieve a high level of preliminary compaction with the paver (between 75 and 95% of Marshall density, depending on the setting of the compaction systems), the remaining compaction increase required in order to achieve the final compaction level (depending on the requirements) is obtained via the use of rollers. Studies have shown that while the compaction systems of the paver subject the material primarily to impact compaction, the combination of pressure and shear stresses under the roller drum allows the aggregate structure to rearrange horizontally, which affects the stiffness and therefore the overall properties of the material [4, 55].

For asphalt compaction, there are several types of rollers. The most commonly used type of roller is the tandem roller with two smooth drums. Alternatively, it is possible to replace one or both drums with rubber tires. While the drums are defined in shape and create a variable stress distribution in the contact area dependent on the diameter of the drum and the material resistance, the rubber tires will partially adjust themselves to the surface of the material depending on the tire pressure and also on the material resistance [39, 41, 49, 50].

Furthermore, smooth drums can either work as static or as dynamic compactors. Static drums use the weight of the roller in order to achieve a compaction effect, while dynamic drums are equipped with one or several unbalance exciters, creating a dynamic movement in the drum according to three different principles: vibration with one circular exciter, vibration with two exciters rotating in opposite directions (thus creating a directed vibration, which can be adjusted), or a rotational oscillation where two exciters rotate in the same direction but with a phase shift of 180°. The advantage of the use of dynamic drums lies in the superposition of different compaction effects. While static compactors are mainly used to stabilize the material during the first roller passes, the use of dynamic drums allows for the material to overcome the internal friction between the aggregates and therefore achieves a higher compaction effort [8, 23, 25, 49, 50].

In recent years, computational-aided methods, such as the Discrete Element Method (DEM) and Finite Element Method (FEM) have been extensively used to simulate and study the construction process of asphalt pavements and the mechanical properties of asphalt mixtures [1, 26, 35, 38, 65]. DEM has attracted significant interest in the simulation of the mechanical responses of granular materials since it was introduced in the 1970s [13, 14]. This simulation method can provide an innovative and effective approach to enhance the understanding of building materials' properties [9, 31, 36, 44]. In DEM simulation, bulk materials are usually treated as an assembly of two-dimensional (2D) disks or three-dimensional (3D) spheres [13, 14, 35, 46], or else as clumps of these shapes made by rigidly connecting and overlapping multiple disks or spheres [19, 20, 46].

When the material flow inside the paver is described with the use of DEM, the computational effort increases with the number of generated particles [36, 35]. Furthermore, describing the compaction behavior of the asphalt mixture during the roller compaction phase requires the definition of a sufficiently large model in order to take the rolling movement of the drum itself into account. For modeling the macroscopic behavior of the asphalt layer in its preliminary compacted state and during a roller pass, continuum mechanics (i.e. FEM) can be used, as seen in [56].

However, it is necessary to use an adequate constitutive law in order to describe the complex behavior of asphaltic material during the compaction process. In general, the mechanical behavior of asphalt mixtures can be considered as elastic-visco-plastic. The elastic properties are mainly determined by the properties of the single aggregates, the plastic flow by the combined aggregate pile, and the temperature-dependent combined visco-elastic and visco-plastic components by the asphalt mortar. In its initially stacked state, the aggregates possess few contact points and are mostly connected by the asphalt mortar. When subjected to external loads, the aggregates begin rearranging, which causes plastic as well as time-dependent visco-plastic deformations, the latter being caused by the deformation resistance exerted by the mortar.

While the number of air voids is reduced, the number of contact points between the aggregates increases, thus enabling the material to support higher loads. As a result, plastic deformations will be fewer with every additional load applied while the percentage of elastic deformations with respect to the overall deformation will increase [21, 27, 43]. One possible method to take into account the effects described above is the use of the theory of bounding surface plasticity combined with the theory of the critical state, commonly used in geo-mechanics. Such constitutive laws have already been used in the past to model the plastic behavior of soils as well as asphaltic materials under cyclic loading conditions [24, 28, 30, 52].

After the compaction process, the mechanical properties of the asphalt mixtures should be further investigated by experimental and numerical methods. Asphalt mixtures are a typical heterogeneous composite material consisting of aggregates with an irregular shape and random distribution, asphalt binder, and air voids at the microscale. In many microscale simulations, the numerical model explicitly represents the individual components of the heterogeneous internal material structure of the asphalt mixture. As a result, specific material models can be assigned to each component in the microscale model. Since the physical effects such as the propagation of microcracks or failure of the interface between aggregates and asphalt mortar are considered separately in microscale simulations, rather simple material formulations can be used for each material phase to represent the complex macroscopic material behavior of asphalt mixture.

In this chapter the compaction process, considering the interaction between the materials and the equipment, is simulated, and the influence of different compaction methods on the mechanical properties of asphalt mixtures is investigated. Before the simulations, important microstructure characteristics (such as morphology of aggregates) are investigated, which have a significant influence on the compaction process and, thus, damage resistance and the durability of asphalt pavements. The details of research strategy and methods can be found in chapters "Computational Methods for Analyses of Different Functional Properties of Pavements" and "Characterization and Evaluation of Different Asphalt Properties Using Microstructural Analysis". Afterward, a pre-compaction model was developed in a DEM simulation, and the model of both materials and the paving machine were generated separately. Selecting the parameters of materials was determined via the laboratory tests, while the setting of the paver's working operations was set based on the real conditions at the field construction site. After the pre-compaction simulation, the theory of bounding surface plasticity was combined with the theory of FEM as well as with a kinematic model of a roller drum, in order to simulate the asphalt mixture behavior during a roller pass and assess the increase in compaction with respect to the roller operation. In order to calibrate the model, laboratory tests were conducted, in which the material behavior was analyzed under compaction temperature conditions as well as under different types of loading (constant loading as well as cyclic loading). In order to ensure consistency both in the laboratory compaction and in field compaction, a new standardized laboratory compaction method has been developed, namely the Aachen compactor. The computer-generated model was developed to explicitly model the

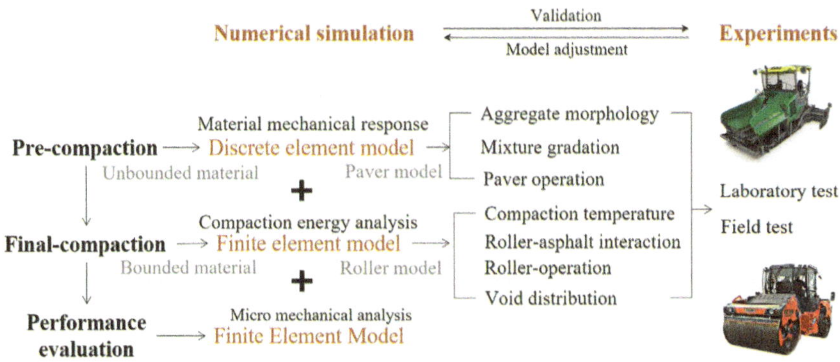

Fig. 1 Research methodology

different material components of the asphalt mixtures. Compared to the homogeneous model, the heterogeneous model is more consistent with reality and thus yields more reliable results. The related testing methods of the material properties used in the heterogeneous model can be found from the chapter "Experimental Methods for the Mechanical Characterization of Asphalt Concrete at Different Length Scales: Bitumen, Mastic, Mortar and Asphalt Mixture". The effect of different compaction methods (Field, Aachen and Marshall Compactions) on the asphalt specimens with regard to the internal structure, mechanical response, and fracture behavior is then compared and evaluated using microscale FEM. Some more advanced microscale models are also introduced at the end of this chapter, and such microscale models, which can be coupled to a macro-model, as shown in the chapters "Multi-physical and Multi-scale Theoretical-numerical Modeling of Tire-Pavement Interaction" and "Multi-scale Computational Approaches for Asphalt Pavements under Rolling Tire Load" to allow for multi-scale analysis. The research methodology of this chapter can be seen in Fig. 1.

2 Simulation of Asphalt Paving Compaction at Mesoscale

The microstructure characteristics of the aggregates are characterized and investigated in this section first. The results can provide basic information for selecting the optimal materials for asphalt compaction. A 3D model based on the DEM was utilized to simulate the pre-compaction of asphalt pavement. The application of DEM in engineering is based on the granular investigations carried out by Cundall and Strack [13, 14]. In DEM simulations, bulk materials are usually treated as an assembly of granular materials interacting at contact points, which can be applied for characterizing the behavior of bulk materials under significant deformation. In the simulation of the pre-compaction of asphalt pavement, the aggregates were generated as clumps with defined morphology consisting of overlapped spherical elements. Newton's second law was used as the basic algorithm to calculate the kinetic behavior of aggregates as well as force-law displacement between the contact points. The equations of motion in the simulation are integrated using an explicit central finite difference algorithm.

2.1 Microstructure Characteristics of Aggregates

The microstructure characteristics of an aggregate (also called morphological characteristics in this section), such as the sphericity, the angularity, and the texture, have a decisive influence on the interlocking, force transmission, and compaction of asphalt mixtures. For this reason, there are strong dependencies between the morphological characteristics of the aggregates and the performance properties of asphalt mixtures. In addition, the morphological characteristics are the most important input values for pavement modeling with numerical methods, e.g., DEM.

The morphological characteristics of 11 types of aggregates were investigated by X-ray Computed Tomography (X-ray CT) in two- and three-dimensions [60, 61, 63, 64]. In order to simulate the polishing process for the road surfacing aggregates, Micro-Deval test (MD) polishing was carried out. The aggregate's morphological properties, such as texture index (TI), 2D Sphericity (2DS), gradient angularity (GA), and 3D angularity (3DA), before and after MD testing were characterized by various parameters calculated from aggregate imaging system (AIMS) and X-ray CT [63]. The Digital Image Processing (DIP) techniques based on X-ray CT images and the detailed description of the four morphological properties are introduced in chapters "Computational Methods for Analyses of Different Functional Properties of Pavements" and "Characterization and Evaluation of Different Asphalt Properties Using Microstructural Analysis".

The GA and 3DA decrease significantly after the MD procedure. The absolute value of change exhibited by the sphericity development is relatively small. The comparison of these four morphological parameters of aggregates before and after MD testing indicated that the AMD state is not dependent on the BMD state. A log–normal function is ideally suited to describe the analyzed morphological characteristics before and after MD. The changes to both the 2DS and the TI only have an ancillary influence [63].

The knowledge obtained from these studies represents an important step regarding the assessment of the polishing resistance of road surfacing aggregate. Furthermore, the morphological properties of the road surfacing aggregate also have a significant influence on the transferable friction between the road surface and the tires, which is analyzed in chapters "Multi-physical and Multi-scale Theoretical-numerical Modeling of Tire-Pavement Interaction" and "Experimental and Simulative Methods for the Analysis of Vehicle-Tire-Pavement Interaction". In addition, the results contribute to the further development of DEM approaches for pre-compaction in this section and also explain the variation of performance characteristics of the roads.

2.2 Development of the Pre-compaction Model in EDEM

There are five steps in the procedure to generate the pre-compaction model in the DEM software EDEM: (1) The geometry sketch of the paver was finished in the computer-aided design (CAD) software AutoCAD, (2) the output from AutoCAD was imported into EDEM as the geometry of paving equipment, the physical properties of the geometries are separately defined with several physical parameters, (3) clumps were generated in EDEM as the templates of particles, the geometrical properties (inertia moment etc.) of the aggregates were calculated via their morphologies, the physical and mechanical properties of them were defined according to laboratory tests, (4) a factory was created in the model for aggregates generation, which was used to define where, when, and how particles appear in the simulation, (5) after defining the parameter that determined the interaction between the paving equipment and bulk material, the aggregates were randomly generated according to the templates, and the size distribution followed the self-defined gradation [66]. Figure 2 illustrates the model of pre-compaction in EDEM. As seen in the figure, during the simulation, the paver geometry (auger and paver screed) moves along the negative direction of the x-axis, to compact and pave the material generated. The materials located in the green box are monitored during the pre-compaction process.

Some detailed information about the modeling in EDEM will be provided in this section. In the particle simulation, a template was generated first. The shape of the template was simplified as a clump which consisted of three spheres; their relative spatial locations formed the basic shape of a particle model with its sphericity. The basic properties of the template, such as density, Poisson's ratio, and shear modulus were defined in the bulk material feature assembled table. The size distribution is then defined for the particles' generation. The particles are generated randomly from the minimum size to the maximum size of coarse aggregate. Materials can be created directly in the simulation or imported from the materials database.

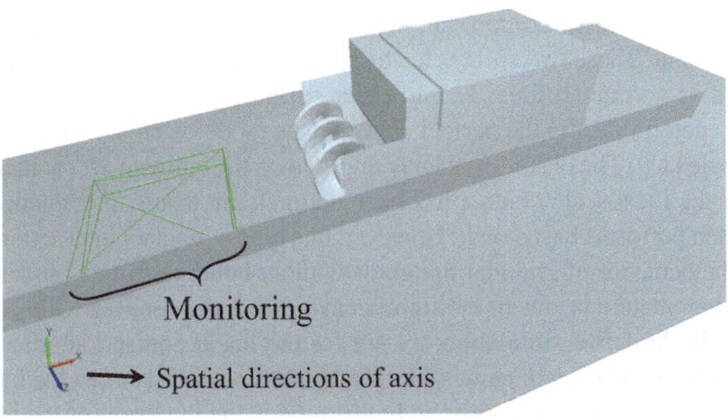

Fig. 2 Pre-compaction model of paver screed in EDEM and its spatial directions of axis [66]

A factory is created in the model for aggregate generation after the templates have been defined. Particle factories are used to define where, when, and how particles appear in a simulation. Any virtual surface or volume (physical or virtual) can be turned into a particle factory. Factories can only be created if a bulk material has been defined. A simulation can have any number of particle factories. The shape of the geometry which generates particles can be designed by the user; there are also several default geometries such as sphere, box, or facet. There are two types of particle factories which can be used to generate particles: static and dynamic. Static factories produce particles at a specified time. The simulation is paused during particle creation. Dynamic factories produce particles over the duration of a simulation, and the simulation continues as the particles are created.

The mechanical properties of aggregates are determined by a set of mechanical parameters (stiffness and friction between contact points) and geometric characteristics (special assemble, size distribution, and morphology) of particles. As the initial purpose of this study is to model the pre-compaction of asphalt pavement at full-scale, the aggregates are not generated with their real shape due to its requirement for a prohibitively long calculation period. More capacity of calculation can therefore be imposed for simulating the mechanical behavior of bulk material with large deformation. In future research, the microstructure characteristics of aggregates will be considered in the aggregate generation.

2.3 Contact Model and Parameters

A contact model describes how elements behave when they have contact with each other, which is greatly important in DEM simulation. Therefore, it is explained in detail in this section.

The interaction between bulk material and equipment is defined after the contact model has been generated. The interactions between aggregates can be simplified as a pair of elastic springs with constant normal and shear stiffness properties acting at the contact point. These two springs have specified tensile and shear strengths, which are defined as normal stiffness k_n and shear stiffness k_s. In addition, the frictional behavior is determined by the perfect elastoplastic model via the micro-scaled index friction angle [13]. The elastic contact is imposed on the entire model at macroscopic scale after the local scale contact has been defined, the macro deformation of bulk material then performs based on it. However, this type of contact interaction can only describe the behavior of granular materials without bonding from a binder.

For asphalt materials, one of the significant issues is the bonding with viscoelastic properties. In the DEM simulation, except for the linear contact, another interface between particles is bonded, namely the binder phase, which is modeled by adding a viscoelastic film around each particle. In this model, the thickness of the binder film wrapped on the surface of the aggregates is assumed to be a constant value, which means the binder is assumed to be uniformly attached to the surface of the aggregate. Furthermore, the thickness of the film obtains the correct volume of the binder phase

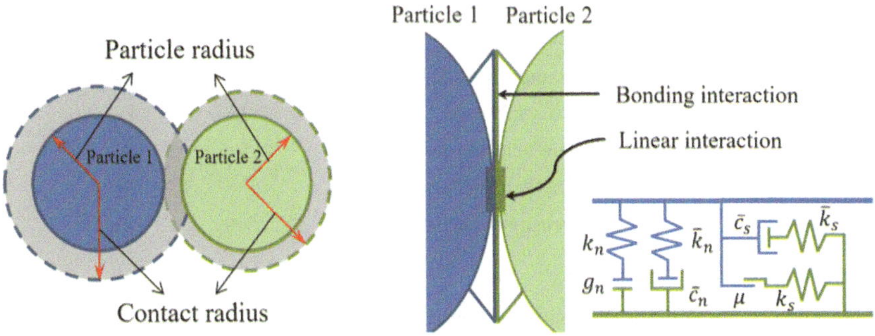

Fig. 3 Visualization of contact between two particles with binder phase [66]

according to the simulated bitumen content. This film can be generated by defining the particle radius and contact radius of a particle, respectively (see Fig. 3).

The viscous bond between two particles acts in parallel with the linear elastic contact. This contact can only be activated when the surface gap of two particles is zero or less than the sum of the film thicknesses of the two particles. Otherwise, this bond will break and the particles will not stick to each other. Where g_n is the contact gap, k_n and k_s are normal and shear stiffnesses, μ is the friction coefficient, and $\overline{c}_s$ and $\overline{c}_n$ are shear and normal critical-damping ratios. $\overline{k}_n$ and $\overline{k}_s$ are normal and shear stiffnesses of the bond load.

The coefficient of friction and rolling resistance between particle and geometry is defined. The Hertz-Mindlin with Johnson-Kendall-Roberts (JKR) model is selected for particle-to-particle interaction, which can add cohesive behavior to the particles' interaction. The cohesive interaction of this model is defined by the parameter surface energy.

Hertz-Mindlin with JKR cohesion is a cohesion contact model that accounts for the influence of Van der Waals forces within the contact zone and allows the user to model strongly adhesive systems, such as wet materials or asphalt mixtures. In this model, the implementation of normal elastic contact force is based on the JKR theory [53]. Hertz-Mindlin with JKR cohesion uses the same calculations as the Hertz-Mindlin (no slip) contact model for the following types of force: tangential elastic force, normal dissipation force, and tangential dissipation force.

JKR normal force depends on the overlap δ and the interaction parameter, surface energy γ in the following way [53, 66]:

$$F_{JKR} = -4\sqrt{\pi \gamma E^* a^{\frac{3}{2}}} + \frac{4E^*}{3R^*} a^3, \tag{1}$$

$$\delta = \frac{a^2}{R^*} - \sqrt{4\pi\gamma \frac{a}{E}^*}. \tag{2}$$

Table 1 Material parameters in simulation [66]

Property	Value	Unit
Young's modulus	10	GPa
Poisson's ratio	0.25	–
Coefficient of restitution	0.5	–
Coefficient of friction	0.7	–
Density of particles	2500	kg/m^3

Here, a is the contact radius, E* is the equivalent Young's modulus and R* is the equivalent radius, which are defined as

$$\frac{1}{E^*} = \frac{\left(1 - v_i^2\right)}{E_i} + \frac{\left(1 - v_j^2\right)}{E_j}, \tag{3}$$

$$\frac{1}{R^*} = \frac{1}{R_i} + \frac{1}{R_j}. \tag{4}$$

Here, a is the contact radius, E_i, v_i, R_i, and E_j, v_j, R_j are Young's modulus, Poisson's ratio and radius of each sphere in contact.

In EDEM, the Hertz-Mindlin mode was applied for calculating the normal and tangential forces, and the JKR here is responsible for the cohesion force calculation. The parameters in the simulation can be seen in Table 1, which were derived from literature studies and laboratory tests, which are introduced in Sect. 2.5.

2.4 DEM Simulation of the Pre-compaction

In order to study the influence of paver working operations on the quality of pre-compaction during the paving process, different working parameters were selected in this research to simulate the paving process. During the simulation, the paving speed of the paver, the paving thickness of the pavement, and the paving angle of the machine are altered in different models. The paving speed is selected as 5, 6, and 7 m/min. The paving angles are chosen as 0°, 1°, and 2°. As for paving thickness, which is the height of pavement after paving, it was chosen as 4, 5, and 6 cm.

During paving compaction in the simulation, different parts of the model geometry have distinct types of movement. The type of geometry can be chosen as physical or virtual, a physical section is an actual surface or volume that particles can interact with. A virtual section (used to create the particle generator) is a surface or volume of interest that does not actually exist and does not interact with anything in a simulation. Sections of paver geometry can be static or dynamic. Static sections remain in a fixed position during the course of a simulation whereas dynamic sections move. Sections can move under translation, rotation, or vibration with a defined frequency. For instance, the test track remains static during the whole process of paving, but other parts such as the paver screed and particles have independent movement depending on the time step.

The tamper and screed plate of the paver both have a vibration in the vertical direction but with different frequencies, which are used for pavement material pre-compaction and surface smoothing. So, the tamper and screed plate are added with "Sinusoidal Translation" kinetic movement, this type of movement can be defined by loop duration, frequency, offset, displacement magnitude, and movement direction. For the auger of the paver, which is used to divide and distribute material evenly on the road surface, this function is achieved by its rotational movement, which is defined by a rotation kinetic option. Besides, all parts of the paver also have a horizontal movement which drags the machine to move forward, which is defined by a stable horizontal velocity.

2.5 *Validation of the Simulation Model Based on Experimental Testing*

Most of the existing research related to pavement focuses on the final compaction of asphalt mixtures, these approaches cannot produce useful knowledge about changes in arrangement or aggregates interlocking inside materials before the mixture was well compacted. Therefore, a device named Smart Rock (SR) is used in this study, which works based on real-time sensing mechanisms, to provide a scientific basis for enhancing the understanding of asphalt particle rearrangements and flow during the pre-compaction phase. The SR can be used for monitoring the rotation and contact force of an aggregate in real-time. A representative material flow test using SR is introduced in this section, which allows for a closer look at the pre-compaction phase (see Fig. 4).

Different gradations can be considered, in which the aggregates are already evenly mixed with alternated oil. The materials after the mixing test are loosely loaded into a round mold. Before each test, a SR is embedded together with the mixed material,

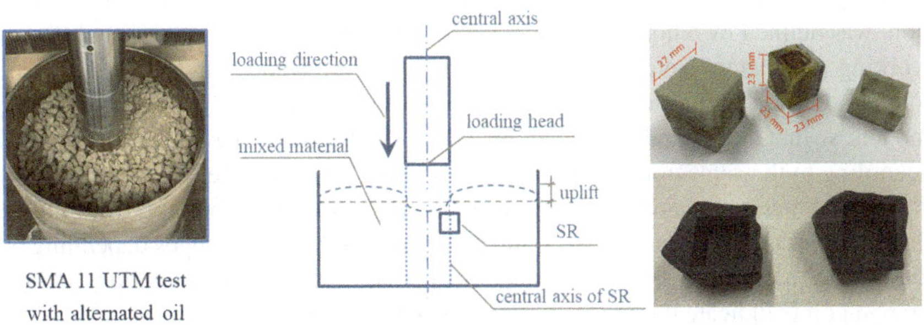

Fig. 4 Material flow test for studying the initial stage of pre-compaction using SR

and it is buried under the loading head, at a position slightly to the right of the middle line. And the right edge of the loading head is aligned with the central axis on the SR, which can be seen in Fig. 4. The location of the SR here is selected because during the compacting process, the movement of the SR, especially the rotation is obvious when it is on the eccentric shaft under the loading head. It is no doubt that the well-compacted materials have a smaller content of air voids compared to materials in which the compaction degree is not enough, and the compaction process is finished via the rotation and re-arrangement of aggregates during compaction. Therefore, the rotation behavior of aggregates during compaction indicates the re-arranging condition. So, the movement of the SR during compacting can be easily monitored when it is at the location mentioned above. When the compaction test starts, the loading head of the machine keeps moving downward with a constant speed until it goes at a pre-set distance. The measured movement of the SR can be used for validation of the DEM model in a future study.

Considering variation between simplified or ideal experiments in the laboratory and field tests, the real condition of pavement pre-compaction was accomplished in the test track. The investigation was carried out on the test track of the MOBA Company in Germany. During the paving process, the paver was driven at a speed of around 5 m/min, and the tamper was operated at 30 Hz. First, the bulk materials were loaded into the paver machine. Secondly, sensors were installed on the paver screed to measure the traction during the paving compaction. The following stage was to study the paving behavior via several working parameters of the paver, namely paving velocity, paving thickness, and paving angle. In the end, the paving condition variation in different working operations was compared based on the derived data from sensors [66]. The field test for paver traction monitoring and paving process investigation can be seen in Fig. 5.

In the DEM paving model, the validation was finished by comparing the relationship between the traction of the paver machine and its vertical supported force. The derived data of the paver from the field test was compared to the results generated in the DEM model with different working parameters. Finally, the model which was closest to real conditions was selected for further study.

The model of pre-compaction uses the real physical parameters of the paver machine and the bulk material in order to simulate the physical conditions of the paving process. The movement of the paver, moving forward with a vertical vibration, was defined by a horizontal velocity per second. Several models with different working parameters were simulated for parameter adjustment and validation. The total amount of the particles with their parameters was kept the same in all of the models, so that the mechanical behavior and the kinetic properties could be compared according to the same conditions. For instance, the contact force between the paver screed and material was analyzed first in this research. As can be seen from Fig. 6, the contact force between the paver and material significantly changes depending on the paving thickness. The blue points indicate the results from simulation, and the orange points indicate the results obtained from field tests.

The relationship between the contact force and the paving thickness yields the distribution function according to mathematical statistics analysis. The vertical

Fig. 5 Traction validation of paver in field test [66]

Fig. 6 Test track validation for simulation [66]

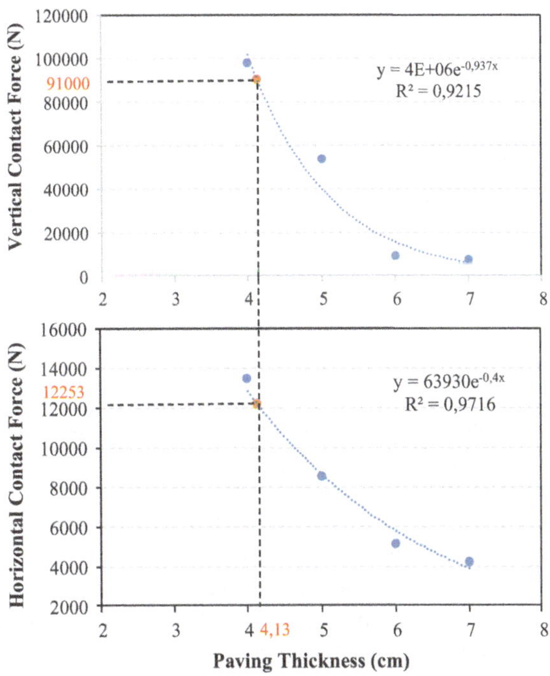

contact force, which is also the direction of gravity, dramatically increases as the paving thickness decreases. When the supporting force from materials to the paver equals to the gravity of the paving machine, then the paving thickness with this compressive force distribution corresponds to the real condition of the paver moving on the road surface, at which the horizontal force of the device is close to the real traction offered by the engine. The gravity of the paver has been calculated to 91,000 N. From this diagram, when the generating volume of particles remains constant in all models, the real mass of the paver is close to the result from the model with a paving thickness of 4 cm (see Fig. 6). Therefore, the model with a paving thickness of 4 cm can be used for kinetic and mechanical analysis. The horizontal force of the device can be derived from this image, which is 12253 N [66].

2.6 Influence of Paving Angle on Paving Compaction

The angle of the contacting plane between the material and the paver screed is determined by the special orientation of the paving machine, namely the paving angle of the screed. Therefore, the force distribution can be changed due to the angle of the contacting plane, also because gravity maintains the same directionality. Therefore, the normal stress, shear stress, and friction between the materials and the screed plate are consequently influenced by this angle.

For the DEM simulations, the reference coordinate system and monitoring area can be found in Fig. 2. It is known that the average angular velocities of aggregates have different levels of fluctuation due to the vibration of the paver screed during paving compaction. Furthermore, it is difficult to obtain the overall rotation of aggregates, namely the effective rotation of aggregates. Paving compaction is also a procedure, during which the aggregates can be pressed and compacted by their relative rotation. The compaction degree therefore can be evaluated by the relative rotational behavior of aggregates. Specifically, the evaluation dominantly considers the rotational properties of aggregates along with the driving (z-axis) direction. Figure 7 illustrates the trend of rotation movement from aggregates during paving. It can be seen from these two figures that the aggregate movement follows a similar trend with the same paving speed and thickness, although with different paving angles [66].

In order to derive the increase in aggregates' effective rotation, this model used a definite integral calculation method. The area enclosed by the curve of angular velocity and the horizontal axis shown in Fig. 7 is calculated and regarded as the average effective rotation of aggregates. In other words, the area calculated for each model is the average effective rotation of aggregates during paving. From the calculated results, the area of 1° paving model is 0.51, and which of 2° paving model is 0.80. It is assumed that the power of the paver screed to the bulk materials remains constant, a larger paving angle indicates more horizontal components (for the paver moving forward) of paving power are used for rotating the aggregates, and there will be fewer components used for vertical compacting. This result means that a smaller paving angle is better for vertically vibrating compaction.

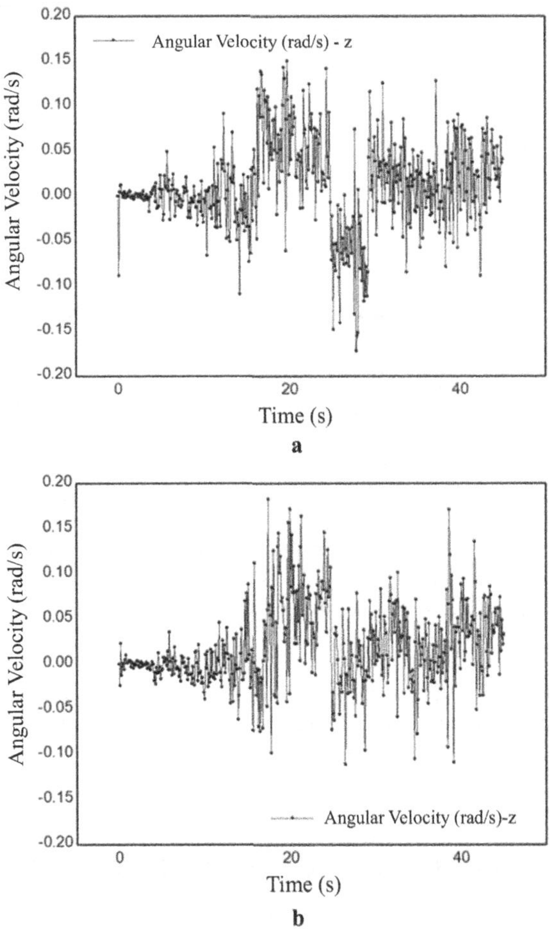

Fig. 7 The trend of average angular velocity around z-axis: paving angle **a** 1°, **b** 2° [66]

Figure 8 shows the comparisons of models with different paving angles; the vertical axis value indicates the final absolute angle of particle rotation, namely the aggregates' effective rotation angle, which was obtained through definite integral calculation. It can be seen from the graph that the effective rotational angle of aggregates performs differently in different rotational directions. However, it should be emphasized that the fluctuation of the rotational angle around both x- and y-directions is not obvious and fluctuated at around 0 rad regardless of the paving angle. On the other hand, the rolling angle following the z-axis changes dramatically when the tamper of the paver starts to contact the particles, fluctuating at a level of more than 0.5 rad. From the results, it can be found that the particles moved obviously around the z-axis, which is parallel to the driving direction, but seldomly move in the other two directions. In other words, during paving compaction, the materials are pre-compacted with a rotation movement of particles around the z-axis; the other

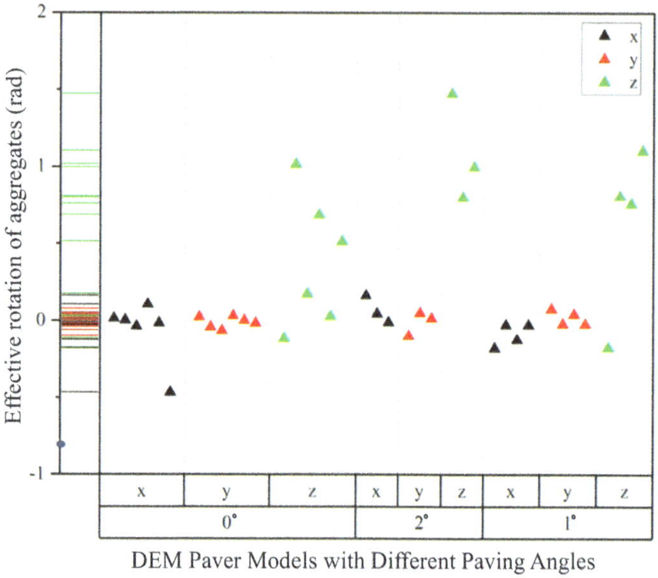

Fig. 8 Effective rotational angle of particles due to different paving angles [66]

two directions do not have obvious rotation. From the results of the simulation, it can also be seen that when the paving angle is 1° or 2°, the increase in compaction density is higher than at an angle of 0° [66].

3 Simulation of Asphalt Roller Compaction at Macroscale

After being pre-compacted, it is necessary to use roller compactors in order to reach the requested final compaction level (depending on the requirements formulated in the technical guidelines) of the asphalt mixture layer. Roller compactors apply a combination of pressure and shear stresses to the asphalt mixture material, which contribute to the increased compaction in the material by rearranging the aggregate structure horizontally [4, 55]. As initially stated, compaction is usually described by the density of the pavement, which increases, while the air-void content of the material decreases.

In this section, the effect of the roller operation on the compaction increase within the asphalt mixture layer is analyzed. For this purpose, a macroscale model of the roller layer interaction is presented, based on previous research [45]. The model needed to fulfill certain requirements, such as representing the movement of the drum (i.e., the rolling movement at different roller speeds as well as the dynamic movement due to different excitations), the contact conditions between the roller and the pavement surface, as well as the behavior of the asphalt mixture when subjected to different types of loading. For this matter, a study on the material behavior under compaction temperatures is presented. This study was conducted through literature analysis and theoretical considerations as well as through laboratory tests, conducted on the material mix. In addition to the reached increase of compaction, an analysis of the void distribution within the layer after the roller pass was conducted as well.

In this section, the pre-compacted asphalt mixture model was described by an FEM model combined with a nonlinear constitutive model (in this case the bounding surface plasticity theory coupled with the critical-state theory), in order to take into account, the material behavior in a loosely stacked state and at compaction temperature conditions. The model is designed in such a way, that the void content in the mineral aggregates (VMA) has a direct impact on the mix properties and therefore the layer properties. Theoretically, the initial conditions of the model (i.e., void distribution before the roller pass) could be defined using the results of the DEM simulations in the previous section. In the underlying study, assumptions regarding the initial state of the layer were made based on practical observations in pavement compaction practice.

Furthermore, the properties of the pavement were determined by laboratory tests. For this purpose, a special laboratory device was designed, which is able to subject the material mix to a variety of triaxial stress states and measure the deformation—including the compression—at compaction temperatures up to 150 °C. The material was subjected to monotonous and cyclic loading in order to capture the effects of time-dependent and repetitive loading. With the help of these tests, the parameters for the constitutive model could be derived and used in order to carry out the simulations of the roller passes at different roller operation modes.

3.1 *Basics of Roller Compaction*

Usually, several different rollers are used, which differ in their sizes, their total weight, and the type of compaction technology they provide. The number and type of rollers depend on the performance of the paver, the type of asphalt mixtures, and the time window available for compaction. The compactability of the material is highly temperature-dependent since it is affected by the viscosity of the asphalt binder, which itself increases with the cooling of the asphalt mixture [43]. Therefore, the binder will evolve a higher resistance against deformation and compaction, so that a higher amount of compaction energy will be required in order to achieve the requested degree of compaction. Studies from [11] have shown an optimum temperature window of 135–155 °C for initiating the roller compaction, outside of which it is still possible to reach the required density with a higher compaction effort, however with a negative effect on the performance properties, such as the indirect tensile strength.

It is, therefore, necessary for the rollers to keep up with the paver, in order to avoid allowing the material to cool off too fast after pre-compaction. The cooling rate itself is dependent on several factors, including meteorological factors as well as the thickness of the layer [11, 56, 67]. Higher pre-compaction values and a lower paving speed can simultaneously decrease the number of required rollers but slow down the construction process. Lower pre-compaction values, on the other hand, require lighter types of rollers for the first roller passes in order to stabilize the material and avoid cracking due to high shear stresses and material displacement. Once the material is stabilized, heavier rollers or rollers equipped with a dynamic drum can be used for breakdown rolling. Furthermore, with lower pre-compaction the number of required roller passes needs to be increased while the roller speed will be reduced in order to achieve a higher increase of compaction with every single pass. Eventually, this will lead to a higher number of required rollers altogether, in order to keep up with the paver [3, 8, 25, 29, 39, 57].

As initially stated, the most common roller type used in pavement compaction is the tandem roller with two smooth steel drums. These drums are defined by the Nijboer number, which is defined as the drum load divided by its width and its diameter. This number should not exceed a certain value, as this could lead to a drum causing more material displacement and cracking instead of compaction. Additionally, the drums can be equipped with unbalance exciters, which create a dynamic movement of the drum (see Fig. 9), allowing for it to overcome the internal friction

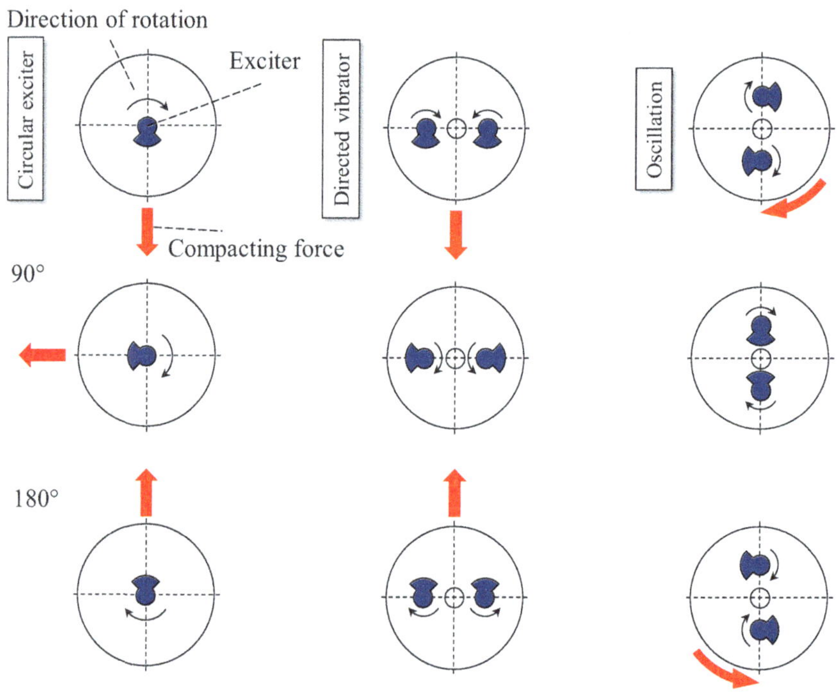

Fig. 9 Principle of different types of dynamic compaction depending on the arrangement of the exciters [50]

between the aggregates and therefore achieve a higher compaction effort [8, 23, 25, 49, 50].

As previously stated, the mechanical behavior of asphalt mixtures can be described as a superposition of the properties of its components, which leads to a combined elastic-visco-plastic behavior. The asphalt binder itself manifests combined visco-elastic—thus time-dependent reversible—characteristics at service temperatures, while when reaching the paving and compaction temperature area, the properties can rather be characterized as visco-plastic (time-dependent, irreversible). According to [27], the reversible deformation properties can be neglected at compaction temperatures.

When subjected to an external load, the asphalt mixture will manifest a resistance against deformation, which can be decomposed into initial resistance, internal friction, and the viscosity of the mixture. The initial resistance can be described as an effect resulting from cohesion and interlocking of the aggregates, while internal friction is mainly defined by the current compaction state, the grain size distribution, and the angularity of the aggregates. The viscous resistance of the mixture is defined by the viscosity of the mortar, which is especially active at loose compaction, since the load distribution inside the material happens through the mortar. As the compaction increases, so does the number of contact points, as well as the frictional and initial resistances [27, 43].

It can be derived that the mechanical behavior of asphalt mixtures at compaction temperatures is comparable to the behavior of soils or similar granular materials. Therefore, it is necessary to take certain effects into account, such as the compression and dilation effect, depending on the combination of hydrostatic pressure and shear stresses. These effects, as well as the phenomenon of the stresses tending towards a specific state when the material is continuously sheared, can be described by the use of the critical-state theory [51].

3.2 Development of the Roller-Asphalt Layer Interaction Model

In this section, the roller compaction is described by using a model, which was developed within the works of [45]. The model consists of three basic parts: a kinematic model of the drum including an excitation unit for oscillation, a contact model (which itself consists of contact conditions for normal contact as well as for dynamic friction between the two contact partners), and the macroscale FEM model of the asphalt mixture layer itself.

All movements are described in a plane coordinate system and on a time-dependent basis. The external loads acting on the system (i.e., the dynamic excitation of the drum) are divided into several time-steps, for each of which, the drum movement response and the corresponding asphalt mixture layer deformation are being computed. This method allows consideration of the continuously changing contact conditions between the drum and the layer due to the translational and rotational drum movement, as well as the oscillatory movement.

The drum itself is assumed as a rigid body and, furthermore, dynamically decoupled from the frame of the roller, as the excitation frequencies of the drum are significantly higher than the natural frequency of the coupled drum-frame system. For each time step, the contact conditions between the drum and the FEM layer model are determined via a radial detection algorithm.

The FEM model is described using a nonlinear constitutive law with plane strain and spatial stress conditions. While conventional plasticity models only take into account plastic deformations when the stress state reaches a defined yield surface, which leads to an abrupt change in the material behavior, the theory of bounding surface plasticity allows a smooth transition to plastic yielding by projecting the stress state onto a defined bounding surface. In this theory, the plastic deformations are calculated with respect to the consistency conditions at the projection point, while at the same time scaling down these plastic deformations dependent on the distance between the projection point and the actual stress state. The definition of an adequate projection rule is one of the key elements in such a model. The theory has been applied in the past for the cyclic loading of unbound granular materials, as well as for asphaltic materials [24, 28, 30, 52]. Furthermore, in [52] the effect of time-dependent behavior due to the temperature-dependent visco-plastic influence of the asphalt binder was considered by implementing consistency based visco-plasticity.

The plastic deformations are calculated, based on the consistency conditions on the bounding surface and the distance of the actual stress point from its projection equivalent. When the actual stress point is identical to the projection center, the deformation is purely elastic. When the stress point reaches the bounding surface, the model behaves similarly to the conventional plasticity theory.

In the underlying study, the theory from [52] combined with the bounding surface formulations from [28] and the stress integration algorithms from [24] were used. In order to differentiate between first-time loading, unloading, and reloading, the projection rules were expanded by additional criteria; Figs. 10 and 11 demonstrate these rules within the coordinates of the hydrostatic pressure p and the deviatoric equivalent Von-Mises stress q.

For first time loading, the projection center is located in the origin of the p-q coordinates, from which the projection point $\bar{\sigma}$ is determined via the current stress point σ. From the location of $\bar{\sigma}$ on the bounding surface, a loading surface is derived with the variable $\hat{p}_0$ as a size-parameter, which then is used for determining the hardening parameters. Furthermore, the vector in the normal direction to the bounding surface $\vec{n}$ needs to be determined as well for the calculation of the plastic deformations.

When a stress reversal takes place, the current loading surface will become a local bounding surface, taking into account the previous loading history of the material and the resulting kinematic hardening. The point of stress reversal is now used as a new primary projection center from which a new point $\bar{\sigma}$ and vector $\vec{n}$ on the local bounding surface are determined and forwarded to the global bounding surface via the series of local bounding surfaces, depending on the previous number of stress

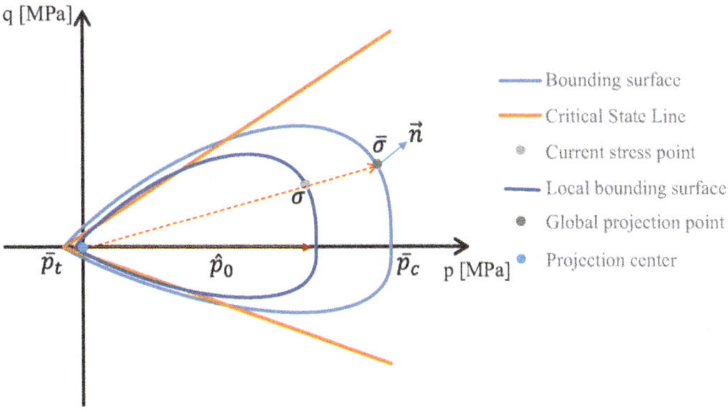

Fig. 10 Projection rules for first time loading [52]

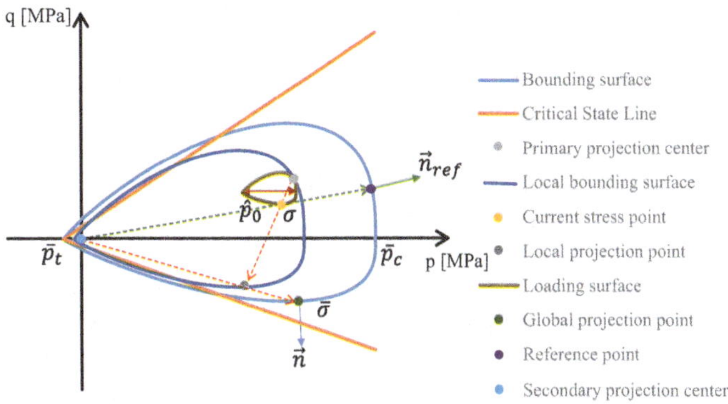

Fig. 11 Projection rules for unloading and reloading [52]

reversals having taken place. At the same time, the origin of the p-q coordinates remains a secondary projection center, from which a reference point and a reference vector $\vec{n}_{ref}$ will be determined.

For first time loading, both the primary and secondary projection center are identical, which means the respective reference points and the vectors $\vec{n}$ and $\vec{n}_{ref}$ will be identical as well, while for unloading and reloading, $\vec{n}$ and $\vec{n}_{ref}$ will point into different directions. Therefore, the following definition for the different loading states based on the cosine function of the angle between the two vectors were introduced

$$\text{first time loading: } \frac{\vec{n} \cdot \vec{n}_{ref}}{|\vec{n}| \cdot |\vec{n}_{ref}|} = 1, \tag{5}$$

$$\text{unloading: } -1 \leq \frac{\vec{n} \cdot \vec{n}_{ref}}{|\vec{n}| \cdot |\vec{n}_{ref}|} < 0, \tag{6}$$

$$\text{reloading: } 0 \geq \frac{\vec{n} \cdot \vec{n}_{ref}}{|\vec{n}| \cdot |\vec{n}_{ref}|} < 1. \tag{7}$$

With these criteria, the hardening parameters can be computed in a way that takes into account the different loading and unloading paths and control the material stiffness accordingly.

3.3 Calibration of the Developed Material Model

In order to calibrate the material model presented in Sect. 3.2 and determine the required model parameter, laboratory tests are needed. In [30], triaxle tests were conducted on sand, while in [52] the asphalt mixture models were determined among other tests with uniaxial tensile strength tests. For this study, a special triaxle testing device was developed, in order to be used at compaction temperatures.

The device, as shown in Fig. 12, consists of a metal base-plate in which heating elements were introduced in order keep the material temperature at a constant level during the test. A confinement frame, which is divided into four mobile segments, is mounted on top of that base-plate. Each segment is fixed with the help of a horizontally arranged metal segment, which in turn is equipped with a force transducer, as well as two displacement sensors. The cubic shaped material sample with an initial edge length of 10 cm, is inserted into this confinement frame. The sample can either be prepared in the laboratory (e.g., in a roller segment compactor) or cut from an existing pavement. The initial mass and exact size of the sample need to be determined prior to the test being carried out, so that the initial compaction state is known.

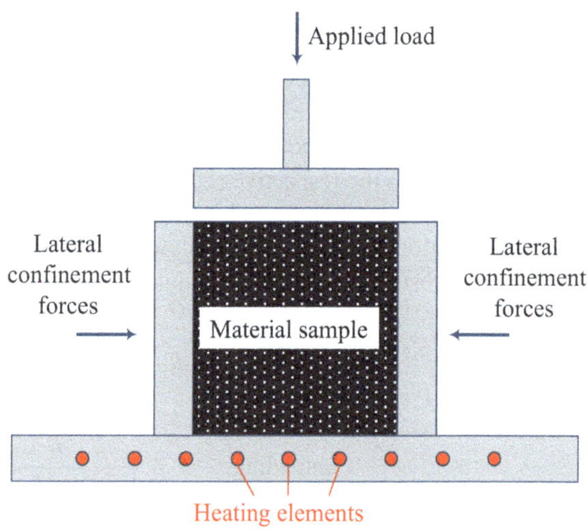

Fig. 12 Schematics of the triaxle testing device used for calibration

The sample can be pre-heated in an oven and brought up to the required testing temperature within the testing device itself, using the heating elements. Once the pre-heated sample is introduced into the frame, a load can be applied via the top plate. The material deformation and the resulting stress states are determined with the help of the displacement sensors and force transducers. Furthermore, the increase of compaction can be continuously monitored during the test.

In this study, two different types of testing configurations were carried out at varying degrees of initial compaction. In the first configuration, the vertical load was applied instantaneously and held for a defined time period after which it was released again. This configuration was aimed at determining the time-dependent behavior of the material under constant loading. In the second configuration, the material was subjected to a cyclic load, in order to analyze the accumulation of plastic deformation as well as the hardening behavior of the material with increasing compaction. The two different testing configurations were carried out using a stone mastic asphalt (SMA) 11 mix, with a polymer-modified asphalt binder. The asphalt binder content was set to 7.0 M.-%. Due to the high number of independent model parameters, assumptions for certain parameters (such as the tensile strength due to cohesion) were made, in order to fit the testing results. The void content VMA at lowest possible stacking was set to 38.7 Vol.-%, which corresponded in the chosen material mix to 75.7% of bulk density relative to the reference Marshall Density. The tests were carried out at material temperatures between 90 and 130 °C in order to determine the temperature-dependent visco-plastic properties.

An example of test results for constant loading can be seen in Fig. 13, where the volumetric strain of the material ϵ_{hyd} is shown as a function of the testing time. The material was subjected to a constant vertical load during the first 30 min of the test. During this time span, the material was instantly compressed by 3.1 ‰ during the phase in which the load was built up, followed by another supplemental 2.0 ‰ during constant loading. At t = 1800 s, the top load was removed, while the horizontal loads remained active. While parts of the deformations observed at the time point of unloading can be attributed to elastic deformations, switching from

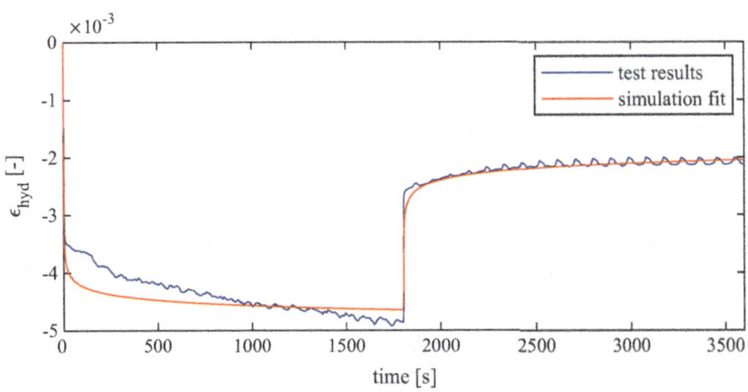

Fig. 13 Test results and parameter fitting of the material model for constant loading

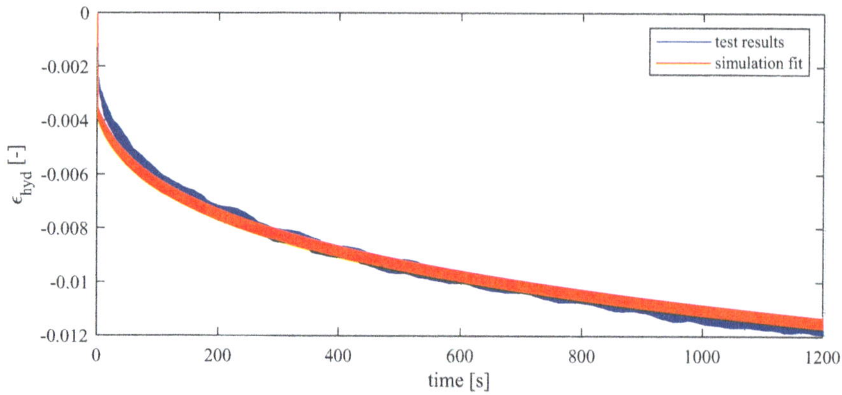

Fig. 14 Test results and parameter fitting of the material model for cyclic loading

a triaxle compressive stress state to a triaxle extensive state will lead to secondary plastic deformations during the unloading phase as well.

The effect of cyclic loading was tested in the second testing configuration, of which exemplary test results and the model parameter fitting can be seen in Fig. 14. The material was subjected to a cyclic threshold load with constant amplitude and a frequency of 2 Hz for a duration of 20 min. While at the first loading an instantaneous compression of 2.5 ‰ was measured, it can be seen that with each additional loading cycle the cumulated deformation increases, reaching a value of almost 12 ‰ after 20 min (corresponding to 2400 loading cycles).

3.4 *FEM Simulation of Asphalt Roller Compaction*

With the material model presented in Sect. 3.2 and the material calibration in Sect. 3.3, roller compaction FEM simulations were run at different roller operation modes in order to study the impact of one roller pass on the increase of compaction and the decrease of the air void content in the resulting asphalt mixture layer. For this purpose, a kinematic model of a smooth steel drum, equipped with a dynamic unit for oscillatory compaction, based on the information from [45], was used in the simulation.

The asphalt mixture layer was defined with a length of 1.5 m in order to allow a sufficiently large enough rolling time at the maximum chosen rolling speed without the roller reaching the end of the layer-model during the simulation. The width of the layer was defined as 1.0 m, which is the exact width of the roller drum itself. The lateral behavior of the material was taken into account via the definition of a spatial stress-state model combined with a plane strain model. The thickness of the layer was chosen as 4 cm, which is a standard value for a typical SMA surface course.

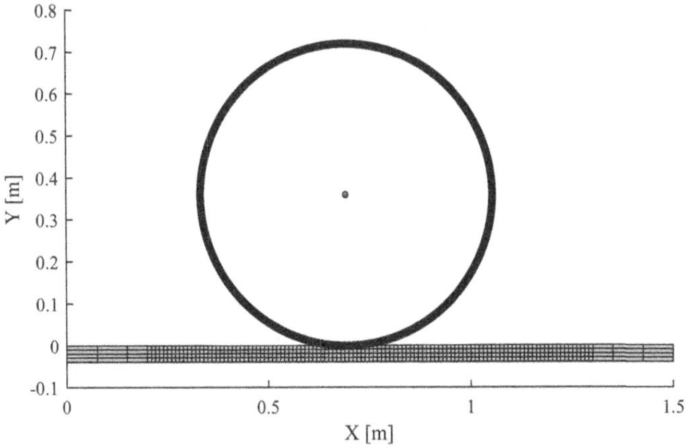

Fig. 15 Roller model with FEM-Asphalt-Mixture-layer

The rigid body model of the oscillatory drum was coupled via the contact model to the FEM-asphalt mixture layer model as described in Sect. 3.2. The simulation time was set to a total duration of 0.8 s, divided into equally sized time-steps with a duration of 0.2 ms, leading to 4000 time-steps. An example of the coupled model can be seen in Fig. 15.

For simplification purposes, the material within the layer model was assumed to be isotropic and fully homogenous in its properties. As such, for every material point in the layer model, the initial content of total voids was set to 26.4 Vol.-%, which corresponds to a density of 90.9% relative to the Marshall density for the reference mix from the tests in Sect. 3.3. Also, the material temperature was set to be overall equal to a temperature of 130 °C, so that identical visco-plastic parameters in every material point could be defined, derived from the parameters that were determined at that same temperature in the tests of Sect. 3.3.

3.5 Compaction Results and Void Distribution

For roller operation parameters, three different modes (static, oscillating at 27, and 42 Hz) and two different roller speeds (0.5 and 1.0 m/s) were chosen for this study, leading to a total of 6 simulation runs. For each simulation run, the resulting void distribution within a chosen section of the FEM layer model was determined after one roller pass. As the compaction results in the field are usually determined using a sample (e.g., a core drilled from the layer), this value only represents an average result for the core. As such, with the information of the material mix setup, the resulting average density value relative to Marshall over the layer section was determined. The difference between this value (after one roller pass) and the initial value (before the

Table 2 Compaction increase after one roller pass at different roller operation modes

Operation mode	Compaction increase at 0.5 m/s (%)	Compaction increase at 1.0 m/s (%)
Static	+1.6	+1.5
Oscil. 27 Hz	+2.4	+1.8
Oscil. 42 Hz	+3.0	+2.4

roller pass) was then calculated and is shown in Table 2 with respect to the different roller operation modes and rolling speeds.

The results show that the dynamic movement of the drum contributes to the increase of compaction in this study. Furthermore, it can be seen that high oscillation frequencies and low roller speeds have a stronger effect on the compaction than lower frequencies at high speeds. This effect can be attributed to the number of loading cycles or impacts to which the layer is being subjected during one roller pass, as this number is dependent on the contact time between the roller drum and a single contact point in the layer surface. Higher frequencies and lower rolling speeds will lead to a higher number of loading cycles, which will lead to a higher compaction, as long as the material stress states remain on the compressive side of the critical state line.

Furthermore, the simulation results allow for determination of the void distribution within the layer. In Fig. 16 the total void (VMA) distribution results after one roller pass at different depths under the layer surface can be seen for a roller speed of 0.5 m/s and a frequency of 27 Hz. The figure shows variations in the void distribution as much in the horizontal direction (X) as in the vertical direction. Reaching almost the bottom of the layer (at −3.5 cm), the total void content is at an average level of 26.4 Vol.-%, which corresponds to the initial value set at the start of the simulation. At levels closer to the surface, a decrease of voids can be observed, up to an average total void

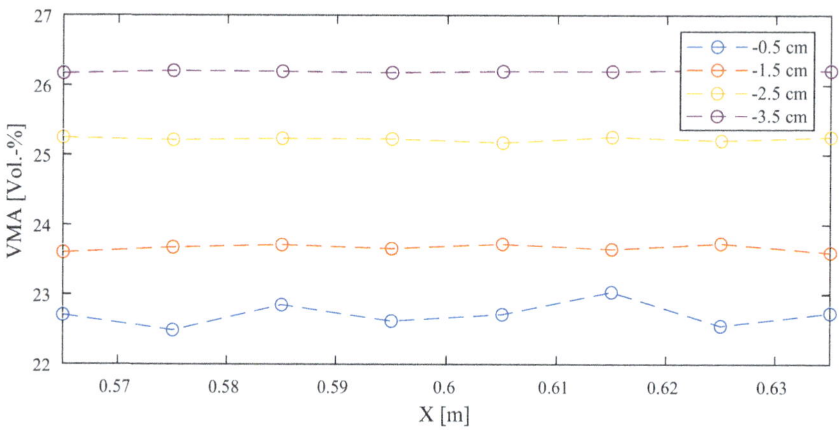

Fig. 16 Total void (VMA) distribution at different levels below the asphalt-mix layer surface (at 27 and 0.5 m/s)

Table 3 Scattering of total void contents in the surface area of the simulated layer at different roller operation modes

Operation mode	VMA scattering at 0.5 m/s (Vol.-%)	VMA scattering at 1.0 m/s (Vol.-%)
Static	±0.002%	±0.002%
Oscil. 27 Hz	±0.166%	±0.841%
Oscil. 42 Hz	±0.399%	±0.923%

content of 22.7 Vol.-% (at a level of −0.5 cm). This can be explained by the fact that the stresses introduced by the roller are highly concentrated in the contact area on the surface and spread over the depth of the layer resulting in overall lower stresses on the bottom of the layer. As a result, the roller possesses a limited depth effect, which results in a stronger compaction effect of the surface zone of the layer.

At the same time, it can be seen that the surface area of the layer also shows a higher scattering of values than the bottom area. The results shown in Table 3 indicate that for static roller passes, the scattering of total voids in the horizontal direction of the layer are comparatively low and will produce an even distribution of voids. With the activation of the oscillation, the voids start to show a higher spreading, which increases with the frequency, as well as with the roller speed.

These results indicate that while the dynamic compaction mode leads to a greater increase of compaction with one single roller pass, it also causes more spreading in the void distribution inside the layer. This effect is connected to the multiple different time-dependent stress states to which the different contact points between the roller and the layer surface will be subjected to. For each contact point, the stresses will be dependent on the contact time, the loading frequency, the loading amplitude, and the phase at which the loading starts and will end. Furthermore, these factors are a function of the size of the contact area between the roller in the rolling direction and the stress distribution within this area. Depending on the constellation of these parameters, the different contact points in the layer surface will be subjected to a different time-dependent constellation of stresses, which in turn will lead to different compaction results, especially on the surface area. The result is an inhomogeneous distribution of voids which can affect the performance properties of the resulting asphalt mixture layer.

It should be pointed out, that this study was only conducted on one single roller pass and with certain simplifying assumptions regarding the distribution of the initial state of material parameters within the layer. For future studies, the effect of several roller passes should be investigated in order to observe the evolution of the void distribution. Furthermore, the initial material parameters within the model should be set with regards to the void and temperature distribution after pre-compaction. For this, the studies carried out with the DEM simulations in the previous section could be used.

4 Simulation of Asphalt Mixtures Manufactured by Different Compaction Methods at Microscale

The mechanical performances of the asphalt mixtures can be influenced by applying the different compaction methods. An innovative standardized laboratory compaction method, named the Aachen compactor, was developed to maintain consistency both in the laboratory and the field compaction [5]. With the Aachen compactor, the cylindrical asphalt samples of two different diameters, namely 100 mm or 150 mm, can be manufactured (by adjusting the setup, the production of other diameters is technically applicable). As the simulation of the field compaction is the primary aim, the use of two conical rotating steel rollers with a smooth banding, compacting the asphalt mixture in a cylindrical mold achieves the roller compaction principle. Figure 17 shows the whole compaction procedure with the Aachen Compactor, and this procedure can be generally divided into three stages. Further details can be found in [34].

In this section, the computer-generated approach was used to reconstruct the microstructures of asphalt mixtures with different material components and produced by different compaction methods. Following, the finite element (FE) software ABAQUS was implemented to develop the microstructure-based FEM models. The influences of the different compaction methods on the asphalt specimens by using microscale FE simulation were investigated with comprehensive analyses, which were concerned with the internal structure, mechanical response, fracture behavior etc.

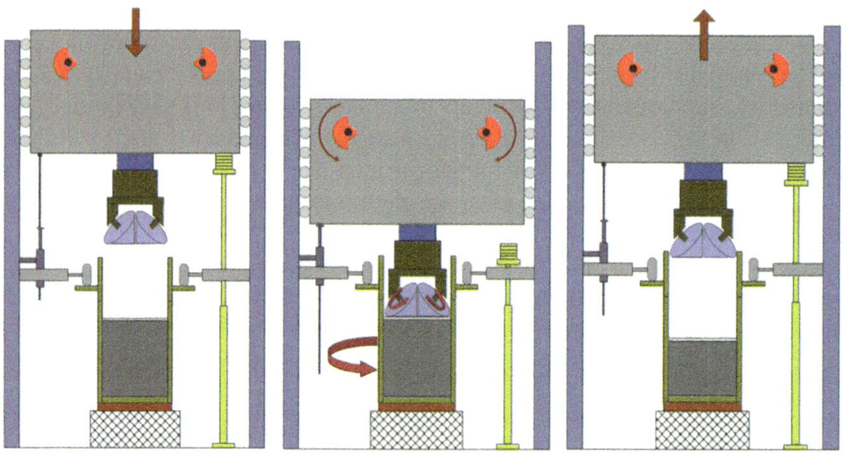

Fig. 17 Operating stages: **a** lowering and pre-rolling stage, **b** main compaction stage, **c** lifting stage [34]

4.1 Preparation of Asphalt Samples and Digital Image Acquisition and Processing

When focusing on a commonly applied asphalt mixture, SMA 11 DS was used as the samples. In this study, the used binder is Bitumen 50/70. More details about SMA 11 DS can be found in the chapter "Experimental Methods for the Mechanical Characterization of Asphalt Concrete at Different Length Scales: Bitumen, Mastic, Mortar and Asphalt Mixture". Three different compaction methods were applied to the asphalt samples. 100 mm (diameter) field-compacted cores were drilled from a test track with an approximate 72 mm construction thickness of the paver, which was compacted with a tandem vibration roller.

The Aachen compactor (Aachen specimens) and Marshall compactor (Marshall specimens) were applied to compact the laboratory samples. The field cores' average density was used as a reference for the volume density of the Marshall and Aachen specimens. German guidelines were referred to manufacture the Marshall specimens [15, 16].

Before the surface grinding, storage of all specimens at room temperature lasted for 48 h. Afterward, they were analyzed with the digital image analysis and numerical simulation. In order to generate the microscale FEM model and investigate the internal structure of the asphalt specimens, image analysis was first applied. Before image capturing, the specimens were sawed into two sections, where the horizontal cut proceeded 30 mm below the surface. For each of the three compaction methods, 3 replicates (100 mm diameter) were prepared, for a total of 9 specimens. A high-resolution optical camera was used for all asphalt specimens to capture the cross-section photographs; in order to obtain adequate image quality, the resolution of the images was set to 450 dpi. For further improvement of the image quality, the captured images were conducted with several procedures in digital image processing (DIP): filter noise, conversion to greyscale images, and binary images to identify objects of interest from the images. To conduct these procedures, the MATLAB Image Processing Toolbox was applied. The selected specimen images are shown in Fig. 18.

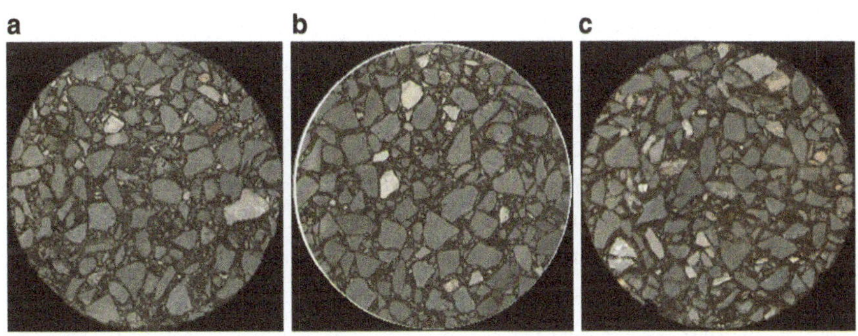

Fig. 18 Cross-section images of **a** field cores, **b** Aachen specimens, **c** Marshall specimens [34]

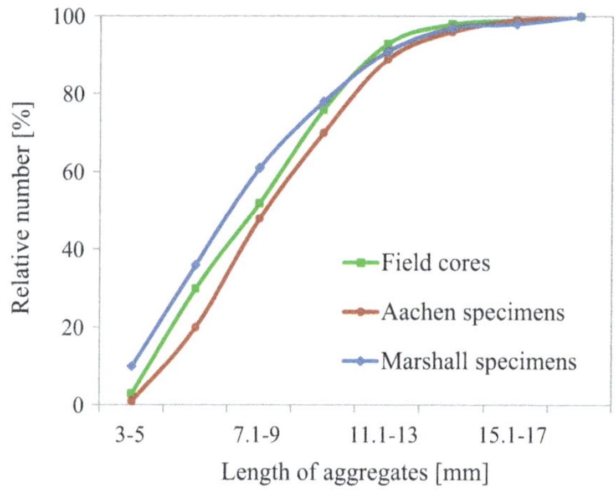

Fig. 19 Relative number of aggregate grains in the cross section (mean value curves) [34]

The DIP techniques were applied to derive the long axis distribution of aggregate grains by averaging each replicate samples' parameters. Figure 19 shows the long axis distribution of aggregate grains for a cross-section. It can be seen that more aggregates with smaller lengths existed the Marshall specimens compared to the other two specimens.

4.2 Development of Microscale FEM Model of Asphalt Mixtures

In order to approach the modeling, the material properties for the aggregate and the asphalt mortar needed to be determined. It is assumed that the asphalt mortar shows linear viscoelastic properties, the results using ABAQUS for a combination of a Prony series are shown in Table 4. A Young's modulus of 55,000 MPa and a Poisson's ratio of 0.20 was assumed for the linear elastic material behavior of the basalt aggregate [69].

Table 4 Prony series of asphalt mortar at 15 °C

Item	ρ_m (s)	E_m (MPa)	Item	ρ_m (s)	E_m (MPa)
1	1.77×10^{-5}	244.262	7	0.06571	446.172
2	6.95×10^{-5}	739.554	8	0.2587	115.784
3	0.0002735	437.066	9	1.0184	119.603
4	0.001077	400.972	10	4.0091	16.708
5	0.004239	487.970	E_∞ (MPa)	3.3521	
6	0.01669	245.953			

The crack propagation may pass through aggregate, asphalt mortar, or the interface layers under consideration of the onset and evolution of damage (i.e., cracking). The material properties of asphalt have a high dependence on temperature. When the temperature is very low, the asphalt mortar is stiffer and some cracks which pass through weak aggregate grains can be seen. As the temperature exceeds 0 °C, very few cracks pass through the aggregate, which leads to more complex failure modes. However, the failure modes are determined by the aggregate morphology, bitumen film thickness, loading rate, and water content [31, 42]. A clear viscoelastic behavior of the asphalt mortar can be observed at 15 °C. This leads to a "self-healing" of micro-cracks or no onset at all because of the relatively low loading speed. Thus, in this investigation, it is assumed that at the interfaces between asphalt mortar and aggregate, the damage and the respective cracks have occurred [42].

Dugdale [18], Barenblatt [10], Rice [48], and others conceived the concept of the cohesive zone model firstly so that crack initiation and propagation could be simulated. According to this concept, resisted by the presence of cohesive forces, fracturing is viewed as a phenomenon where separation occurs between two adjacent virtual surfaces across an extended crack tip (cohesive zone). The implementation of the cohesive fracture behavior in numerical models required the determination of a few parameters. Thus, a static three-point semi-cylinder bending test (3PSCBT) [33] with asphalt specimens complying with the DIN EN 12,697-44 [17] was carried out. A high-speed camera (FastCam SA 5) was used during the test to record the specimen. The applied force and the crack tip opening displacement (CTOD) were used for the determination of the required parameters.

The implementation of the adhesive fracture behavior in ABAQUS was considered as linear softening. The constitutive response of the cohesive element was determined by using a bilinear traction–separation law. The initial values for the parameter T° and G_f under consideration of the concept of dissipated fracture energy were determined using the results of the static 3PSCBT [54, 58]. For proper results to fit from the experiment to the numerical simulation, the optimizing of the material parameters was done. They are listed in Table 5.

Figure 20 shows the 2D models: the aggregates and asphalt mortar are colored in grey and black respectively. For a better illustration of the aggregates and asphalt mortar, the mesh was hidden and due to the small size of the air voids, the air voids are invisible. 4.0 mm was chosen as the minimal grain size of the coarse aggregate and, thus, the asphalt mortar contains the fine aggregate and the coarse aggregates whose size was smaller than 4.0 mm. The cross-section consists of asphalt mortar, air voids, and aggregate with an assumed circularity between 0.7 and 0.8 [40]. To generate the 2D microscale FEM models randomly, Neper, an open-source software package for the polygon and polyhedron generation and meshing, was used [47]. A

Table 5 Material parameters for cohesive elements

Phase	K (MPa)	T° (MPa)	G^f (mJ/mm^2)
Mortar-aggregate interface	14,200	3.56	0.344

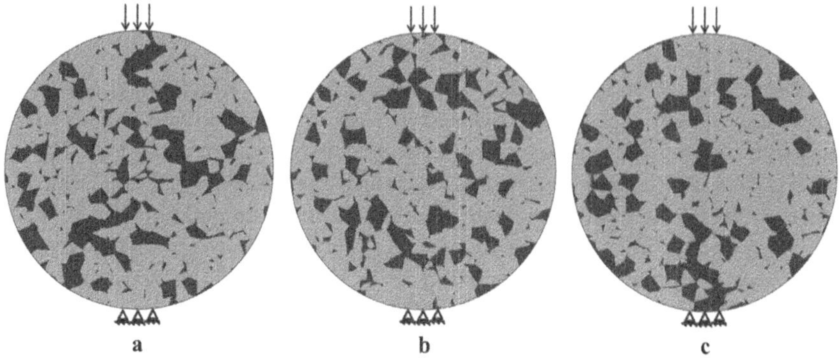

Fig. 20 2D FEM models: **a** field cores, **b** Aachen specimens, **c** Marshall specimens [34]

hard contact relation between aggregates as well as in the interface aggregate and asphalt mortar was characterized. Table 5 shows an adhesive failure potential. To discretize the asphalt samples, which contain asphalt mortar and coarse aggregates, linear triangle 3-node plane strain elements (CPE3) were used. For the discretization of the interfaces between aggregate and asphalt mortar, 4-node 2D cohesive elements (COH2D4) were used. After extensive research on mesh, 0.7 mm was selected as the average mesh size of elements in the FEM model. 12.7 mm was selected as the width of the loading and support strips for the simulation with the indirect tensile test. Until failure of the specimens, the loading distribution was uniform at a constant deformation rate of 50 mm/min. Horizontal deformation was possible, while the vertical displacement was fixed at the support strip.

4.3 Load-Bearing Capacity of the Specimens Manufactured by Different Compaction Methods

The load-bearing capacity of the specimens can be described by the evolution of the applied load and the corresponding displacement of the loading strip. The comparison of the load-bearing capacities of asphalt specimens, which were manufactured with different compaction methods, is shown in Fig. 21.

All three types of asphalt specimens show similarities in the initial stiffnesses (curves' slope) until the displacement reaches 0.1 mm. As can be seen, the load-bearing capacity of the Marshall specimens is the highest, while the field cores have the lowest value. The Aachen specimens' load-bearing capacity curve is closer to that of the field cores with displacements below 1.0 mm, when compared with the Marshall specimen. During the production process of the Marshall samples, higher compaction energy was applied. Thus, an expectation of a more uniform stress distribution in these specimens, which lead to higher load-bearing capacities, is claimed.

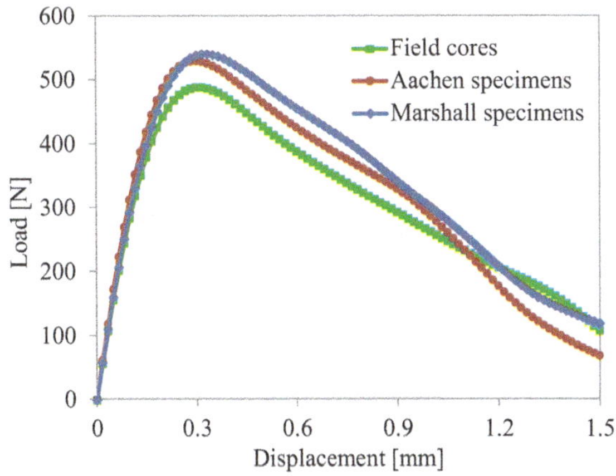

Fig. 21 Comparison of load–displacement curves between models under different compaction methods [34]

Table 6 Displacement of the loading strip and the applied loads at different states

State	Field		Aachen		Marshall	
	Dis (mm)	Load (N)	Dis (mm)	Load (N)	Dis (mm)	Load (N)
Microcrack initiation	0.10	284	0.09	291	0.12	331
Macrocrack initiation	0.31	489	0.30	530	0.33	540
Failure	0.90	294	0.93	318	0.94	325

As the slope of the curve strays from its linear path, the micro-cracking begins. When the curve reaches its maximum, the macro-cracking occurs. Once 60% of the maximum load is reached, complete specimen failure is expected. Table 6 shows the displacements of the loading strip and the corresponding loads. Comparing the values derived from the Aachen specimens and the Marshall specimens, both values of the Aachen specimens fit better to that of the field cores, and this is in accordance with the results observing the curves in Fig. 21.

4.4 Fracture Patterns of Asphalt Specimens

Figure 22 shows the fracture patterns at the moment of macro-crack initiation in asphalt specimens. Different distributions of the macro-cracking in the asphalt specimens can be observed; in the field specimen, most macro-cracks cumulate in regions of the loading and support strips. However, in the Marshall specimens, this can be found throughout the entirety of the specimens.

The overall damage of the mortar-aggregate interfaces can be expressed by the damage parameter, i.e., the stiffness degradation. The stiffness of the cohesive elements of the mortar-aggregate interfaces decreases to zero with the increase of the

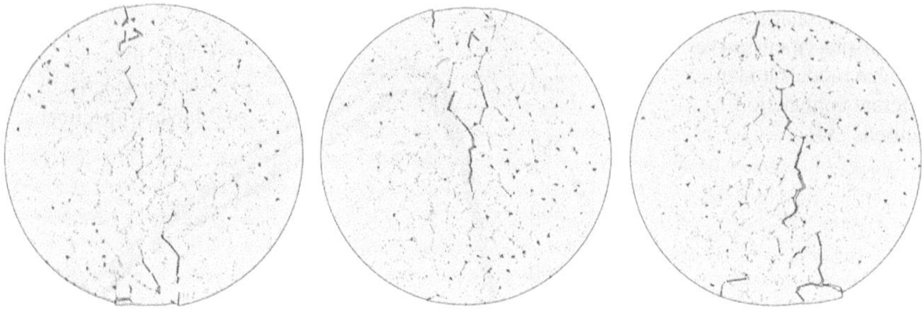

a b c

Fig. 22 Fracture patterns at the moment of macro-crack initiation: **a** field specimen, **b** Aachen specimen, **c** Marshall specimen [34]

stiffness degradation value from 0 to 1. Therefore, fracturing takes place. The stiffness degradation of the cohesive elements between different compaction methods is analyzed by using the cumulative distribution function (CDF), which is shown in Fig. 23. Exceedance of a limit value is not possible, so all CDFs trail off to the right. Compared to the other two compacted specimens at macro-crack initiation, the ratio of the degraded stiffness in the Marshall specimens is the highest. More cohesive elements are needed due to the damage and fracture for the reduction of the Marshall specimens' stiffness. This is in accordance with the fact that the resistance to permanent deformation increases with the increased stiffness of the Marshall specimens. More similarity in the cumulative distributions of the stiffness can be observed in the field cores and the Aachen specimens. Thus, using the specimens with the production of the Aachen compactor can better predict the crack initiation and propagation in practice than the Marshall specimens.

4.5 Further Development of FE Models of Asphalt Mixtures at Microscale

Besides using the aforementioned FEM models of asphalt mixtures to investigate the effect of the different compaction methods on the asphalt specimens, the FEM models with different microstructures of asphalt mixtures are further developed in three more studies using both a computer-generation based approach and an image-based approach.

Facing the current global challenges, hydrological cycle recovery and urban flood risk reduction must be taken into consideration. Porous asphalt (PA) is one promising and effective permeable pavement solution, which is characterized by void-rich pavement materials. The performance of PA mixtures can be significantly affected by the fillers used because of the specific gradation. An investigation with the numerical method for the mechanical responses of the PA mixtures influenced by the

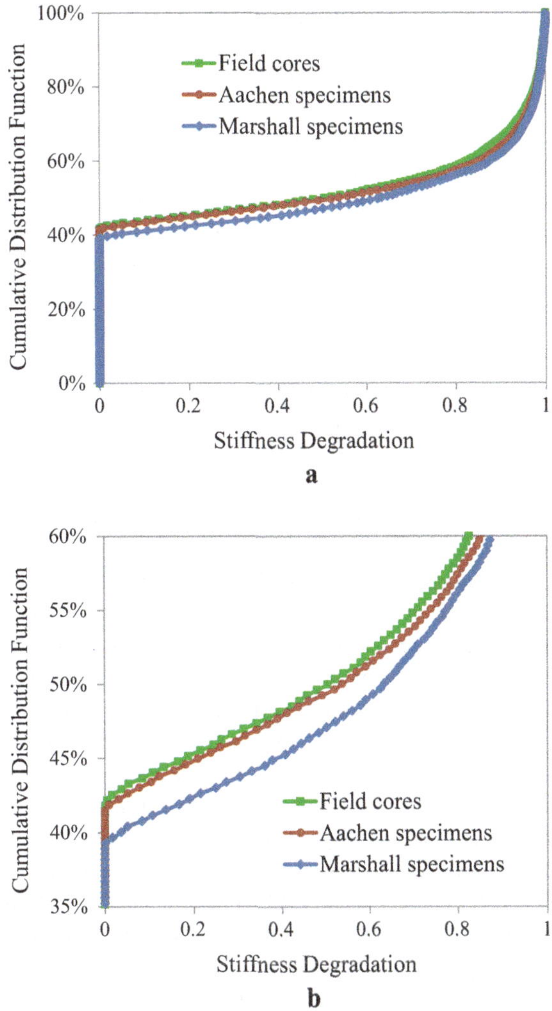

Fig. 23 Cumulative distribution functions of the stiffness degradation for various compaction methods at macro-crack initiation: **a** overall, **b** partial magnification [34]

four different fillers, namely Limestone, Dolomite, Rhyolite, and Granodiorite, was carried out in the first study [37]. The microstructure of PA specimens was detected and reconstructed by using X-ray CT scanning and DIP techniques. An indirect tensile test was simulated by using the 2D FEM model to compute and compare the mechanical responses of the PA mixtures (load-bearing capacity, von Mises stress, and creep strain). Based on the results, the mechanical responses of the PA mixtures are considerably affected by the different fillers. A ranking of the performance of the PA mixtures with different fillers was presented. The selection of optimal filler, which improves the permeable pavement design, can be realized based on this ranking.

In the second study, the 3D microscale FE simulations were used to improve understanding of the effect of temperature (−5, 5 and 15 °C) on the mechanical performances of the asphalt mixtures [32]. The cross-sectional images of the asphalt specimens were captured by X-ray CT scanning, which were then processed with DIP techniques for the reconstruction of the 3D microstructure of the asphalt specimen. The simulation of the repeat loading triaxial test (RLTT) on this microstructure and the investigation of the evolution of creep strain, the stress states at the asphalt mortar-aggregate interface, and the variation of energy dissipation of asphalt mixtures were presented. The investigation shows that with the increase of the temperature during the RLTT, the magnitude and amplitude of creep strain increase, and the strain growth rate decreases. Additionally, as the temperature increases, the proportion of larger maximum principal stress at the interface between the asphalt mortar and aggregates increases. The proportion of larger maximum principal stress in both coarser and finer aggregates increases when the temperature increases. Moreover, as the temperature increases, the energy dissipations increase. Before the external load is removed, the dissipation of creep energy is increasing gradually. However, a cyclic variation of the dissipation of strain energy can be observed and it recovers to zero after the external load is removed.

The third study was aimed at studying the effect of aggregate morphology (concentrated on aggregate angularity) on the mechanical response and damage behavior of asphalt mixture [33]. A gradually decreasing aggregate angularity was applied for the creation of four microscale FEM models. Analysis of the mechanical response and damage behavior of the models was presented. Based on the results, the load-bearing capacity of the asphalt mixture is significantly affected by the aggregate angularity. There is no linearly proportional variation correlation between the load-bearing capacity and the aggregate angularity. At a defined temperature, the creep dissipation energy of the asphalt mixtures can be slight influenced by the different angularities of the aggregates. In general, the creep strain in asphalt mortar reduces with decreases in the aggregate angularity. For the most part, damage becomes visible at the interfaces near sharp corners of the aggregate, with the help of the visualization technique. A connected damage network with complicated crack bridging and branching is built by the extension of damage bands to their surrounding areas. Lower stress concentrations are led from a lower aggregate angularity, less damage and crack initiation at the mortar-aggregate interfaces occur, also the release of damage dissipation energy is reduced. Negative correlations are determined between aggregate angularity and some kinds of dissipation energy, e.g. damage, friction, and creep dissipation energy.

5 Conclusions and Outlook

To simulate the pre-compaction of pavement via paver machine at the mesoscale, a DEM model was developed in this chapter. The models of both materials and the paving machine were generated separately. Selecting the parameters of materials was determined based on the laboratory test results, while the setting of the paver's working operations was selected based on the real conditions at the field construction site. The chosen parameters in this model that can be adjusted for model validation are the paving speed, paving angle of the paver, and the paving thickness of the road surface.

The generated and presented data indicate much variability in vital parameters of the paver and their influence on pre-compaction. During paving compaction, a larger paving angle causes more horizontal components of paving power to be used for rotating the particles, and thus fewer components used for vertical compacting. This result shows that within a proper range of variation, a smaller paving angle has a better effect on vertical vibrating compaction.

The DEM numerical simulation of pavement pre-compaction in this research is based on simplified granular materials, and the validation of the DEM model is based on general field tests with a paver machine. However, the investigation on the paving compaction is closely related to the morphological distribution of granular materials, and also to the interlocking behavior among aggregates during pre-compaction. In addition, specific lab- and field tests are planned to be conducted for further study of the movement of bulk materials in pre-compaction. For the future study of pavement compaction, the real shape information will be accounted for in simulation, and advanced technology for monitoring the movement of particles will be used during the compaction test.

The simulation of asphalt roller compaction at macroscale was carried out with the help of an FEM model describing the asphalt mixture layer, coupled to a kinematic model of a steel roller drum, which in turn was equipped with an oscillatory excitation unit. Both models were coupled with the use of nonlinear contact law, taking into consideration dynamic friction processes within the contact area. For the description of the elastic-visco-plastic deformation behavior of the asphalt mixture at compaction temperature level, a material model, based on the bounding surface plasticity theory, was applied and calibrated with the use of a laboratory device, which in turn was designed specifically for testing the mechanical behavior of asphalt mixtures at compaction temperature level.

The information obtained during the laboratory tests was then used to set up the mechanical properties for an asphalt mixture layer model with a defined initial compaction and temperature distribution. This layer was subjected in different simulations to a roller pass, each being set to different roller operation modes and different rolling speeds. The effect of the different operation modes was then analyzed by the resulting increase of compaction and total void distribution within the layer after one roller pass. The results show that dynamic compaction can contribute to the compaction increase, but at the same time will cause strong local variations in the void content distribution within the layer.

The laboratory tests in combination with the theoretical considerations in the literature study have shown that the used material model fulfills the requirements to describe the mechanical behavior of asphalt mixes at compaction temperature level. For a more thorough study on the compaction progress during different roller passes at different compaction modes and their impact on the final result, the initial state of the layer with respect to the initial void and temperature distribution due to the pre-compaction process needs to be considered. For future studies, the effect of several roller passes should be investigated in order to observe the evolution of the void distribution. Furthermore, the initial material parameters within the model should be set with regards to the void and temperature distribution after pre-compaction, which could be derived from the DEM simulations.

A laboratory compaction device, which is called the Aachen compactor, has been developed to attain a higher correspondence between samples compacted in the laboratory and field compaction. Different compaction methods including the field compaction, Aachen compaction, and Marshall compaction were considered in this research. The indirect tensile test was simulated with 2D FEM models to investigate the influence of compaction methods with respect to the mechanical response and fracture behavior. Almost all values derived from the Aachen specimens are closer to those from field cores as compared to the Marshall specimens. The different performances of the asphalt specimens manufactured by the various compaction methods may be a result of higher compaction energies in the Marshall compaction. The uniform impact loading of the Marshall hammer and the absence of a kneading effect are most likely the major reason behind the poor correlation to the field cores. Both the compaction energy and kneading effect have an influence on the aggregate orientation and therefore on the related performance.

In future research, more experimental testing should be carried out with regard to different testing types and different testing temperatures to further prove the advantages of the Aachen compactor. More comprehensive analyses using DIP techniques should be performed to consider the characteristics of the air voids. The 3D numerical models should be reconstructed based on the DIP technique and thus deliver more detailed insights into the mechanical responses and deformation properties of the asphalt specimens than the current 2D models.

References

1. Abbas, A., Masad, E., Papagiannakis, T., Harman, T.: Micromechanical modeling of the viscoelastic behavior of asphalt mixtures using the discrete-element method. Int. J. Geomech. **7**, 131–139 (2007)
2. Ahmed, T.M., Green, P.L., Khalid, H.A.: Predicting fatigue performance of hot mix asphalt using artificial neural networks. Road Mater. Pavement Des. **18**, 141–154 (2017)
3. Angst, C.: Der Einfluß der Verdichtung auf die mechanischen Eigenschaften bituminöser Schichten. Mitteilungen ETH Zürich (1981)
4. Arand, W.: Zur Wirksamkeit von Hochverdichtungsbohlen. Straße und Autobahn **41**, 14–22 (1990)
5. Arnold, P.: Aachener Verdichter–praxisadäquate Laborverdichtung von Walzasphaltprobekörpern. Mitteilungen des Lehrstuhls und Instituts für Strassenwesen, Erd- und Tunnelbau, RWTH Aachen (2009)
6. Axel, F., Ronald, S.: VÖGELE – Operators Guide. Joseph Vögele AG, Ludwigshafen (2016)
7. Axel, F., Ronald, S.: VÖGELE Einbaufibel. Joseph Vögele AG, Ludwigshafen (2012)
8. BOMAG GmbH: Grundlagen der Asphaltverdichtung. BOMAG GmbH, Boppard (2009)
9. Bardet, J.P., Huang, Q.: Numerical modeling of micropolar effects in idealized granular materials. Mechanics of Granular Materials and Powder Systems **37**, 85–92 (1992)

10. Barenblatt, G.I.: The mathematical theory of equilibrium cracks in brittle fracture. Adv. Appl. Mech. **7**, 55–129 (1962)
11. Bijleveld, F.: Professionalising the asphalt construction process: aligning information technologies, operators' knowledge and laboratory practices. Ph.D. thesis, University of Twente (2015)
12. Commuri, S., Zaman, M.: A novel neural network-based asphalt compaction analyzer. Int. J. Pavement Eng. **9**, 177–188 (2008)
13. Cundall, P.A., Strack, O.D.L.: Discrete numerical model for granular assemblies. Geotechnique **29**, 47–65 (1979)
14. Cundall, P.A., Strack, O.D.L.: Modeling of microscopic mechanisms in granular material. Stud. Appl. Mech. **7**, 137–149 (1983)
15. Deutsches Institut für Normung: DIN 1996-4: 11/1984, Prüfung von Asphalt - Herstellung von Marshall-Probekörpern. Berlin (1984)
16. Deutsches Institut für Normung: DIN EN 12697-30: 2004, 09/2004, Asphalt - Prüfverfahren für Heißasphalt – Teil 30: Probenvorbereitung, Marshall-Verdichtungsgerät. Berlin (2004)
17. Deutsches Institut für Normung: DIN EN 12697-44: Bituminous mixtures—test method for hot mix asphalt-part 44: crack propagation by semi-circular bending test. Berlin (2014)
18. Dugdale, D.S.: Yielding of steel sheets containing slits. J. Mech. Phys. Solids **8**, 100–104 (1960)
19. Favier, J.F., Abbaspour-Fard, M.H., Kremmer, M.: Modeling nonspherical particles using multisphere discrete elements. J. Eng. Mech. **127**, 971–977 (2002)
20. Favier, J.F., Abbaspour-Fard, M.H., Kremmer, M., Raji, A.O.: Shape representation of axisymmetrical, non-spherical particles in discrete element simulation using multi-element model particles. Eng. Comput. **16**, 467–480 (1999)
21. Figge, H.: Verdichtungs-und belastungsverhalten bituminöser gemische. Ph.D. thesis, RWTH Aachen (1987)
22. Fischer, A., Schug, R.: VÖGELE Booklet on Paving. Joseph Vögele AG, Ludwigshafen (2017)
23. Floss, R., Reuther, A.: Vergleichsuntersuchungen ueber die Wirkung von Vibrierend und Oszillierend Arbeitender Verdichtungswalze. Tiefbau, Ingenieurbau, Strassenbau **33**, 31–34 (1991)
24. Habte, M.A.: Numerical and constitutive modelling of monotonic and cyclic loading in variably saturated soils. Ph.D. thesis, University of New South Wales (2006)
25. Hamm, AG.: Verdichtung im Asphalt- und Erbau. Hamm, A.G. Steyrermühl (2008)
26. Hu, J., Liu, P., Steinauer, B.: A study on fatigue damage of asphalt mixture under different compaction using 3D-microstructural characteristics. Front. Struct. Civ. Eng. **11**, 329–337 (2017)
27. Huschek, S.: Zum Verformungsverhalten von Asphaltbeton unter Druck, Ph.D. thesis, ETH Zürich (1983)
28. Kan, M.E., Taiebat, H.A., Khalili, N.: Simplified mapping rule for bounding surface simulation of complex loading paths in granular materials. Int. J. Geomech. **14**, 239–253 (2014)
29. Kappel, M.: Angewandter Strassenbau: Strassenfertiger im Einsatz. Springer, Berlin (2016)
30. Khalili, N., Habte, M.A., Valliappan, S.: A bounding surface plasticity model for cyclic loading of granular soils. Int. J. Numer. Meth. Eng. **63**, 1939–1960 (2005)
31. Kim, H., Buttlar, W.G.: Discrete fracture modeling of asphalt concrete. Int. J. Solids Struct. **46**, 2593–2604 (2009)
32. Liu, P., Hu, J., Wang, H., Canon Falla, G., Wang, D., Oeser, M.: Influence of temperature on the mechanical response of asphalt mixtures using microstructural analysis and finite-element simulations. J. Mater. Civ. Eng. **30**, 4018327 (2018)
33. Liu, P., Hu, J., Wang, D., Oeser, M., Alber, S., Ressel, W., Falla, G.C.: Modelling and evaluation of aggregate morphology on asphalt compression behavior. Constr. Build. Mater. **133**, 196–208 (2017)
34. Liu, P., Xu, H., Wang, D., Wang, C., Schulze, C., Oeser, M.: Comparison of mechanical responses of asphalt mixtures manufactured by different compaction methods. Constr. Build. Mater. **162**, 765–780 (2018)

35. Liu, Y., You, Z.: Discrete-element modeling: impacts of aggregate sphericity, orientation, and angularity on creep stiffness of idealized asphalt mixtures. J. Eng. Mech. **137**, 294–303 (2011)
36. Liu, Y., Zhou, X., You, Z., Ma, B., Gong, F.: Determining aggregate grain size using discrete-element models of sieve analysis. Int. J. Geomech. **19**, 04019014 (2019)
37. Liu, P., Lu, G., Yang, X., Jin, C.: Influence of different fillers on mechanical properties of porous asphalt mixtures using microstructural finite-element analysis. J. Transp. Eng., Part B: Pavements **147**, 4021004 (2021)
38. Lu, G., Wang, C., Liu, P., Pyrek, S., Oeser, M., Leischner, S.: Comparison of mechanical responses of asphalt mixtures under uniform and non-uniform loads using microscale finite element simulation. Materials **12**, 3058 (2019)
39. McLeod, N.W.: Influence of viscosity of asphalt-cements on compaction of paving mixtures in the field and discussion. Highway Res. Rec. **158**, 76–115 (1967)
40. Mgangira, M.B., Anochie-Boateng, J., Komba, J.J.: Quantification of aggregate grain shape characteristics using 3-D laser scanning technology. In: Southern African Transport Conference (SATC 2013), Pretoria (2013)
41. Milster, R., Emperhoff, W., Graf, K.: Ratschläge für den Einbau von Walzasphalt. Asphalt-Leitfaden **152**, 51 (2004)
42. Mo, L.T.: Damage development in the adhesive zone and mortar of porous asphalt concrete. Ph.D. thesis, Road and Railway Engineering, TU Delft (2010)
43. Nijboer, L.W.: Plasticity as a Factor in the Design of Dense Bituminous Road Carpets. Elsevier, Amsterdam (1948)
44. Olsson, E., Jelagin, D., Partl, M.N.: New discrete element framework for modelling asphalt compaction. Road Mater. Pavement Des. **20**, 604–616 (2019)
45. Otto, F.: Dynamisches Interaktionsverhalten zwischen Oszillationsbandagen und Asphaltoberflächen. Ph.D. thesis, RWTH Aachen University (2020)
46. Potyondy, D.O., Cundall, P.A.: A bonded-particle model for rock. Int. J. Rock Mech. Min. Sci. **41**, 1329–1364 (2004)
47. Quey, R., Dawson, P.R., Barbe, F.: Large-scale 3D random polycrystals for the finite element method: Generation, meshing and remeshing. Comput. Methods Appl. Mech. Eng. **200**, 1729–1745 (2011)
48. Rice, J.R.: Mathematical analysis in the mechanics of fracture. Fracture: An Advanced Treatise, vol. 2, pp. 191–311 (1968)
49. Römer, A.: Oszillationswalzen - Effektiv verdichten ohne Vibrationsbelastung. Tiefbau/BauPortal **7**, 17 (2000)
50. Römer, A.: Verdichtungssysteme für die Asphaltverdichtung. Tiefbau/BauPortal **126**, 23–26 (2003)
51. Schofield, A., Wroth, P.: Critical State Soil Mechanics. McGraw-Hill, New York (1968)
52. Shahbodagh, B., Habte, M.A., Khoshghalb, A., Khalili, N.: A bounding surface elasto-viscoplastic constitutive model for non-isothermal cyclic analysis of asphaltic materials. Int. J. Numer. Anal. Meth. Geomech. **41**, 721–739 (2017)
53. Solutions, D.E.M.: EDEM 2.6 theory reference guide. Edinburgh, United Kingdom (2014)
54. Song, S.H., Wagoner, M.P., Paulino, G.H., Buttlar, W.G.: δ25 Crack opening displacement parameter in cohesive zone models: experiments and simulations in asphalt concrete. Fatigue Fract. Eng. Mater. Struct. **31**, 850–856 (2008)
55. Tappert, A.: Verdichtung Bituminoeser Schichten-Erfahrungen mit Hochverdichtungsbohlen. Teerbau-Veröffentlichungen 43–50 (1985)
56. Ter Huerne, H.L.: Compaction of asphalt road pavements: using finite elements and critical state theory. Ph.D. thesis, University of Twente (2004)
57. Velske, S., Mentlein, H., Eymann, P.: Strassenbau, Strassenbautechnik. Reguvis Fachmedien, Cologne (2009)
58. Wagnoner, M.P., Buttlar, W., Paulino, G.H.: Disk-shaped compact tension test for asphalt concrete fracture. Exp. Mech. **45**, 270–277 (2005)
59. Wan, Y., Jia, J.: Nonlinear dynamics of asphalt—screed interaction during compaction: application to improving paving density. Constr. Build. Mater. **202**, 363–373 (2019)

60. Wang, H., Bu, Y., Wang, D., Oeser, M.: Dreidimensionale Charakterisierung der Kornform und Scharfkantigkeit von Gesteinskörnungen mittels Röntgen-Computertomographie. Bautechnik **90**, 436–443 (2015)
61. Wang, D., Wang, H., Bu, Y., Schulze, C., Oeser, M.: Evaluation of aggregate resistance to wear with Micro-Deval test in combination with aggregate imaging techniques. Wear **338–339**, 288–296 (2015)
62. Wang, H., Wang, C., Bu, Y., You, Z., Yang, X., Oeser, M.: Correlate aggregate angularity characteristics to the skid resistance of asphalt pavement based on image analysis technology. Constr. Build. Mater. **242**, 1–10 (2020)
63. Wang, H., Wang, D., Liu, P., Hu, J., Schulze, C., Oeser, M.: Development of morphological properties of road surfacing aggregates during the polishing process. Int. J. Pavement Eng. **18**, 367–380 (2017)
64. Wang, C., Wang, H., Oeser, M., Mohd Hasan, M.R.: Investigation on the morphological and mineralogical properties of coarse aggregates under VSI crushing operation. Int. J. Pavement Eng. **21**, 1–14 (2020)
65. Wang, H., Wang, C., You, Z., Yang, X., Huang, Z.: Characterising the asphalt concrete fracture performance from X-ray CT Imaging and finite element modelling. Int. J. Pavement Eng. **19**, 307–318 (2018)
66. Wang, C., Moharekpour, M., Liu, Q., Zhang, Z., Liu, P., Oeser, M.: Investigation on asphalt-screed interaction during pre-compaction: Improving paving effect via numerical simulation. Constr. Build. Mater. **289**, 123164 (2021)
67. Wendebaum, J.: Nutzung der Kerntemperaturvorhersage zur Verdichtung von Asphaltmischgut im Straßenbau. Ph.D. thesis, Universität Fridericiana zu Karlsruhe (2004)
68. Xu, Q., Chang, G.K., Gallivan, V.L., Horan, R.D.: Influences of intelligent compaction uniformity on pavement performances of hot mix asphalt. Constr. Build. Mater. **30**, 746–752 (2012)
69. You, Z., Adhikari, S., Dai, Q.: Three-dimensional discrete element models for asphalt mixtures. J. Eng. Mech. **134**, 1053–1063 (2008)

Computational Methods for Analyses of Different Functional Properties of Pavements

Tim Teutsch, Barbara Schuck, Tobias Götz, Stefan Alber, and Wolfram Ressel

Abstract Different computational methods dealing with functional properties of road surfaces are presented. Drainage and skid resistance as functional properties are treated. A special focus is set on the relationship between functional properties and asphalt structures, examples of important connecting aspects are described with regard to drainage of porous pavements and relevant void structures. Furthermore, in that context, analyses of the inner structure of asphalt are performed with XRCT scanning methods in order to develop a better understanding of asphalt structures and their implications on functional properties. In addition, the deformation behavior in before/after comparisons of XRCT images after uniaxial load tests are investigated.

Keywords Drainage · Asphalt structure · Skid resistance · XRCT technology

1 Introduction

A highly occupied road network needs high performance pavements. They have to ensure bearing capacities for high traffic loads—repeated millions of times—without severe deterioration during their lifetime. Durability is therefore of great importance in pavement engineering and, in fact, powerful engineering tools such as analyzing techniques, computational methods and related models are needed to be able to design durable pavement structures.

The Research Unit FOR 2089 "Durable Pavement Constructions for Future Traffic Loads: Coupled System Pavement-Tyre-Vehicle" funded by the German Research Foundation (DFG), deals with these challenges in pavement engineering in different sub-projects. Within the sub-project presented in this section, the research focuses on

Funded by the German Research Foundation (DFG) under grant RE 1620/4.

T. Teutsch · B. Schuck · T. Götz · S. Alber (✉) · W. Ressel
Chair for Road Design and Construction, Institute for Road and Transport Science, University of Stuttgart, Stuttgart, Germany
e-mail: stefan.alber@isv.uni-stuttgart.de

M. Kaliske et al. (eds.), *Coupled System Pavement—Tire—Vehicle*,
Lecture Notes in Applied and Computational Mechanics 96,
https://doi.org/10.1007/978-3-030-75486-0_3

the functional properties of pavements. In addition to the mechanical and structural requirements, which relate to a sufficient durable load bearing capacity without early deterioration effects, the functional properties mentioned above are important for road users (e.g. drainage and skid resistance) and partly also as environmental aspects (e.g. drainage and noise reduction). Functional properties arise from the pavement surface or, in a broader sense, the wearing course layer and need to be guaranteed permanently.

Selected computational methods for the analysis of different functional properties of the pavement surface (drainage and skid resistance) are presented in this contribution.

The inner structure of asphalt can be analyzed using computational methods. Of particular interest are asphalt aggregates and void structures (see Sect. 2), both of which have an influence on various functional properties. The description of the inner structure is based on X-ray computed tomography (XRCT) scans and their analysis. In this context, deformation processes are also studied with regard to the movement of aggregates in the asphalt after loading.

Regarding the functional property of drainage (see Sect. 3), different models for dense and porous road surfaces are compared in terms of simulation decisions (such as mathematical approach, problem dimensions, consideration of capillary effects, numerical techniques). Several approaches are shown and evaluated based on model capabilities for possible applications. Parameters of drainage models are named and distinguished for porous and dense road surfaces. Suitable methods for the determination of leading parameters are briefly discussed, e.g. derived from the inner structure analyses of porous asphalt (see also Sect. 2). The related own research on drainage modeling is presented and placed in the context of the discussion of the state of the art regarding pavement drainage modeling.

In Sect. 4, the development of a (wet) skid resistance modeling approach is shown; different aspects and influencing parameters are discussed. The presented approach depicts a single aspect of skid resistance (hysteresis friction depending on microtexture effects) and helps to understand the overall phenomenon of friction between road and tire in more detail.

Finally, the conclusion in Sect. 5 gives an overview of XRCT analyses possibilities in terms of functional properties and summarizes the individual topics of this contribution.

2 Analyses of Asphalt Structures Using X-Ray Scans and Implications on Functional Properties

The following section describes a method, based on X-ray scans, to analyze the three-dimensional structure of asphalt mixtures. On the one hand, it provides additional insights on the aggregates, such as positioning and orientation, as well as the movement during rutting. The information can be used for modeling this process. On the other hand, it is used to gather insights about the air voids, leading to a better understanding of the drainage process, in open graded pavements.

2.1 Three Dimensional Analysis of Asphalt Drill Cores

Asphalt basically has three components that have to be analyzed for a better understanding of the internal structure. The components are aggregates, mastic (bitumen and filler) and air voids. This study focuses on the examination of aggregates and air voids. The following describes which materials are examined and which experiments were carried out during the research.

Materials In this study two physical mechanisms are examined, for which two principally different materials are used. On the one hand, to obtain information on the structural changes in the aggregates under loading, a stone mastic asphalt (SMA 11 S) is examined. The material used is specified in [39]. The aggregate size distribution of the material is also shown in Fig. 9 marked as "mix design".

On the other hand, the air voids of porous asphalt will be analyzed for drainage, wherefore the porous asphalt (PA11), which is described in [5], is used.

Experimental tests In order to reproduce realistic loads on the asphalt structure, a dynamic load test is typically performed. A creep test was chosen, since the asphalt deformations are most relevant for the calibration and validation of the described method used to analyze the rutting process. The unconfined creep test is easily implemented and high plastic deformations can be applied under controlled conditions.

Asphalt specimen A test track was constructed using the stone mastic asphalt described above at the RWTH Aachen by subproject 2, as mentioned in [38]. Then, drill cores were taken from the total pavement layer, which were subsequently reduced to the size of around 75 mm in diameter and 75 mm in height, as a requirement for the load tests. The surfaces on top and bottom of the drill cores were afterwards sanded parallel to each other in order to ensure even load transfer during the test. This reduces the height by about 3–4 mm of each specimen. Three of these reduced drill cores are further analyzed in this study (SMA 1, SMA 2 and SMA 3).

Test procedure In order to give the asphalt time to deform its structure, preventing the specimen from failing too soon and causing large cracks to occur, the stress needs to be applied incrementally. For this purpose there are two different methods to regulate the load. One is the force based control, where the distance is regulated to reach and hold a defined load for a set time. In this case, the force is increased to 5 kN in steps of 0.5 kN and each followed by a holding time of 600 s.

The other one is displacement controlled, where the load is regulated in order to reach and hold a defined distance in a set time. It is increased in 0.1 mm steps and a following holding time of 300 s to a total distance of 3.7 mm. In both cases, this leads to a compression of the specimen by around 5%, to ensure that the results are comparable.

Test results As mentioned before, the test is carried out in two ways. For the force controlled test, the specimens are compressed by around 5% due to the maximal loading. Based on these results, the boundary conditions are set for displacement controlled tests to reach the approximately same compression, for better comparability. The stress-strain diagrams of these tests are shown in Figs. 1b and c. Due to non-parallel top and bottom surfaces, the vertical compression of the specimen "SMA 1" is higher than those of "SMA 2".

X-ray scans The "modular and open micro X-ray Computed Tomography (μXRCT) system", described in [67], is used to acquire the volume images of the asphalt specimen. In order to analyze the deformation process, the drill cores are scanned before and after the deformation by the XRCT.

2.2 Aggregate Analysis

3D processing of X-ray scans The flow chart shown in Fig. 2 shows the entire process of the image processing procedure described in this section.

Pre-processing In order to segment the aggregates in the three-dimensional images (volume image) captured by the XRCT, they must be prepared and processed first. The steps of image pre-processing are shown on an exemplary image section in Figs. 3 and 4. Figure 3a shows the unprocessed image, where only the contrast is enhanced for better visibility. As the drill core is cylindrical and the volume images are cuboid, a mask is first generated to define the region of interest (ROI). All following steps are executed within this mask.

For technical reasons, all X-ray scans have a radial gray value gradient from inside to outside as mentioned for instance by [40], that means the image edges are usually darker or brighter than the center. This gradient is compensated in order to significantly reduce the problems that may occur on the edges during segmentation. A detailed description of the gradient adjustment process is presented in [69]. Figure 3 shows a quarter of the same image, once without (see Fig. 3a) and once with image adjustment (see Fig. 3b). It can be seen, that the edge of the drill core area is darker without than with adjustment, which later has a great influence on the segmentation. The contrast is enhanced on both images for better visibility.

The next step is to apply a gray value transformation to adjust the contrast of the gradient adjusted image (see Fig. 4a). Then a median filter is used to reduce the image noise while preserving the edges in the image (see Fig. 4b). Then, another contrast improvement is applied with an adaptive histogram equalization, followed by a smoothing of the entire image (see Fig. 4c). Since the material contains aggregates, which are represented by different brightness due to their material composition, a morphological filtering is finally performed to further reduce the aggregate texture. Figure 4d shows the pre-processing results.

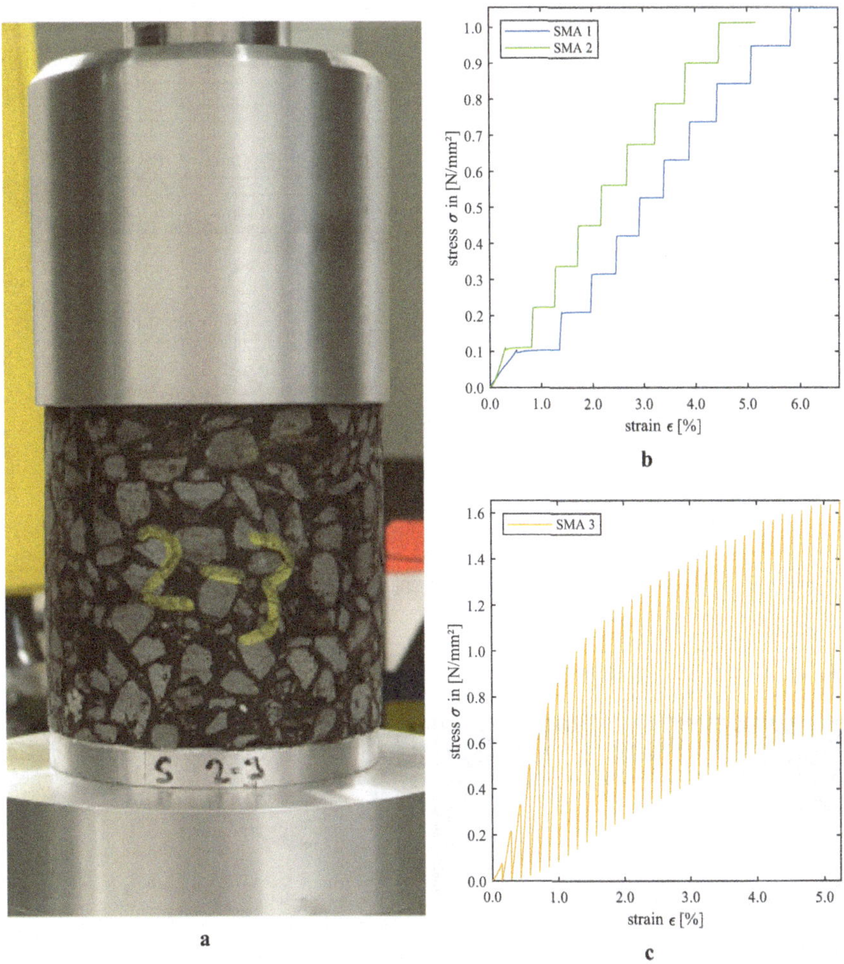

Fig. 1 **a** Asphalt specimen mounted inside the testing machine, **b** stress-strain diagram of the force controlled test, **c** stress-strain diagram of the displacement controlled test

Segmentation As described above, some of the aggregates vary significantly in brightness, so they are segmented by a combination of two methods. This ensures that the aggregates are segmented as accurately as possible.

First, the aggregates are segmented with the adaptive threshold method developed by Bradley [12] in order to reach a high segmentation of the aggregate edges. The result image of the Bradley segmentation is shown in Fig. 5a. However, large aggregates contain holes, which are difficult to fill, so the following method is carried out on the adjusted image (Fig. 4d), to be merged into one image during the post-processing.

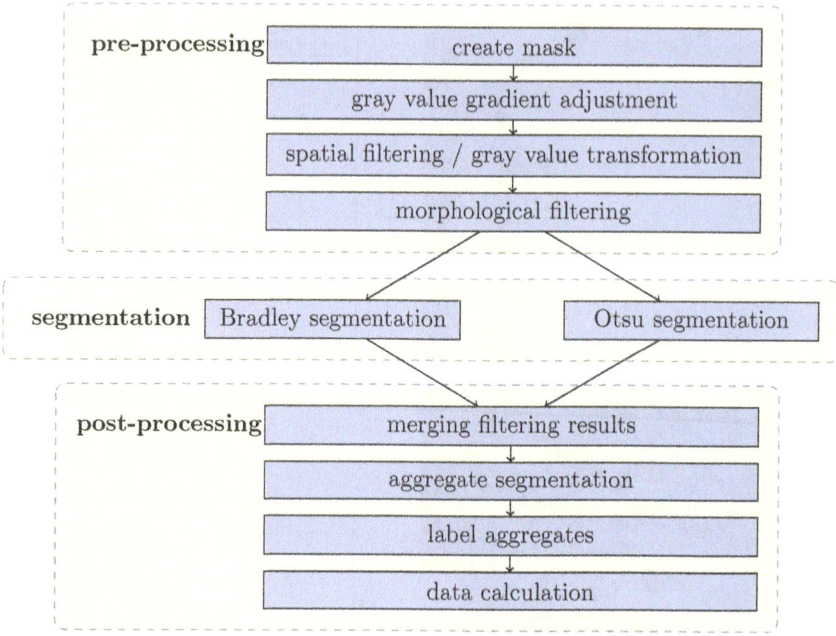

Fig. 2 Image processing sequence

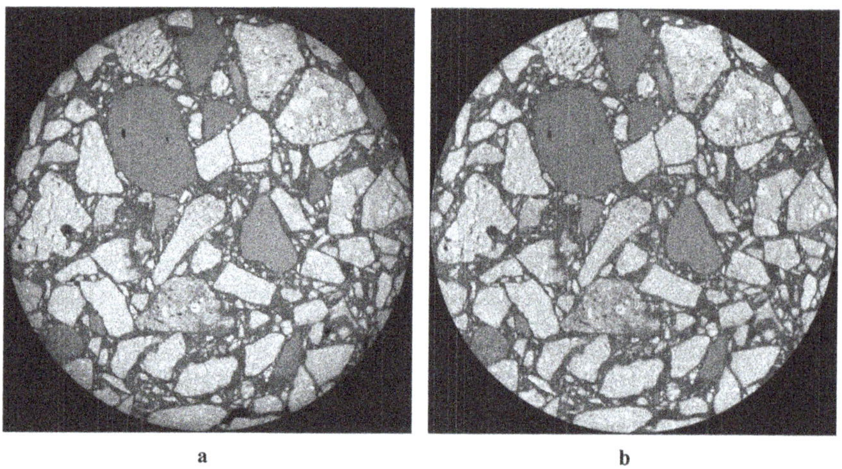

Fig. 3 Radial gray value gradient adjustment: **a** original image without gradient adjustment, **b** original image with gradient adjustment

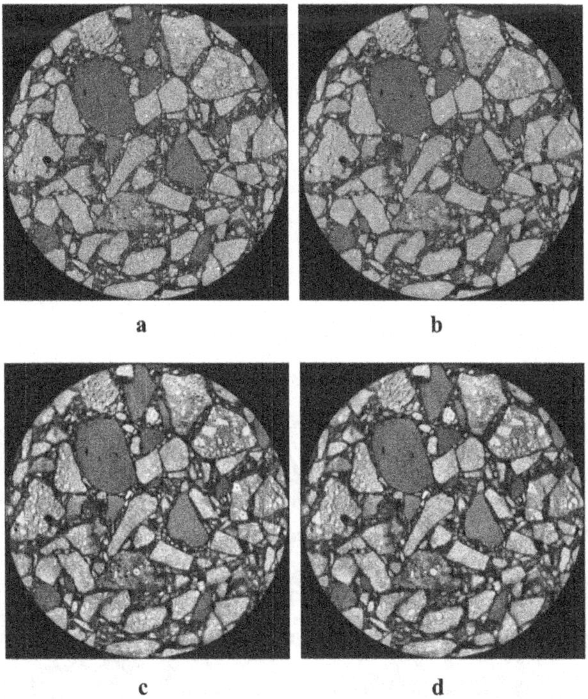

Fig. 4 Image filtering process: **a** first contrast adjustment, **b** noise reduction, **c** second contrast adjustment, **d** aggregate texture reduction

Second, the aggregates are segmented using the threshold method developed by Otsu [56], to segment the inner area of the aggregate area. Since this method calculates a global threshold for the entire image, which leads to poor results due to the brightness variations of the aggregates, a watershed transformation [32, 33] is applied before, to pre-segmented images. Then, the Otsu method is carried out on each segment. Figure 5b shows the result of the Otsu segmentation.

Post-processing Afterwards, the resulting images of the two segmentation methods are merged and the aggregates edges are smoothed (see Fig. 6a). However, this may cause connections between some individual aggregates. With a final watershed transformation [32, 33], the aggregates are distinctly separated from each other. As Fig. 6b shows, in some cases this can lead to over-segmentation, which means that some of the larger aggregates are fragmented, so they are manually identified and reassembled semi-automatic.

The segmentation outputs are binary volume images containing ones representing the aggregates and zeros defining the mastics and air. Now, objects of connected voxels are detected and numbered to generate a three-dimensional labeled matrix. In this step, the aggregates smaller than 2 mm are removed for the mentioned reasons. Figure 6c shows the final result of the segmentation process, where every single

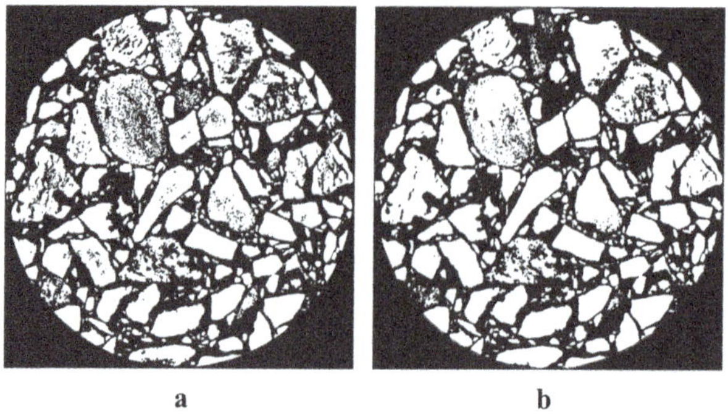

Fig. 5 Image segmentation process: **a** Bradley segmentation, **b** Otsu segmentation

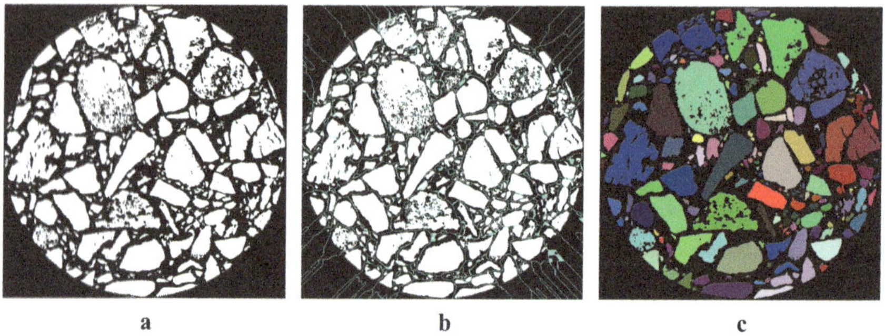

Fig. 6 Image post processing: **a** merged segmentation results, **b** aggregate segmentation, **c** labeled aggregates

aggregate is represented by one color. For later analysis, the aggregates are stored in tables containing their properties. For each aggregate, a volume image is stored, with a position inside the drill core defined by a bounding box and the aggregate volume is calculated. The result binary volume image of the segmentation process is shown in Fig. 7a.

Furthermore, a simplified mathematical definition of the aggregate as ellipsoid is calculated based on the volume image and its position inside the drill core. An ellipsoid is defined by a center with three position coordinates, the size with three radii which are orthogonal to each other and the rotation with three angles around the three coordination axes (see Fig. 8). Their calculation method is described by Legland in [50]. Figure 7b shows all ellipsoids representing the aggregates of one drillcore.

3D analyses of aggregates This section describes the different possibilities of aggregate analyses by examining real asphalt structures provided by this method. As described above, aggregates, which are smaller than 2 millimeters, can be considered as mastic and are therefore not included. The aggregates on drill core edges are also excluded from the analysis because they are incomplete and would therefore falsify the results.

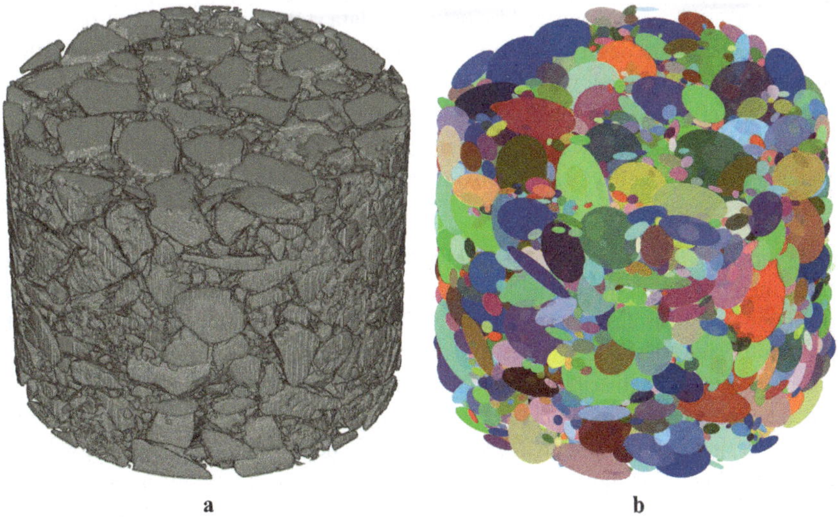

Fig. 7 Segmentation results: **a** segmented volume image, **b** volume displayed as ellipsoids

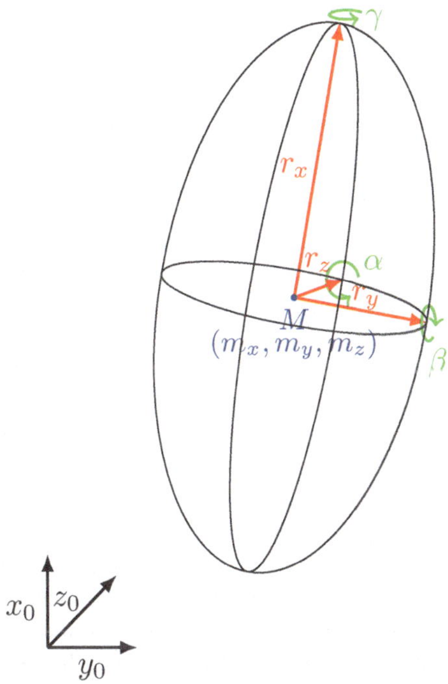

Fig. 8 Definition of an ellipsoid to characterize an aggregate

The aggregates are characterized by using the data stored in the tables of properties during the previous described segmentation process. Below the aspects and their parameters are described, which are examined in this study. The parameters, which describe the ellipsoid, are shown in Fig. 8. The center is defined by the three position coordinates m_x, m_y, m_z, the spatial expansion by the radii r_x, r_y, r_z and the orientation by the angles α, β and γ. This is also the order in which the parameters are stored in the tables of properties.

Aggregate size distribution One aspect to evaluate quality and reliability of the introduced method is to compare the aggregate size distributions resulting from the data and sieving analysis described in Sect. 2.1. The aggregates or representing ellipsoids are digitally sieved and sorted into the aggregate classes, which are defined to determine the typical size distribution of stone mastic asphalt (SMA 11 S), as described in [26]. The aggregate size is defined by the second largest radius (r_y) of the ellipsoid, because in sieving analyses, an aggregate does not need to fit through a sieve with the maximum elongation, but with its width. This is used to classify the aggregates into size classes of 2, 2.8, 4, 5, 5.6, 8, 11.2 and 16 mm. Then, the classes percentage of the specimen's total volume is determined for each size class, from which afterwards the aggregate size distribution is accumulated. The size distributions of the specimen are shown in Fig. 9 in comparison to the control test size distribution of the used material.

As shown in Fig. 9, there are some variances from the values given in some size classes of the size distribution. This is primarily caused by the sample size. The fact, that a drill core contains considerably less aggregates than a sample typically used for sieving analysis, inhomogeneities in the distribution of aggregates inside the paved asphalt lead to these variances.

Aggregate positioning and orientation One of the method's greatest advantages is the ability to analyze the aggregates positioning. Due to the simplification as ellipsoids, they are accurately mathematically defined. As described above, the ellipsoid center is specified by three coordinates.

The aggregate size is defined by the radii of its corresponding ellipsoids, where the longest radius r_x represents the length (L), r_y the width (B) and r_z the thickness (E). Besides its usage to determine the size distribution, they are used to characterize the aggregate shape. By calculating the ratio between length and thickness (L/E), the shape is characterized. So if the shape ratio of an aggregate equals 1, it is more or less spherical and the higher the ratio gets, it becomes more elongated. According to [20], it is considered as unfavorably or non-cubic shaped, when the ratio is higher than three (L/E $>$ 3). The German guidelines for aggregate characterization [27] specifies an "aggregate shape index", which limits the maximum amount of such non-cubic aggregates. Figure 10 shows a distribution of the aggregate shapes within a drill core. The vertical line at a L/E-ratio of three indicates the limit above which the aggregates are considered non-cubic.

The aggregates orientation inside the drill core is described by the three rotation angles of its corresponding ellipsoid as shown in Fig. 8. The final orientation is defined in the following order of the Euler angles. α is specified as counterclockwise

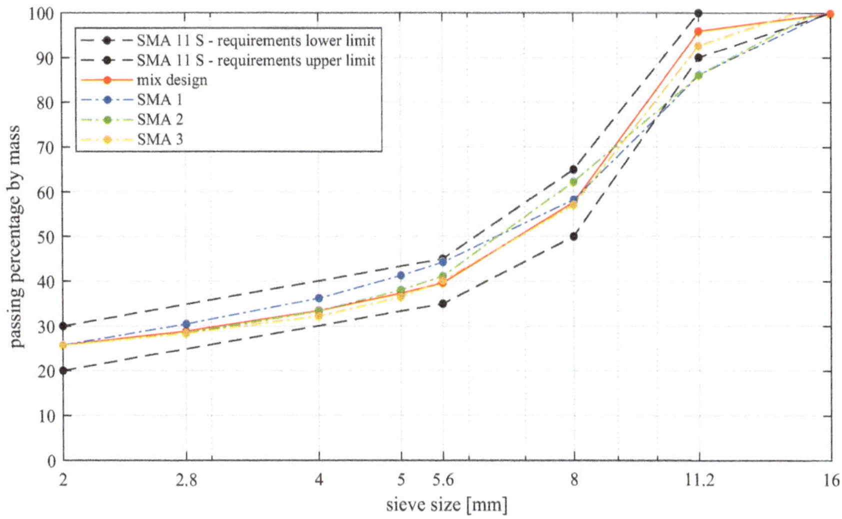

Fig. 9 Aggregate size distributions of the digital analyzed specimen compared to the mix design

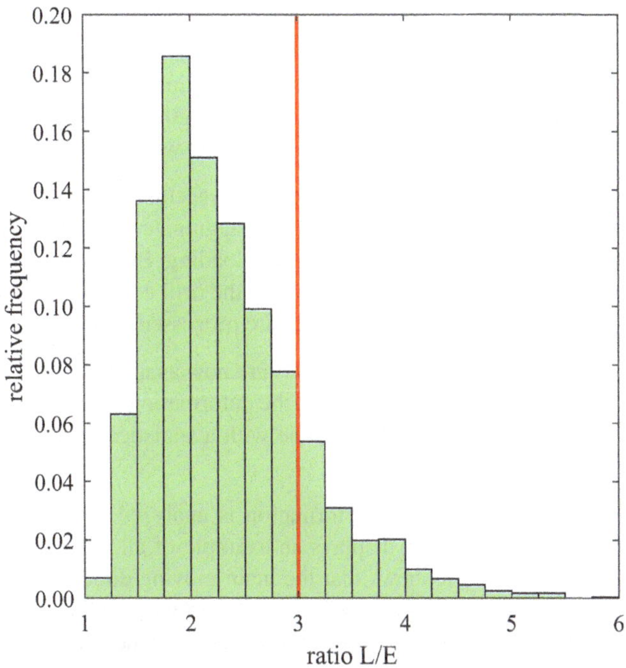

Fig. 10 Distribution of aggregate shapes

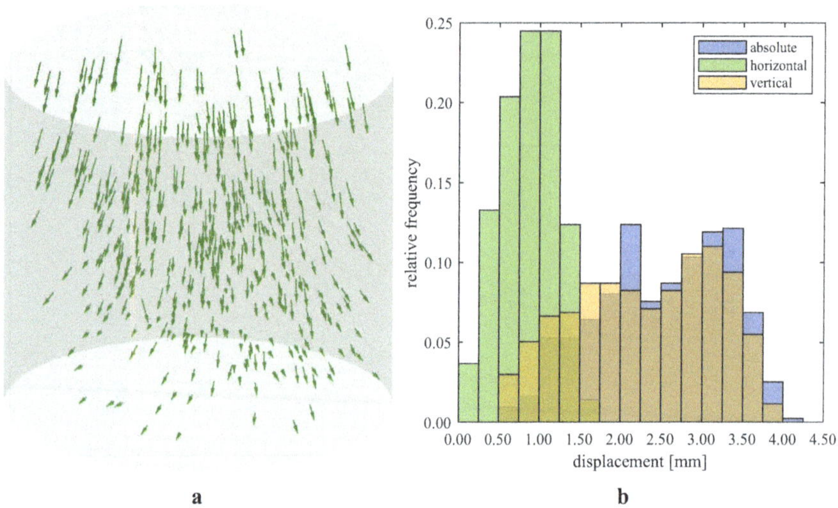

Fig. 11 **a** Visualization of the aggregate movement, **b** relative frequencies of displacements

rotation around the z-axis and, therefore, the horizontal direction of the aggregate length. This is followed by the clockwise rotation around the y-axis by the angle β, which sets the vertical alignment of the aggregate length. Finally, a rotation around the x-axis by the angle γ specifies the direction of the width and thickness.

Analyses of deformation processes Since the analysis method can be used to determine the position and orientation of individual aggregates in drill cores, it is also able to analyze their transformation as a result of loading. Before and after the deformation by performing the unconfined creep test, the drill cores are scanned with the XRCT [67] and the resulting volume images are processed, as described above.

Aggregate assignment Data on the aggregates are now available for the original and the deformed condition. In order to analyze the deformation process, the aggregates must be assigned to each other. This is done with a statistical analysis of the size, position and orientation of the ellipsoids.

Aggregate movement analysis The deformation is analyzed by characterizing the aggregate movement. Figure 11a displays an example of all aggregate movements tracked within a drill core. It shows that the main moving direction is the same as the load input direction, but the aggregates are also moving radially away from the center. The diagram in Fig. 11b shows the displacements percentage in horizontal and vertical direction, as well as the absolute displacement.

2.3 Air Void Analysis

XRCT images and their analysis can also be used to assess the functional properties influenced by air voids. The size, distribution and shape of the air voids are of special interest here and will be discussed briefly in this contribution. Furthermore, these approaches can be used in more materialistic analyses such as determining the contact points between aggregates to describe the structural properties of the road (see e.g. [45]). The bitumen itself could also be of interest, as the bitumen film is in contact with water for a long time during the whole of the evaporation process.

The above described method can be used to recognize and analyze air voids. This is done both for dense road surfaces, such as hot-mix asphalt, as well as for porous ones. The geometry and distribution of air voids for pavement specimens can be done relatively simply with air void analyses. Connected air voids (also called effective air voids) can be distinguished from isolated or dead-end pores. Especially the estimation of a material's permeability or hydraulic conductivity is the most common use case of such air void analyses.

Influence of air voids on drainage Permeability and hydraulic conductivity are equivalent to each other and differ only in the way they take fluid properties, like the viscosity, into account. From here on, the term permeability will be used.

Permeability is important as it is a measurement of the possible flow velocities through a porous medium. It can therefore describe the drainage capacity of a porous pavement and its change through clogging. Thus, the permeability (both modeled homogeneously and heterogeneously) is an important input parameter for pavement drainage modeling of porous pavements (see Sect. 3).

Studies show that the permeability in porous pavement structures in the horizontal and vertical direction are different. The horizontal permeability was found in experiments and models to be 2 [60], 10 [54] or even 50 [48] times higher than the vertical one. This is a result of the aggregate orientation and the anisotropic distribution of air voids, with bigger and fewer on top and bottom and smaller, but more in the middle of the porous layer.

The permeability of a pavement specimen can either be measured in an experiment [53], in-situ [46], can be derived from XRCT analyses [45] or from hydrodynamic modeling (see [48] and Sect. 3).

In addition to this, a closer look at pore size distributions and possible constrictions to fluid flow could help to understand and assess the drainage process in porous pavements. Figure 12 shows separated pores in an XRCT image analysis of a porous asphalt specimen. A unique color is assigned to every pore. The pores are separated by their location as well as at points where small constrictions appear. This approach is especially useful for recognizing zones with a high clogging risk. Additionally, these constrictions also represent zones with high flow velocities. These high flow velocities lead to high pressure gradients and shear stresses at this point and increase the damaging potential of the water [48].

A more comprehensive look at air void and pore throat analysis is published in [69].

Fig. 12 Separated pores in an XRCT image of a porous asphalt specimen [69]

3 Drainage Modeling of Dense and Porous Pavement Surfaces

Understanding the drainage processes of pavements is important for a number of reasons. The drainage process (with a focus on outflow and water film depths) itself is important for designing outlets, layer thicknesses and texture depths, for example. For porous pavements, the water infiltration is an important functional property, but still damages can occur (e.g. stripping of bitumen [48]). In addition to these models that describe only the drainage process as such, other system processes of the road can also be modeled. Evaporation, is one such property and is especially relevant for porous pavements, as this process takes in the order of days [41, 49] until the porous structure is dry again. Furthermore, clogging processes in porous structures reduce permeability and, thus, the infiltration capacity and are therefore an important use case for drainage modeling.

This section will present a thorough literature review of pavement drainage modeling. Thereafter, a model for dense drainage modeling and its extension towards porous modeling will be presented and compared to the existing models. Model outcomes and conclusions will then be presented that give a deeper understanding of asphalt material and its performance. This was a goal of this Research Unit FOR 2089 funded by the German Research Foundation (DFG).

3.1 Modeling Decisions

In the following, the basics of drainage modeling both for dense as well as for porous road surfaces will be explained. A review of existing models will be presented and the most important modeling decisions will be summarized, compared and are depicted in Table 1 for dense drainage modeling and in Table 2 for the porous case. Considered modeling decisions are the solution type (analytical, empirical or numerical), the problem dimension (1D-3D), the underlying mathematical equations and their physical meaning, the flow regime (laminar, transient or turbulent) and the temporal state (steady-state or time-dependent modeling). Specifically for porous models, the underlying characteristics are further the coupling of the surface and subsurface flows and the handling of unsaturated flow conditions.

Model types Basically, existing models can be distinguished in empirical and hydrodynamic models. Empirical models try to link flow behavior to experimentally observed relationships. Often, this is modeled as a 1D flow problem and power functions are used to describe the relationships.

Hydrodynamic models are based on the conservation of fluid properties and on momentum equations of the flow [80]. Although some hydrodynamic problems can be solved analytically, this has seldom been done in the field of pavement drainage modeling (see Charbeneau and Barrett [16] for an analytical solution of the porous pavement drainage problem). Most of the hydrodynamic equations, used in pavement surface modeling, require a numerical solution approach.

Table 1 Overview of surface drainage models

Name	Type	Dimension	Equation	Flow type	State
Gallaway [30]	Empirical	1D	Power function	Laminar	Steady
PAVDRN [11]	Analytical	1D	Kinematic SWE	Laminar	Steady
PSRM [80]	Numerical (FVM)	2D	Dynamic SWE	Laminar	Unsteady

Table 2 Overview of porous drainage modeling

Name	Type	Dimension	Equation surface	Equation subsurface	Flow type	State	Saturation
Charbeneau and Barrett [16]	Analytical	1D	Darcy-Weisbach	Darcy with Dupuit-Forchheimer	Laminar	Steady	Saturated
Pratico and Moro [59]	Analytical	1D	unspecified	Richards equation	Laminar	Steady	Saturated
Ranieri [62]	Empirical	2D	unspecified	Adapted Darcy	Laminar-turbulent	Steady	Saturated
Alawi [3]	Numerical (FEM)	1D	unspecified	Darcy's law	Laminar	Steady	Unsaturated
1D-HYDRUS [13]	Numerical (FEM)	1D	unspecified	Richards equation	Laminar	Unsteady	Unsaturated
2D-HYDRUS [75]	Numerical (FEM)	2D	unspecified	Richards equation	Laminar	Unsteady	Unsaturated
Cortier [19]	Numerical (FEM)	2D	unspecified	Richards equation	Laminar	Unsteady	Unsaturated
Tan [73]	Numerical (FEM)	3D	unspecified	unspecified	unspecified	Unsteady	Unsaturated
Hsieh and Chen [37]	unspecified	unspecified	Back-calculated from subsurface flow	Simplified Navier-Stokes equations	unspecified	unspecified	Saturated
Chen [17]	Numerical (FEM)	2D	Back-calculated from subsurface flow	Richards equation	Laminar	Unsteady	Unsaturated
Liu [52]	Numerical (FVM)	2D/3D	Diffusive SWE	Richards equation	Laminar	Unsteady	Unsaturated
PERFCODE [21]	Numerical (FVM)	2D	Diffusive SWE	Darcy with Dupuit-Forchheimer	Laminar	Unsteady	Saturated

Mathematical formulation The mathematical description of the drainage process is a big difference of the models. To describe the complex drainage process mathematically, simplifications (for example reduced problem dimension or simpler flow type) are necessary. These simplifications should be adapted to the envisaged model use case.

Problem dimension Besides the mathematical problem formulation and the model type (empirical, analytical or numerical), models can be distinguished by their problem dimension (1D, 2D or 3D). Most models use a 1D or 2D approach. The reduction of the problem to 1D can be used both for dense and porous pavements. The resulting slope combining longitudinal and cross slope is treated as the water flow path (both on dense surfaces and in the porous structure). Thus, this dimensional reduction is still quite close to the physical problem, but can fundamentally simplify the underlying mathematical setup.

Flow regime The flow regime, characterized by the Reynolds number, can have a great effect on the quality of the model results and on the allowable simplifications. Flow can be described either as laminar or turbulent. In laminar flow, the fluid flow takes place in parallel layers and without any changes perpendicular to the flow direction. Chaotic changes in the flow direction and the pressure mark turbulent flow. There is no hard jump between laminar and turbulent flow. Transient flow, as a third flow regime, is therefore often introduced. In transient flow, a mixture of laminar and turbulent flow conditions occurs. Assuming a laminar flow usually allows a simplified mathematical formulation, see for example the derivation of the Shallow Water Equations for the surface flow in [80] or the use of Darcy's [3, 16] or Richards equation [52, 59] in porous pavement modeling. However, a certain error in the model results will be the consequence of this simplification. According to Charbeneau et al. [15] and Herrmann [36], turbulence in surface flow will only occur in unusually strong rain events [80], which might allow for the laminar simplification for practical purposes. In pavement subsurface drainage, turbulent flows were calculated as well as measured [81]. Lu et al. [53] also observed non-linear flow behavior in permeameter tests. Ranieri et al. even registered mostly transient flow regimes [62]. In porous pavement modeling, a simplification to laminar flow conditions might therefore reduce the quality of the model results.

Stationarity Depending on the temporal consideration, the model complexity will show differences. The rain event in all models, so far, is assumed as spatially uniform. With the beginning of the rain event, the surface texture will start to get filled with water (in surface flow) or the water will start to infiltrate into the porous structure and will drain along the underlying dense layer and, therefore, the water table will rise inside of the porous layer. After some time has passed and wetting has fully taken place, runoff or outflow will take place. Usually, a certain balance between the incoming rain (when modeled as spatially uniform with a constant intensity) and the drainage state of the pavement (water film depth, outflow etc.) will be achieved. This state can be classified as the steady-state solution of pavement drainage. For road engineers and planers, the steady-state solution of the drainage process will be enough in most cases, as it allows to check the drainage capacity of an existing or planned road under consideration of a certain design rain event. However, a temporal resolution can be implemented to model variable rain intensities. In addition, a non-steady-state model can, for example, provide information on which pavement surface area is affected with critical drainage situations first or how fast this happens.

Saturation Another modeling decision is how unsaturated flow conditions in porous pavements are treated. A neglect of unsaturated flow would, again, cause an error in the model results, while, again, facilitating the mathematical formulation and its solution. The water saturation level of a porous structure has a huge influence especially on the infiltration process and the permeability of the porous structure, as it influences the capillary forces. This relationship can be described via the water retention curve. To find a relationship here is, however, relatively difficult [17]. If only saturated conditions are considered, the infiltration rate is assumed not to change with the water saturation in the structure and the water will always infiltrate in its given rate to the water table at the bottom of the porous layer. Eck [22] stated that unsaturated flows can be neglected as the permeability (and therefore the infiltration rate from the surface to the water table inside the porous layer) is much greater than the rain fall intensity. While this is true, at least for not overmuch clogged porous media, this neglects already wetted states. This could cause problems as drying can take up to days (again see [41, 49]) and, therefore, residual water contents prior to a rain event are overlooked.

Anisotropy of permeability In most models, so far, the porous medium is assumed as homogeneous. However as stated in Sect. 2, there is a strong difference in the permeability in the vertical and horizontal direction. This could lead to a rather recognizable error in modeling. So far, the heterogeneity, if included, is only modeled as a factor of horizontal versus vertical permeability (see [17] and [73] for an example). This is already a big step towards modeling the anisotropic permeability values, needed when clogging or cleaning processes of porous pavements should be included. It should also be noted again, that the permeability varies not only with the pore structure (material parameters and state of clogging), but also on the state of saturation, as has been explained above.

3.2 Basics for Modeling Drainage of Dense Pavements

In surface modeling, the investigated system boundaries differ significantly and different model types can therefore be distinguished. There is a number of models (the first of these models was by Gallaway [30]), that calculate water film depths and mostly use empirical relationships for this. Furthermore, there are skid resistance models that derive aquaplaning speeds (for example the *PAVDRN* model [11]). More complex skid resistance models take both the pavement surface and properties of the tire into account (see e.g. [18]) and can only be solved with up to date computational numerical solution approaches. The water film depth is in both cases treated rather simple and is an input into the model. Lastly, there also exists a splash and spray assessment tool [25]. Table 1 presents one pavement surface runoff model for each of the described types.

Runoff over a dense pavement surface corresponds to sheet flow in more intuitive hydraulic research areas such as groundwater flow and watershed modeling. The most common assumption in dense pavement modeling is that shallow water flow occurs. This can be assumed when the horizontal spread of the water film is much larger compared to the vertical water film thickness.

The Shallow Water Equations (SWE) are derived from the Navier-Stokes equations. They can be modeled at three complexity levels. A kinematic version, where only 1D flow can be modeled. The diffusive approach can model 2D flow with simplified flow properties. The full, dynamic derivation is more complex in its flow properties. However, it is computationally expensive. Nearly all pavement runoff models with numerical solution approaches solve for one of those Shallow Water Equations set-ups. This is also the case when surface runoff is coupled to a porous pavement model.

3.3 Modeling Permeability of Porous Pavements

Models of porous pavement drainage can best be distinguished by their model results. For this contribution, approaches that actually model flow in the porous medium coupled or non-coupled to surface flow will be the focus and will be discussed in Sect. 3.4.

There are other models (empirical, analytical or numerical) that focus on the flow behavior of porous pavement and not on the water table height or the water film depth on the surface. These models (such as [10]) calculate the flow velocities and pressure differences in the porous medium, often using XRCT images as basis. The flow characteristics are then used to calculate the permeability in terms of spatial and time variation.

XRCT analysis The advantage of using XRCT images as basis is, that image processing tools can also be used to calculate the permeability values. It is relatively simple to gain the parameters for the Kozeny-Carman equations (most common are porosity and average particle size as input, see e.g. [1] or [53]), the Ergun equation (mostly using porosity and diameter of the spheres that represent aggregate dimension, see [10]) and the Berryman-Blair permeability limit (with porosity and average radius of the (assumed) spherical particles [10]). Studies have introduced a variety of factors in these equations to refine the calculations, such as nominal maximum aggregate size [48] or the layer thickness [74]. All these empirical solution approaches are used to estimate the permeability on the basis of simple characteristics and empirical coefficients.

Computational fluid dynamics With an improvement, not only of XRCT and image analysis techniques, but also of computational modeling, solution approaches using computational fluid dynamic (CFD) models have been applied to porous pavement structures more and more.

Lattice-Boltzmann method The so called Lattice-Boltzmann (LB) method can be used to model flow velocities and permeability values in porous pavements. Space is here discretized as lattice nodes (each node represents aggregates or air void space) and inside each node velocity vectors of fluid flow define the velocity [77]. The velocity of fluid particles traveling along these vectors is determined by satisfying the Navier-Stokes equations [47]. The Lattice-Boltzmann method can be used to gain insight into hydraulic parameters of pavement at a lower computational cost and simulation difficulty instead of modeling the whole drainage process (including a coupled surface-subsurface flow problem).

Umiliaco and Benedetto [77] have used the LB method to calculate the velocity and the flow path inside of a virtually generated Marshall-type asphalt specimen. The results were compared to Darcy permeability measurements (so only laminar flow regimes were assumed). Because flow paths were identified, tortuosity could also be calculated. Here, the LB model could handle a 2D unsteady flow simulation.

SIMPLE scheme Like the Lattice-Boltzmann scheme, there exists a solution process in CFD, called SIMPLE (semi-implicit method for pressure-linked equations) that is used by Al-Omari and Masad [10]. Here, as in the LB models, the basis can be XRCT images, that are digitized and 3D porous flow through this structure is modeled. The model result is the permeability of the specimen.

3.4 Modeling Drainage in Porous Pavements

A list of the above mentioned permeability models could be extended significantly. The focus however, of this research work and this contribution will be on flow modeling in and over porous pavements. As the flow itself is the model result, the permeability will be simplified to a mere input data without complex modeling techniques as the above described LB models. Porous flow can be imaged as shown in Fig. 13. Water infiltrates across the surface into the porous structure. It builds up at the bottom of the underlying impermeable layer and forms a curved water table. The water then drains along the impermeable interface and flows out of the system. The area below the water table can be described as saturated flow while the area above shows unsaturated flow behavior.

The porous structure is therefore able to absorb the water. Outflow will be retarded, as first the void structure is wetted (see e.g. experiments by [68]). In addition, the water retention capacity of the porous pavement will retard peaks in outflow and often even prevent them [22]. A drawback of this ability is the long evaporation times of porous pavements (see Sect. 3.5).

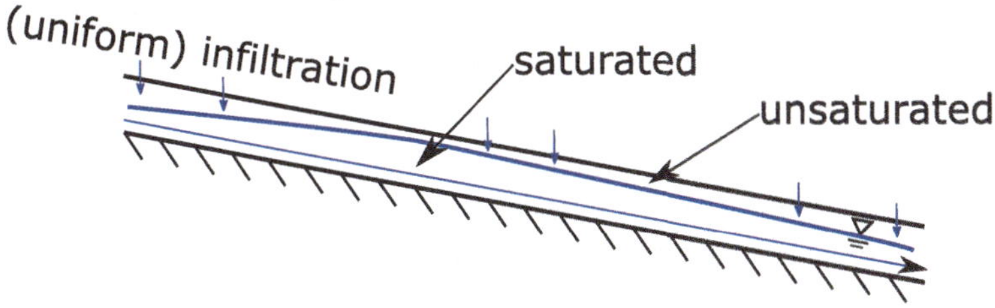

Fig. 13 Subsurface flow in porous pavement

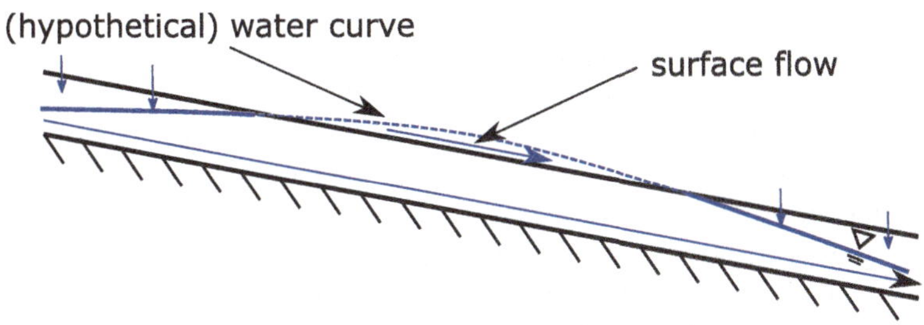

Fig. 14 Subsurface and surface flow in porous pavement

Especially porous pavement surfaces are usually able to drain incoming rain events completely. However, due to clogging or an unfavorable combination of road design parameters, the drainage capacity of the pavement could be exceeded. Surface pavement flow can occur in these cases and manifests itself in ponding. Here, the curved water table rises above the layer thickness (see Fig. 14) and surface runoff will start from this point. Depending on the circumstances, this ex-filtrated water can either infiltrate into the porous layer again or appear as runoff across the rest of the flow path to the road embankments or outlets. Only certain areas of the pavement surface are therefore affected and pose an aquaplaning risk. This could even be more dangerous than aquaplaning on dense pavements, as the drivers experience a mostly dry surface without splash and spray and do not reduce their velocity according to the rain event. With this excessive speed, a sudden appearance of ponding could be particularly dangerous.

If surface flow is included in porous pavement modeling, a coupling of the two types of flow (surface and subsurface) is needed as different equations have to be used for each flow type. This coupling ensures the dependencies and parameter exchanges.

Solutions for this drainage problem are typically based on a physical understanding rather than simpler empirical approaches.

In the following, analytical solution strategies of hydrodynamic equations will be presented first. Models that consider porous pavements as part of Low Impact Development (LID) techniques will then be described. These LID-models have their own direction, assumptions and possible use cases. Lastly, numerical solutions for this hydrodynamic problem will be shown. These very complex porous models (be it coupled or uncoupled to surface flow) are able to give a detailed flow description of the drainage process. A short overview of all the presented models can be found in Table 2.

Analytical modeling Charbeneau and Barrett [16] provided an analytical solution to the surface drainage problem. It is intended both for the pavement design process and for validation of more complex models. It is based on equations of the flow process in porous media. The problem was simplified to a 1D problem (as can be done for dense surfaces, where only the maximum flow path length is investigated). Steady-state and unsaturated conditions were assumed and the flow was solved with the Dupuit-Forchheimer assumptions to Darcy's equation. This is applicable to horizontal porous media flow in an unconfined aquifer (this can be assumed for lower slopes [16]). The outflow was then modeled as proportional to the thickness of the layer. Analytical solutions can be provided then. The solutions can distinguish between subsurface and subsurface/surface flow. Results were water depth in the porous layer and residence time depending on rainfall intensity, layer thickness, permeability, slope and drainage path length (depending on slopes and width).

Pratico and Moro [59] based their model on the Richards equation. This is also an extension of Darcy's equation and allows a consideration of unsaturated flows in a very simple way. Water discharge and permeability can be calculated with the model. The model was able to reproduce experimental permeability tests.

Ranieri et al. [62] developed an analytical solution for the design process of porous pavements. Starting from Darcy (so laminar) flow, they adapted the flow equations to the other flow regimes, as most flow conditions inside the porous medium were found to be in the transient regime [61]. This was done with the Lindquist-Kovács equations. Similar to the Kozeny-Carman equation (see Sect. 3.3), this allows to predict the permeability and resulting seepage velocity based on pore characteristics (such as porosity and the shape coefficient of the aggregates). Furthermore, this attempt takes the range of Reynolds number (representing the flow regime) into account and provides a discontinuous equation for four flow regimes. The required empirical values were derived from experiments. Additionally, in [63] the approach was used to also provide design charts and dependencies of the inner-related parameters permeability, slope (resulting slope from longitudinal and cross slope), space between sub-drains, pavement thickness and rain intensity.

Modeling of subsurface flow Schlüter and Jefferies [68] were one of the first to propose a porous modeling technique. They configured the Stormwater Software Package Erwin with modules that could at least in part represent the porous pavement outflow rates. They recognized how important the water content of the porous layer prior to the rain event is and, therefore, modeled cycles of rain events. The model was able to predict the outflow rates from a porous parking area surface and to validate the results against measurement data and to quantify the storage capacity of the porous layer of water.

Alawi [3] can integrate even two-phase flow into their 1D model, so transport of fine particles can be included. Thus change in permeability and porosity due to clogging can be modeled. The model domain consisted of different porous layers, such as on parking lots. An overflow and, thus, a coupling with a surface flow was not done.

Low Impact Development The HYDRUS software (see [70]) is widely used in literature for simulating water, heat and solute transport processes. It was also used for modeling flow behavior of porous pavements. HYDRUS uses Richards equation (so laminar, unsaturated flow modeling) for subsurface flow. This modeling technique has been called computationally expensive and unpredictable [75]. However, due to its commercial availability and comparably easy handling, it has been widely used in porous pavement modeling, especially in the field of Low Impact Development (LID). This is an approach in land and engineering design to reduce human impact into the environment especially in inner-city areas. Chosen methods are, for example, green roofs (see e.g. [14]) or parking lots that allow an infiltration of the rain water (by using for example interlocking concrete pavement) and, therefore, a reduction of the accruing surface runoff. This explains why porous structures that could only symbolize parking lots rather than highway structures (which would be the obvious use case for porous pavement modeling for road engineers) are included in these HYDRUS models.

Brunetti et al. [13] used a 1D-HYDRUS model to investigate an overlay of different porous materials, such as built for parking lots. This work was extended by Turco et al. [75] into a 2D-HYDRUS model for a multi-layer porous pavement model. Here, interlocking concrete pavement was modeled over a pervious sub-base. The outflow at the vertical end of the pavement and the water saturation in the middle of the system are the main model outputs. As in the 1D-HYDRUS model, a coupling with surface flow was not accomplished and the separate layers were assumed to be homogeneous with regard to their permeability and aggregate or air void characteristics.

Cortier et al. [19] used a commercial partial differential equation solver called FlexPDE for LID modeling. They also considered a multi-layer, porous construction, here even with a drainage pipe inside the base layer. Outflow rates, as well as, the infiltration at the layer boundaries can be calculated. Unsaturated flow conditions can also be handled here.

Complex modeling of subsurface flow The cited LID modeling techniques cannot be used for an in-depth flow consideration or to review the road design parameters of a planned or existing road.

Tan et al. [73] used a commercial software called SEEP3D. They assumed a constant infiltration rate and did not couple with a surface flow. Here, the permeability anistropy was modeled as a simple factor (horizontal to vertical permeability as a ratio of two). The model was applied to a couple of use cases and to generate a set of design curves, similar to the analytical approach of Ranieri et al. [61]. These curves can help to choose the interconnected values of longitudinal/cross slope, pavement width and porous layer thickness. The model data was validated roughly by experimental data. While the software is able to model 3D unsaturated flow, the publications did not show the used flow equations, flow conditions and problem dimenisonality. Therefore, it is hard to compare this model to the others.

The model proposed by Tan et al. [73] and the LID models show the continuous development of commercial software to solve computational fluid problems.

Modeling of coupled surface and subsurface flow Hsieh and Chen [37] presented the theory for a full model of the surface drainage mechanism of porous pavements. They proposed the simplified Navier-Stokes equations for the subsurface flow, with the surface flow inferred indirectly as a simple method to couple surface and subsurface flow. While the mathematical equations are displayed, the numerical scheme (type, grid, boundary conditions etc.) was not mentioned. A short case study was included with a variation of transverse slope, road width and rainfall intensity. While this looks like promising work, the results are not validated or compared to experimental data or to other models.

The model by Chen et al. [17], a 2D unsaturated modeling approach (closely related to the work by Gaia Ferreira et al. [29]), is another one to take the anisotrpy of permeability into account. Their subsurface flow assumed that permeability in the horizontal direction is constantly 20 % higher than the vertical one. However, as this study aimed to model the surface flow occurring on porous pavement surfaces, this anistropy was not assumed to have a huge influence [17]. The surface flow in [17] was assumed as an extension of the subsurface flow. The shape of the water curve inside the pavement was just extended over the free surface (see the dashed line in Fig. 14). This approach allows a simple but effective coupling between surface and subsurface flow. Chen et al. [17] focused especially on the conditions when surface flow occurs. The area affected by surface flow was found to be highly dependent on transverse slope, rainfall intensity and, of course, on porous layer thickness (which had, as expected, the highest influence on the drainage capacity of a porous pavement). The surface water spread could cover different traffic lanes or parts of traffic lanes during one rain event. Curvature of road and a changing slope profile cannot be regarded in this model.

A 2D (surface)/3D (subsurface) model was presented by Liu et al. [52]. For the surface, the diffusive approach of the Shallow Water Equations was used. This is a simplified flow behavior compared to [80]. They used the open source platform OpenFOAM to solve the partial differential equations. The model was applied on two benchmark tests and compared to the analytical solution proposed by Charbeneau and Barrett [16]. Furthermore, a case study with a realistic porous pavement thickness over a dense sub-layer with curvature, longitudinal and cross slopes on a two-lane road and drainage inlets was included.

Eck et al. [21] have developed the most complex coupled surface-subsurface model so far for modeling infiltration, porous drainage and subsurface runoff. The open-source model, the *permeable friction course code (PERFCODE)*, is applicable to a wide range of road orientations and other design parameters. Rain was modeled as spatially uniform, but can be varied over time. However, unsaturated flow conditions and heterogeneous porous properties were neglected.

3.5 Modeling Complex Drainage Processes

While the above described models aim at explaining the water movement in drainage problems and help in designing pavements with adequate drainage capacities, some models rather focus on other processes. They constrict (or rather simplify) the flow problem and, thus, can concentrate on the water flow process as a whole. Sun et al. [72] used a 3D FEM model (laminar, unsteady, unsaturated flow conditions without a coupled surface flow) to simulate the dynamic response of a wet porous asphalt and the corresponding pore water pressure under load. When comparing saturated and unsaturated states and their dynamic response, the importance of an adequate drainage is made clear [72]. Thus, even a saturated state not yet showing signs of surface flow is to be avoided and unsaturated states are more desirable for our road environment.

Other examples for drainage process modeling are evaporation models (see e.g. [2, 41, 49, 58]). Furthermore, clogging influences the change of drainage capacity of porous pavements. So far, no model has been able to incorporate clogging into flow modeling.

3.6 Development and Application of Drainage Modeling of Dense and Porous Pavement Surfaces

In a work prior to this research group, a computationally inexpensive model of dense pavement drainage was developed. The *PLANUS* model [36, 64] calculates slope lines that depend on longitudinal and cross slope as well as on the transition geometry of the road. The water flow is simulated along these slope lines. Flow paths are defined by parameters of the resulting slope and the flow path length along the slope lines. Thus, the model can calculate water film depths along the course of these lines, leading to a 2D distribution of water film depths. However, this methodology is based on 1D considerations. It is very reliable in many cases of standard geometries and is validated with experiments [64]. The methodology with slope lines, nevertheless, shows certain problems, e.g. at the end of slope lines, at pavement edges or in areas with slopes close to 0 %. For more complex geometries, a 2D model approach would therefore be preferable.

Model of dense pavement drainage Subsequently, a drainage model called *Pavement Surface Runoff Model (PSRM)* was developed [80]. As can be seen in Table 1, this 2D finite volume model uses the dynamic solution to the Shallow Water Equations for modeling dense pavement runoff.

The simulation is performed with the DuMu^x software toolbox [24]. It is a simulator for flow and transport processes in porous media.

The model assumes laminar flow conditions [80]. In comparison to many other models [11, 30, 36], unsteady state modeling is possible. The gradual spreading of water films over a pavement surface can, therefore, be displayed. The underlying

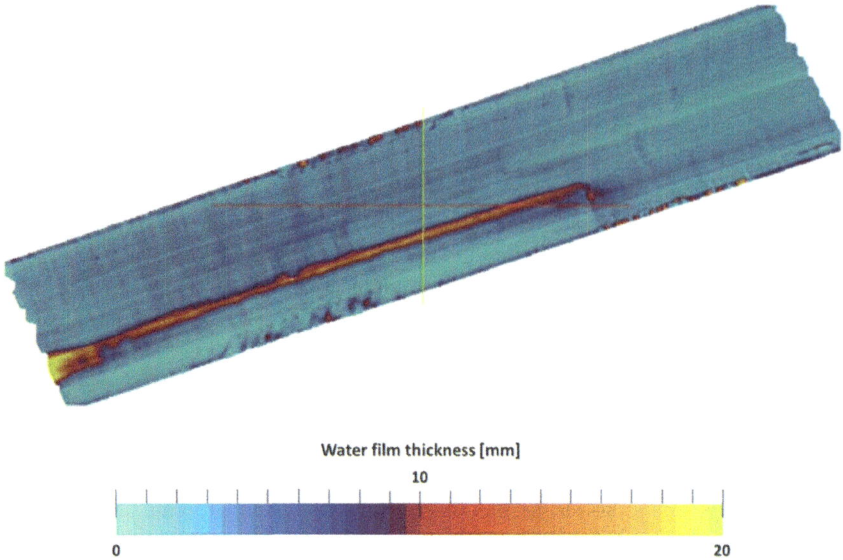

Fig. 15 *PSRM* simulation of water film thickness on a pavement surface with a deep rut [65]

numerical grid allows a simulation of nearly arbitrary road geometries. In addition to slopes, curves and transition zones, even individual or small irregularities such as ruts or road markings, can be included in the analysis (see e.g. [9]).

Figure 15 shows an exemplary simulation result from *PSRM*. Here, the water film thickness over a dense pavement surface with one marked rut is visualized with a very fine grid (see also [65]). Further model results of the *PSRM* in connection with this research group are presented in the chapter "Simulation Chain: From the Material Behavior to the Thermo-mechanical Long-term Response of Asphalt Pavements and the Alteration of Functional Properties (Surface Drainage)" and are published in [9].

Rain events can be modeled with chosen duration and intensity. However, as with for example the *PERFCODE* model [21], the rain is modeled as homogeneously in time and distribution.

The model is validated with the same texture input data as in [36]. In pavement drainage, the influence of pavement surface roughness is contradictory. It enables contact between road and tire even when the roughness valleys start to fill with water. An increased roughness could thus lead to a higher drainage capacity. Furthermore, roughness acts as resistance to flow. It retards the water movement and could thus lead to a reduced drainage capacity. This relationship is explained in more detail for example in [4, 65].

The roughness of the modeled pavement surface is represented by the mean texture depth and several depth values that represent realistic pavement surfaces are implemented in the *PSRM*. One of the these is a porous pavement surface. This approach will be discussed in the following.

Uncoupled model of porous pavement drainage As described above, the *PSRM* [80] can be extended to model porous pavement drainage albeit in a simplistic way. Each grid cell is assigned a drainage capacity with a constant infiltration rate. This approach allows to model pavement drainage over the porous pavement without the need to couple a separate model for the subsurface. It is therefore similar to the model by Chen et al. [17] presented in Sect. 3.4. Simulations with this simplified porous *PSRM* model show the important contribution of porous pavements to reduce aquaplaning conditions even in critical road sections. This was done, for example, in a systematic way for different infiltration rates and road geometries (simulating high and low drainage capabilities of the porous asphalt pavement) in [51].

However, with this approach, the complexities of fluid flow inside the porous pavement layer cannot be modeled.

Firstly, as has been described in Sect. 3.4, the high permeability and slope of the layer lead to an uneven rise of the water table inside of the porous pavement. Surface flow or ponding on the surface of the pavement will therefore only take place in certain areas under certain conditions. As has been argued in Eck [22], the subsurface flow controls the surface flow conditions, not the other way round as in most precipitation-infiltration processes. This coupling cannot be taken into account with the described approach.

Secondly, other factors such as anisotropic permeability and turbulent flow conditions (see Sect. 3.1) or complex processes such as clogging or evaporation (see Sect. 3.5) have to be neglected.

Thirdly, an evaluation of the drainage capacity, retention capabilities and even contaminant reduction in stormwater cannot be made. This is especially important in the context of Low Impact Development (see Sect. 3.4).

The modeling of porous pavement drainage can be improved by using a finite-volume method or pore-network model. Both approaches can couple surface with subsurface flow and will be described shortly.

Another approach to simply assess the drainage capabilities will also be presented first.

Describing model of porous pavement drainage To explain drainage processes and capabilities of porous pavements empirically, a linear reservoir model [8] was applied to data from runoff experiments over soiled porous asphalt specimens.

The aim is to infer time-dependent runoff, retention and discharge rates by defining a so-called storage constant. The infiltration into the porous structure is imagined as a reservoir that gets filled with water.

In the model representation, an overflow of the reservoir would signify pavement surface runoff in reality. In this way, the retention can be mathematically described by the reservoir properties. With simple mass and storage equations, the discharge rate and runoff can be calculated. To model more complex runoff behavior, imagined reservoirs can be connected in series.

Calculations show that higher rainfall intensities and coarser mix design diminish the retention capacity (illustrated by a lower reservoir storage constant) as outflow rates rise [8]. A prior wetting of the porous structure can also be modeled here, as this could be modeled as lower storage constants leading to the expected lowered retention capacity.

Coupled model of porous pavement drainage The DuMu^x software toolbox [24] also allows simulating a coupled finite-volume surface-subsurface pavement drainage model (see [66]). The input parameters have first to be gained, see e.g. [69] for modeling or calculating permeability values. The numerical grid and resolution of the rain event can be implemented based on the approaches in the *PSRM*, as a finer resolution or more simulated details seem not to be relevant for pavement engineering purposes. 2D modeling of the subsurface flow with Darcy's equation allows taking the porous structure into account in detail. Even clogging processes could then be applied (see [7] for the influence of soiling on the functional properties of pavements). This allows to model temporal changes, however, assumes only laminar flow conditions with saturated states. The mathematical formulation for this model can be found in [65].

Outlook on further modeling This contribution tries to present a deep and comprehensive literature analysis of pavement drainage modeling. It can be seen, that there is still no model with a broad application to porous pavement design. In this research group, the basis for such a model were developed (see [65] and [66]) and will be extended into a full numerical model of porous pavement drainage. An approach using a pore-network model, as presented in [79], could improve porous drainage modeling even more. Here, the advantages of models with a basis on XRCT images (realistic pore structures, easy modeling of permeability anisotropy and clogging processes) can be combined with the fast computational numerical solution approaches (see also Sect. 2 and [69]).

4 Contribution to Skid Resistance Modeling Under Wet Conditions Based on Micro-texture Data

4.1 Introductive General Remarks About Skid Resistance Modeling

The phenomenon of skid resistance depends on a complex interaction between tire and road surface. It is widely accepted that both the micro-texture and the macro-texture of a road surface influence friction.

With respect to road surfaces, macro-texture typically describes roughness wavelengths between 0.5 and 50 mm originating from the coarse aggregates at the surface, while micro-texture refers to wavelengths below 0.5 mm and represents the roughness on individual coarse aggregates (e.g. described in [44]).

The two texture scales contribute to the overall skid resistance to different extents, depending on e.g. speed and/or slip conditions (see e.g. [55]). Moreover, in the case

Fig. 16 Macro-texture (valleys) filled with water [43]

of wet friction as the more critical case, the macro-texture influences the drainage behavior of water from the top surface of the aggregates, where the contact between tire and road mostly takes place, into macro-texture valleys (see Fig. 16). This has an important effect on skid resistance because the more water remains on top of the aggregates, the less effective is micro-texture friction. The drainage effect of the macro-texture—in principle also quantitatively calculated with the PSRM model (see Sect. 3)—and its implication for remaining friction under wet conditions was presented exemplarily in [43]. The influence of residual water in the micro-texture (on top of the coarse aggregates) is another important impact on friction in this context and will be discussed in more detail in Sect. 4.4 with respect to a possible modeling approach. Friction effects between tire rubber and the pavement surface are based on hysteresis and adhesion effects (e.g. described in [44]). In addition to the road surface properties, the behavior of the tire—as the second part of the friction process—is based on viscoelastic rubber properties as well as on tire tread patterns and can be described with different rubber models.

In recent years, quite a number of different models have been developed and published, dealing with different approaches to rubber friction, texture description and calculation methods for wet and dry conditions, e.g. [23, 42, 71, 76].

With regard to the many different influencing parameters and effects, which in many cases are not even independent of each other, a partial examination of different friction effects can help to understand the complex phenomena of skid resistance more comprehensively.

Therefore, an analytical hysteresis friction model was developed in [78] that only accounts for micro-texture effects using a basic rubber model (see Sect. 4.2) under dry conditions. The neglect of macro-texture and adhesion effects in this approach are intended to separate different effects contributing to skid resistance as a whole. This approach is described in Sect. 4.2. The model has been enhanced with a more profound rubber model (see Sect. 4.3). Moreover, wet friction approaches have been integrated in the model (see Sect. 4.4).

4.2 Hysteresis Friction Model for the Micro-texture

In [78], a friction model was developed that describes the influence of the micro-texture (without considering macro-texture).

Therefore, it is necessary to separate the different scales/wavelengths of the texture in a filtering process. According to the focus on the micro-texture, only small areas of an asphalt surface (about 10 mm to 10 mm) were measured using a fringe projection method with a very high solution. Thus, approximately only one single coarse aggregate is considered in each (micro-texture) measurement sample. Using a Gauss filtering method with measured texture profiles, the micro-texture is separated from the macro-texture and outliers are eliminated [78].

This measured and pre-processed 3D texture data can be reduced to one-dimensional texture profiles with approximately comparable contact properties and, thus, friction behavior. This reduction method, which generates simplified equivalent profiles (considering certain boundary conditions) was developed by Popov [31, 57]. The implementation within the presented model is described in [35, 78].

Depending on this (micro-texture) profile, hysteresis forces are calculated with respect to a simplified rubber block representing the tire and its basic viscoelastic properties. In fact, considerations of the tire pattern are neglected, which might be acceptable for the micro-texture scale.

The hysteresis forces f_x are calculated in [78] from the deformation process of the rubber activated by the (micro-)texture, as shown in Fig. 17. They are influenced by the normal force f_i, where the resulting penetration in z-direction depends on the texture profile contact points.

A simulation of a moving rubber block in contact with the reduced one-dimensional texture profile is performed in several time discretization steps. The contact points are determined and the horizontal forces are summed up for each discretization step (in time) in order to calculate the friction coefficient—depending only on micro-texture and hysteresis friction for the presented model. An exemplary texture profile as well as the simulation of the moving tire rubber penetrating into the micro-texture are shown in Fig. 18.

Adhesion effects are neglected, only dry surface conditions are considered, and a simple viscoelastic model (Kelvin-Voigt model) is used to describe the rheological behavior of the tire rubber.

In fact, the description of a single effect (hysteresis, depending only on micro-texture contact) of friction [78] separately from the various influencing parameters of skid resistance is a strength of the presented model [78]. It can help to understand the combination of effects that determine the whole phenomenon in more detail. Weise [78] also provided a general approach for generating virtual artificial stochastic textures with defined parameters (e.g. based on the power spectrum), which was later enhanced and used for 3D prints of artificial (micro-)textures [28]. This combination is quite a helpful tool to generate realistic but defined textures, since it is not possible to create (real) asphalt samples consisting only of micro-texture elements. Real samples always represent a more or less undefined mixture of micro- and macro-texture.

4.3 Enhancement of the Rubber Model

As described above, the model developed by [78] assumes a simple Kelvin-Voigt model consisting of a spring and a dashpot in parallel to describe the viscoelastic rubber properties of the tire. In [35], a comparison of different skid resistance models

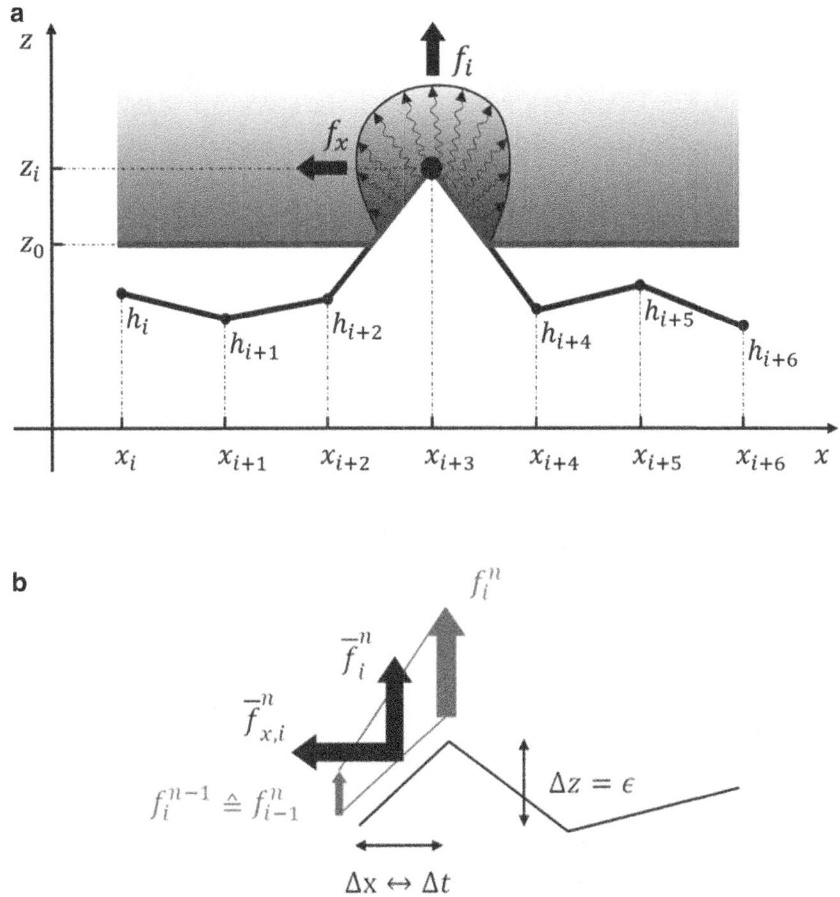

Fig. 17 **a** Hysteresis interaction between rubber and texture [35, 78], **b** calculation of hysteresis friction force f_x [28]

is described. The model with two parameters (shear modulus G for the spring and viscosity η for the dashpot) has been enhanced by Götz using a Zener model approach, where the dashpot is replaced by a Maxwell element (spring and dashpot in series). Thus, the enhanced model used for the calculations in [35] is a model with three parameters. The parameters were derived from model comparison with a multiscale model in which the material is described based on a generalized Maxwell model by fitting the material response of both models as well as possible for different speeds.

Afterwards, the model was developed further and a generalized Maxwell approach was also implemented by Götz in [28] in the hysteresis micro-scale model from Sect. 4.2. Thus, a more realistic material description with Prony series (using 15 Maxwell elements in this case) has been realized. Using the enhancement of the

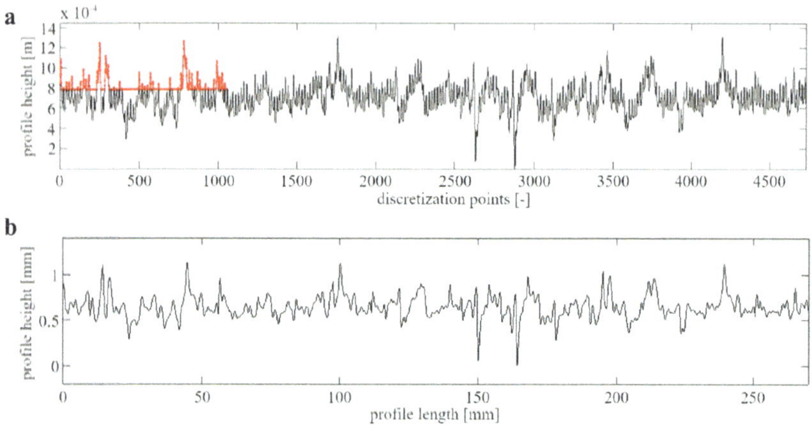

Fig. 18 **a** Exemplary reduced one-dimensional texture profile, **b** simulation of contact points of a moving rubber block in the model, after [28]

model in [28], a basic adhesion formulation was added, which is considered at each contact point in the discretization.

4.4 Wet Friction Approaches

A further development of the (enhanced) model approach described in Sects. 4.2 and 4.3 addresses the integration of wet friction approaches, which has not yet been considered in the earlier stages of the model. In [34], an algorithm that gradually fills the (micro-)texture with water to simulate different wetness states and degrees of filling of the texture valleys (see Fig. 19) is being developed. The simulation of a moving rubber block changes in a way that the penetration depth is influenced and only some texture elements can contribute to hysteresis (and adhesion) friction [28]. An adhesion formulation has been added to the model for that purpose (compare Sect. 4.3). The water thus generates a new effective micro-texture, as the water surface in the texture valleys reduces the depth of penetration and thus the hysteresis effects. Since the filling mechanism is performed with the reduced equivalent one-dimensional texture profile, the real filling degree of the 3D-texture and the corresponding water volume have to be calculated backwards. This approach was also applied in [28] when comparing real dry and wet friction measurements on 3D-prints of defined artificial micro-textures generated virtually, as described in principle in Sect. 4.2. In a backwards calculation, the amount of water in the texture is determined by the friction loss (by comparing wet to dry state) in the measurement.

Thus, this step of model enhancement gives an idea of how water in the micro-texture can decrease hysteresis and adhesion friction. Nevertheless, it must be stated that it can neither describe viscous effects of water on wet friction nor water displacement by the rolling tire.

In addition to the model enhancement steps already described, and in comparison with different model approaches and measurements, this is another important contribution to a more comprehensive understanding of friction as a whole by looking at a single effect in more detail.

5 Conclusions and Outlook

3D imaging methods using XRCT technique (see Sect. 2) open a wide field in asphalt technology and the analyses of functional properties of pavements.

In addition to material characterization and description of material behavior under different load states, properties of the inner structure related to functional properties such as drainage (see Sect. 3) can also be determined in a suitable way. Pore structures, pore connections, pore sizes, clogging effects and derived parameters such as permeability are just a few important aspects of drainage in porous pavement structures that can be reliably described using three dimensional analyzing methods such as XRCT scanning.

Besides drainage, sound absorption is another functional property of porous pavements that leads to noise reduction. Sound absorption depends significantly on different pore structure parameters, which is why this can be another use case for XRCT analyses as 3D imaging technique (e.g. [6, 7]).

Noise reduction is also influenced by the (macro-)texture characteristics of the road surface—especially for (common) dense surfaces. XRCT scanning also offers the possibility to determine these related surface macro-texture parameters, also relevant for skid resistance. With very high resolution XRCT methods [67], even the analyses of micro-texture on the aggregates surface—with implications for skid resistance (compare Sect. 4)—might be possible as well.

In fact, XRCT scanning methods can help to understand material behavior and functional properties in a more fundamental sense. This may, after further research in the field, improve our understanding and perhaps eventually offer new design methods and requirements in asphalt technology.

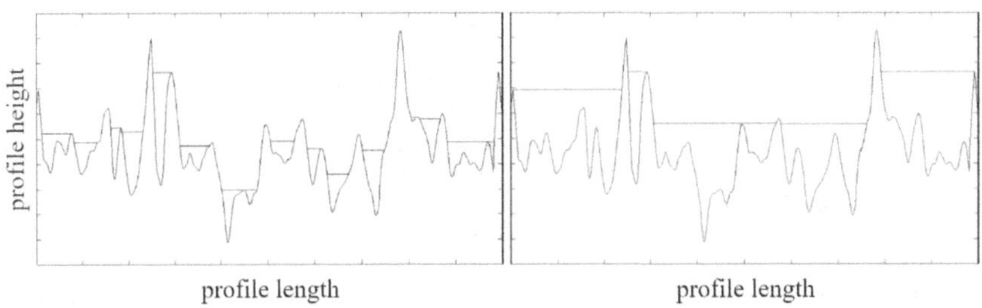

Fig. 19 Exemplary filling of micro-texture with different volumes of water, after [28]

References

1. Aboufoul, M., Garcia, A.: Factors affecting hydraulic conductivity of asphalt mixture. Mater. Struct. **50**, 1–16 (2017)
2. Aboufoul, M., Shokri, N., Saleh, E., Tuck, C., Garcia, A.: Dynamics of water evaporation from porous asphalt. Constr. Build. Mater. **202**, 406–414 (2019)
3. Alawi, M., El-Qadi, M., El-Ameen, M.: Modeling and simulation of flow and formation damage of asphalt-paved roads. Abstr. Appl. Anal. **2013**, 384640 (2013)
4. Alber, S., Schuck, B., Ressel, W.: Importance of pavement drainage and different approaches of modelling, in: Chen, X.; Yang, J; Oeser, M.; Wang, H.: Functional Pavements, Proceedings of the 6th Chinese–European Workshop on Functional Pavement Design (CEW 2020), Nanjing, China, 18–21 Oct 2020, CRC Press, 403–406 (2020)
5. Alber, S.: Veränderung des Schallabsorptionsverhaltens von offenporigen Asphalten durch Verschmutzung. Ph.D. thesis, University of Stuttgart (2013)
6. Alber, S., Ressel, W., Liu, P., Hu, J., Wang, D., Oeser, M., Uribe, D., Steeb, H.: Investigation of microstructure characteristics of porous asphalt with relevance to acoustic pavement performance. Int. J. Transp. Sci. Technol. **7**, 199–207 (2018)
7. Alber, S., Ressel, W., Liu, P., Wang, D., Oeser, M.: Influence of soiling phenomena on air-void microstructure and acoustic performance of porous asphalt pavement. Constr. Build. Mater. **158**, 938–948 (2018)
8. Alber, S., Ressel, W., Schuck, B.: Explaining drainage of porous asphalt with hydrological modelling. Int. J. Pavement Eng. **2020**, 1–11 (2020)
9. Alber, S., Schuck, B., Ressel, W., Behnke, R., Canon Falla, G., Kaliske, M., Leischner, S., Wellner, F.: Modeling of surface drainage during the service life of asphalt pavements showing long-term rutting: a modular hydromechanical approach. Adv. Mater. Sci. Eng. **2020**, 879362 (2020)
10. Al-Omari, A., Masad, E.: Three dimensional simulation of fluid flow in X-ray CT images of porous media. Int. J. Numer. Anal. Methods Geomech. **28**, 1327–1360 (2004)
11. Anderson, D.A., Huebner, R., Reed, J.R., Warner, J., Henry, J.J.: Improved surface drainage of pavements. The Pennsylvania State University, Technical Report (1998)
12. Bradley, D., Roth, G.: Adaptive thresholding using the integral image. J. Graph. Tools **12**, 13–21 (2007)
13. Brunetti, G., Simunek, J., Piro, P.: A comprehensive numerical analysis of the hydraulic behavior of a permeable pavement. J. Hydrol. **540**, 1146–1161 (2016)
14. Brunetti, G., Simunek, J., Piro, P.: A comprehensive analysis of the variably saturated hydraulic behavior of a green roof in a mediterranean climate. Vadose Zone J. **15**, 1–17 (2016)
15. Charbeneau, R.J., Jeong, J., Barrett, M.E.: Highway drainage at superelevation transitions. Center for Transportation Research, The University of Texas at Austin, Technical Report (2008)
16. Charbeneau, R.J., Barrett, M.E.: Drainage hydraulics of permeable friction courses. Water Resour. Res. **44**, W04417 (2008)
17. Chen, X., Wang, H., Li, C., Zhang, W., Xu, G.: Computational investigation on surface water distribution and permeability of porous asphalt pavement. Int. J. Pavement Eng. **2020**, 1–13 (2020)
18. Chu, L., Fwa, T.: Pavement skid resistance consideration in rain-related wet-weather speed limits determination. Road Mater. Pavement Des. **19**, 334–352 (2018)
19. Cortier, O., Boutouil, M., Maquaire, O.: Physical model of hydrological behavior of permeable pavements using FlexPDE. J. Hydrol. Eng. **24**, 04019035 (2019)
20. Deutsches Institut für Normung e. V.: Prüfverfahren für geometrische Eigenschaften von Gesteinskörnungen - Teil 4: Bestimmung der Kornform - Kornformkennzahl (DIN EN 933-4) (2015)
21. Eck, B.J., Charbeneau, R.J., Barrett, M.E.: Drainage hydraulics of porous pavement: Coupling surface and subsurface flow. Center for Research in Water Resources, University of Texas at Austin, Technical Report (2010)

22. Eck, B.J., Barrett, M.E., Charbeneau, R.J.: Coupled surface-subsurface model for simulating drainage from permeable friction course highways. J. Hydraul. Eng. **138**, 13–22 (2012)
23. Falk, K., Lang, R., Kaliske, M.: Multiscale simulation to determine rubber friction on asphalt surfaces. Tire Sci. Technol. **44**, 226–247 (2016)
24. Flemisch, B., Darcis, M., Erbertseder, K., Faigle, B., Lauser, A., Mosthaf, K., Müthing, S., Nuske, P., Tatomir, A., Wolff, M., et al.: DuMu^x: DUNE for multi-{phase, component, scale, physics,...} flow and transport in porous media. Adv. Water Resour. **34**, 1102–1112 (2011)
25. Flintsch, G.W., Tang, L., Katicha, S.W., de Leŏn Izeppi, E., Viner, H., Dunford, A., Nesnas, K., Coyle, F., Sanders, P., Gibbons, R.B., et al.: Splash and spray assessment tool development program. Technical Report, Virginia Tech Transportation Institute, Virginia Tech (2014)
26. Forschungsgesellschaft für Strassen- und Verkehrswesen (FGSV): Technische Lieferbedingungen für Asphaltmischgut für den Bau von Verkehrsflächenbefestigungen (TL Asphalt-StB). FGSV-Verlag, Köln (2007)
27. Forschungsgesellschaft für Strassen- und Verkehrswesen (FGSV): Technische Lieferbedingungen für Gesteinskörnungen im Strassenbau (TL Gestein-StB 04). FGSV-Verlag, Köln (2004)
28. Friederichs, J., Wegener, D., Eckstein, L., Hartung, F., Kaliske, M., Götz, T., Ressel, W.: Using a new 3D-printing method to investigate rubber friction laws on different scales. Tire Sci. Technol. **48**, 250–286 (2020)
29. Gaia Ferreira, W.L., Castelo Branco, V.T.F., Caro, S., Vasconcelos, K.: Analysis of water flow in an asphalt pavement surface layer with different thicknesses and different permeability coefficients. Road Mater. Pavement Des. **2020**, 1–19 (2019)
30. Gallaway, B.M., Rose, J.G., Schiller, R., Jr.: The relative effects of several factors affecting rainwater depths on pavement surfaces. Highw. Res. Rec. **396**, 59–71 (1972)
31. Geike, T., Popov, V.L.: Mapping of three-dimensional contact problems into one dimension. Phys. Rev. E **76**, 036710 (2007)
32. Gonzalez, R.C., Woods, R.E.: Digital Image Processing, 4th edn. Pearson, Uttar Pradesh (2018)
33. Gonzalez, R.C., Woods, R.E., Eddins, S.L.: Digital Image Processing Using MATLAB, 2nd edn. Gatesmark Publishing, Knoxville (2009)
34. Götz, T.: Mechanisch-analytische Modellierung der Nassgriffigkeit von Straßenoberflächen. Ph.D. thesis, University of Stuttgart, unpublished draft version
35. Hartung, F., Kienle, R., Götz, T., Winkler, T., Ressel, W., Eckstein, L., Kaliske, M.: Numerical determination of hysteresis friction on different length scales and comparison to experiments. Tribol. Int. **127**, 165–176 (2018)
36. Herrmann, S.R.: Simulationsmodell zum Wasserabfluss- und Aquaplaning-Verhalten auf Fahrbahnoberflächen. Ph.D. thesis, University of Stuttgart (2008)
37. Hsieh, P.C., Chen, Y.C.: Surface water flow over a pervious pavement. Int. J. Numer. Anal. Methods Geomech. **37**, 1095–1105 (2013)
38. Hu, J., Liu, P., Steinauer, B.: A study on fatigue damage of asphalt mixture under different compaction using 3D-microstructural characteristics. Front. Struct. Civ. Eng. **11**, 329–337 (2017)
39. Hu, J., Liu, P., Wang, D., Oeser, M., Tan, Y.: Investigation on fatigue damage of asphalt mixture with different air-voids using microstructural analysis. Constr. Build. Mater. **125**, 936–945 (2016)
40. Hu, J., Qian, Z., Wang, D., Oeser, M.: Influence of aggregate particles on mastic and air-voids in asphalt concrete. Constr. Build. Mater. **93**, 1–9 (2015)
41. Jerjen, I., Poulikakos, L.D., Plamondon, M., Schuetz, P., Luethi, T., Flisch, A.: Drying of porous asphalt concrete investigated by X-ray computed tomography. Phys. Procedia **69**, 451–456 (2015)
42. Kassem, E., Awed, A., Masad, E.A., Little, D.N.: Development of predictive model for skid loss of asphalt pavements. Transp. Res. Rec. **2372**, 83–96 (2013)
43. Kienle, R., Ressel, W., Götz, T., Weise, M.: The influence of road surface texture on the skid resistance under wet conditions. Proc. Inst. Mech. Eng. Part J J. Eng. Tribol. **234**, 313–319 (2020)

44. Kogbara, R.B., Masad, E.A., Kassem, E., Scarpas, A.T., Anupam, K.: A state-of-the-art review of parameters influencing measurement and modeling of skid resistance of asphalt pavements. Constr. Build. Mater. **114**, 602–617 (2016)
45. Krol, J.B., Khan, R., Collop, A.C.: The study of the effect of internal structure on permeability of porous asphalt. Road Mater. Pavement Des. **19**, 935–951 (2018)
46. Kumar, K., Kozak, J., Hundal, L., Cox, A., Zhang, H., Granato, T.: In-situ infiltration performance of different permeable pavements in a employee used parking lot-a four-year study. J. Environ. Manag. **167**, 8–14 (2016)
47. Kutay, M.E., Aydilek, A.H., Masad, E.: Laboratory validation of lattice Boltzmann method for modeling pore-scale flow in granular materials. Comput. Geotech. **33**, 381–395 (2006)
48. Kutay, M.E., Aydilek, A.H., Masad, E., Harman, T.: Computational and experimental evaluation of hydraulic conductivity anisotropy in hot-mix asphalt. Int. J. Pavement Eng. **8**, 29–43 (2007)
49. Lal, S., Prat, M., Plamondon, M., Poulikakos, L., Partl, M.N., Derome, D., Carmeliet, J.: A cluster-based pore network model of drying with corner liquid films, with application to a macroporous material. Int. J. Heat Mass Transf. **140**, 620–633 (2019)
50. Legland, D.: Image ellipsoid 3D—computation of equivalent ellipsoid coefficients from moment matrix (2020). https://www.mathworks.com/matlabcentral/fileexchange/34104-image-ellipsoid-3d
51. Lippold, C., Vetters, A., Ressel, W., Alber, S.: Vermeidung von abflussschwachen Zonen in Verwindungsbereichen-Vergleich und Bewertung von baulichen Lösungen. Berichte der Bundesanstalt für Straßenwesen, Reihe Verkehrstechnik, Heft V 319 (2019)
52. Liu, X., Chen, Y., Shen, C.: Coupled two-dimensional surface flow and three-dimensional subsurface flow modeling for drainage of permeable road pavement. J. Hydrol. Eng. **21**, 04016051 (2016)
53. Lu, G., Wang, Z., Liu, P., Wang, D., Oeser, M.: Investigation of the hydraulic properties of pervious pavement mixtures: characterization of Darcy and non-Darcy flow based on pore microstructures. J. Transp. Eng. Part B Pavements **146**, 04020012 (2020)
54. Masad, E., Al Omari, A., Chen, H.C.: Computations of permeability tensor coefficients and anisotropy of asphalt concrete based on microstructure simulation of fluid flow. Comput. Mater. Sci. **40**, 449–459 (2007)
55. Oeser, M., Ueckermann, A.: A method to predict skid resistance from texture using a rubber friction model. In: 17. Internationales Stuttgarter Symposium, pp. 1263–1280. Springer, Wiesbaden (2017)
56. Otsu, N.: A threshold selection method from gray-level histograms. IEEE Trans. Syst. Man Cybernet. **9**, 62–66 (1979)
57. Popov, V.L., Heß, M.: Methode der Dimensionsreduktion in Kontaktmechanik und Reibung. Springer, Berlin (2013)
58. Poulikakos, L., Gilani, M.S., Derome, D., Jerjen, I., Vontobel, P.: Time resolved analysis of water drainage in porous asphalt concrete using neutron radiography. Appl. Radiat. Isot. **77**, 5–13 (2013)
59. Praticò, F.G., Moro, A.: Flow of water in rigid solids: development and experimental validation of models for tests on asphalts. Comput. Math. Appl. **55**, 235–244 (2008)
60. Qian, N., Wang, D., Li, D., Shi, L.: Three-dimensional mesoscopic permeability of porous asphalt mixture. Constr. Build. Mater. **236**, 117430 (2020)
61. Ranieri, V.: Runoff control in porous pavements. Transp. Res. Rec. **1789**, 46–55 (2002)
62. Ranieri, V., Colonna, P., Ying, G., Sansalone, J.: Model of flow regimes in porous pavement and porous friction courses. Transp. Res. Rec. **2436**, 156–166 (2014)
63. Ranieri, V., Ying, G., Sansalone, J.: Drainage modeling of roadway systems with porous friction courses. J. Transp. Eng. **138**, 395–405 (2012)
64. Ressel, W., Herrmann, S.: Aquaplaning und Verkehrssicherheit in Verwindungsbereichen dreistreifiger Richtungsfahrbahnen-Berechnung der Wasserfilmdicke. Forschung Straßenbau und Straßenverkehrstechnik, Bonn **997** (2008)
65. Ressel, W., Wolff, A., Alber, S., Rucker, I.: Modelling and simulation of pavement drainage. Int. J. Pavement Eng. **20**, 801–810 (2019)

66. Rucker, I., Ressel, W.: A numerical drainage model to simulate infiltration into porous pavements for higher road safety. In: 17. Internationales Stuttgarter Symposium, 1293–1303. Springer, Wiesbaden (2017)
67. Ruf, M., Steeb, H.: An open, modular, and flexible micro X-ray computed tomography system for research. Rev. Sci. Instrum. **91**, 113102 (2020)
68. Schlüter, W., Jefferies, C.: Modelling the outflow from a porous pavement. Urban Water **4**, 245–253 (2002)
69. Schuck, B., Teutsch, T., Alber, S., Ressel, W., Steeb, H., Ruf, M.: Study of air void topology of asphalt with focus on air void constrictions-a review and research approach. Road Mater. Pavement Des. **22**, 425–443 (2021)
70. Simunek, J., van Genuchten, M.T., Sejna, M.: Development and applications of the HYDRUS and STANMOD software packages and related codes. Vadose Zone J. **7**, 587–600 (2008)
71. Srirangam, S., Anupam, K., Scarpas, A., Kasbergen, C., Kane, M.: Safety aspects of wet asphalt pavement surfaces through field and numerical modeling investigations. Transp. Res. Rec. **2446**, 37–51 (2014)
72. Sun, Y., Guo, R., Wang, X., Ning, X.: Dynamic response characteristics of permeable asphalt pavement based on unsaturated seepage. Int. J. Transp. Sci. Technol. **8**, 403–417 (2019)
73. Tan, S., Fwa, T., Chai, K.: Drainage considerations for porous asphalt surface course design. Transp. Res. Rec. **1868**, 142–149 (2004)
74. Tarefder, R.A., White, L., Zaman, M.: Neural network model for asphalt concrete permeability. J. Mater. Civ. Eng. **17**, 19–27 (2005)
75. Turco, M., Kodešová, R., Brunetti, G., Nikodem, A., Fér, M., Piro, P.: Unsaturated hydraulic behaviour of a permeable pavement: laboratory investigation and numerical analysis by using the HYDRUS-2D model. J. Hydrol. **554**, 780–791 (2017)
76. Ueckermann, A., Wang, D., Oeser, M., Steinauer, B.: A contribution to non-contact skid resistance measurement. Int. J. Pavement Eng. **16**, 646–659 (2015)
77. Umiliaco, A., Benedetto, A.: Unsteady flow simulation of water drainage in open-graded asphalt mixtures. Procedia Soc. Behav. Sci. **53**, 346–355 (2012)
78. Weise, M.: Einflüsse der mikroskaligen Oberflächengeometrie von Asphaltdeckschichten auf das Tribosystem Reifen-Fahrbahn. Ph.D. thesis, University of Stuttgart (2015)
79. Weishaupt, K., Joekar-Niasar, V., Helmig, R.: An efficient coupling of free flow and porous media flow using the pore-network modeling approach. J. Comput. Phys. X **1**, 100011 (2019)
80. Wolff, A.: Simulation of pavement surface runoff using the depth-averaged shallow water equations. Ph.D. thesis, University of Stuttgart (2013)
81. Zhang, J., Cui, X., Tang, W., Lou, J.: Approximate simulation of storm water runoff over pervious pavement. Int. J. Pavement Eng. **18**, 247–259 (2017)

Experimental Methods for the Mechanical Characterization of Asphalt Concrete at Different Length Scales: Bitumen, Mastic, Mortar and Asphalt Mixture

Sabine Leischner, Gustavo Canon Falla, Mrinali Rochlani, Alexander Zeißler, and Frohmut Wellner

Abstract This chapter presents a comprehensive characterization of asphalt concrete at different scales of observation. State-of-the-art characterization procedures for bitumen, mastic, mortar and asphalt are described in detail. The procedures were envisaged to provide experimental data for parameter identification and validation of constitutive numerical models. The validation of numerical models against experimental data is a prerequisite for the use in any application. The temperature-frequency dependency of the rheological properties of bitumen and mastic was characterized using temperature sweeps in the dynamic shear rheometer. The results showed that the bitumen provenance and the filler's mineralogy have a major impact on the rheological response of the asphalt. A new rheometer, known as Dresden dynamic shear tester, was developed with the aim of characterizing the thermo-viscoelastic properties of mortar. This novel equipment was also used to identify the stiffening effect of the aggregates by comparing the results of bitumen, mortar and asphalt. Finally, the short and long term behavior of different asphalt mixtures were characterized with the repeated load triaxial tests and with the indirect tensile tests. The results showed that the performance of asphalt is highly affected by the bitumen type and the aggregate gradation.

Keywords Experimental characterization · Rheology · Bitumen · Mastic · Mortar · Asphalt mixture

Funded by the German Research Foundation (DFG) under grant WE 1642/11 and grant LE 3649/2.

S. Leischner (✉) · G. Canon Falla · M. Rochlani · A. Zeißler · F. Wellner
Institute of Urban and Pavement Engineering, Technische Universität Dresden, Dresden, Germany
e-mail: Sabine.leischner@tu-dresden.de

M. Kaliske et al. (eds.), *Coupled System Pavement—Tire—Vehicle*,
Lecture Notes in Applied and Computational Mechanics 96,
https://doi.org/10.1007/978-3-030-75486-0_4

1 Introduction

In recent years, the road structures have been deteriorating at an elevated rate due to the augmentation in service traffic densities, axle loading, changes in weather conditions and reduced maintenance services. The governments have been investing large sums of money in order to attain exceptional, and long-lasting pavements. However, these surfaces keep showing early signs of distresses [22]. In order to minimise the degradation of pavement surfaces and increase the durability, there is a dire need to improve and optimize the construction materials so as to achieve performance related properties such as resistance to fatigue and low temperature cracking along with adequate protection against permanent deformation or rutting.

A better material understanding results in multiple scales of consideration, starting from the microscale which includes the mechanical properties of bitumen, mastic and mortar to the macroscale that deals with the properties of the asphalt mixtures. As a consequence, a significant part of pavement research in the last years has been dedicated to developing and validating new models and test methods in order to assess the performance of asphalt from micro to macro length scales.

In this context, this chapter presents different characterization techniques aimed to better understand the behavior and performance of asphalt concrete from micro to macro level. The research shown was carried out within the subproject number 4 (TP4) of the research group-FOR 2089 "Durable Pavement Constructions for Future Traffic Loads: Coupled System Pavement-Tyre-Vehicle", funded by the German Research Foundation (DFG). One of the main objectives of TP4 was the development of material characterization techniques to provide experimental data to fit and validate the numerical models implemented by the other subprojects. For example, the results of repeated load triaxial test (RLTT) on asphalt were used to determine the material parameters of the short and long term numerical models presented in the chapter "Multi-physical and Multi-scale Theoretical-Numerical Modeling of Tire-Pavement Interaction". The data of the RLTT was also used in chapter "Numerical Simulation of Asphalt Compaction and Asphalt Performance" to determine the Prony series parameters of the micromodel introduced therein.

All experimental data of the project was collected into a database of laboratory test results on asphalt, mortar, mastic, bitumen and aggregates. Table 1 gives a general overview of the laboratory testing data that is available in the database. Some testing results were selected from the database and are presented in this chapter to give a general overview of the state-of-the-art characterization methods for bituminous materials used in road construction.

Outline. Sections 2 and 3 of this chapter present experimental results of bitumen and mastic, respectively. The material characterization was performed using the dynamic shear rheometer (DSR). This study involved a competent and advanced testing plan, such that comprehensive data regarding the performance and rheology related characteristics of bituminous mixtures could be derived. Section 4 deals with the rheology of mortar. A novel testing equipment called Dresden dynamic shear tester (DDST) was developed to determine the temperature dependent viscoelastic

Table 1 General overview of the laboratory testing data that is available in the database of the project

Material	Testing device	Type of test
Bitumen	RLTT	Temperature sweeps
	DDST	Temperature sweeps
	DSR	Temperature sweeps
		Strain sweeps
		Time sweeps (fatigue)
		Cryogenic tests
		Creep tests
Mastic	DSR	Temperature sweeps
		Strain sweeps
		Time sweeps (fatigue)
		Cryogenic tests
		Creep tests
Mortar	RLTT	Temperature sweeps
	DDST	Temperature sweeps
Aggregates	RLTT	Stress sweeps
	DDST	Friction tests
Asphalt	RLTT	Stress-temperature sweeps
		Creep tests
		Time sweeps (permanent deformation)
		Strain sweeps
	DDST	Temperature sweeps (shear load)
	ITT	Time sweeps (fatigue)

DDST: Dresden Dynamic Shear Tester, DSR: Dynamic Shear Rheometer, ITT: Indirect Tensile Test device, RLTT: Repeated Load Triaxial Test device

behavior of mortar. In Sect. 5, the results of short and long term performance tests on asphalt are presented.

2 Mechanical Characterization of Bitumen

2.1 Background

Asphalt concrete is a heterogeneous material whose holistic properties depend on the properties of each of its constituents. Asphalt concrete is made mainly of bitumen and mineral aggregates. The bitumen essentially acts as a binder for the mineral aggregates to form the asphalt mixture. The percentage of bitumen in the mix depends on the type of mixture, however, typically 5 wt% bitumen is blended at high temperature (around 160 °C) with 95 wt% of aggregates to fabricate what is commonly known as hot mix asphalt.

Because bitumen's properties directly influence the macroscopic behavior of asphalt concrete, the characterization of bitumen is currently being employed to better understand the macroscopic behavior of the mixture. Up-scaling techniques have emerged as a possibility to design and engineer asphalts with properties that better adapt to the specific requirements in terms of stiffness, viscoelasticity, fatigue resistance and plasticity [8]. Hence, a new challenge has arisen to optimize bitumen to

- resist cracking due to thermal stresses at low temperatures,
- resist permanent deformation at high temperatures,
- resist fatigue under repeated loading at intermediate temperatures,
- resist aging,
- resist moisture damage and
- increase the adhesion with the granular aggregates.

Optimized bitumen development is possible only through the use of phenomenological test procedures that characterize the rheological behavior of the material. Traditional bitumen tests, such as ring and ball softening point and needle penetration, are empirical in nature. Due to their empirical essence, the information obtained from these tests is minimal, and it cannot be used for the development of new materials neither as input to mechanistic-empirical design methodologies. Thus, a new trend has emerged in Germany in which traditional tests are being replaced by advanced tests that determine performance-oriented properties that may be used not only for quality assurance and material development, but also as input to rheological models.

This section presents a complete testing program to characterize the rheological behavior of bitumen as well as its performance in terms of resistance to permanent deformation, to low temperature cracking and to fatigue cracking. Four bitumen were compared using the results from different tests with the DSR.

2.2 *Dynamic Shear Rheometer*

DSR is usually used to determine the dynamic shear modulus, $|G^*|$, and the phase angle, ϕ, of bituminous binders at different temperatures, stress/strain levels, and frequencies. The standard DSR test arrangement consists of a bitumen sample sandwiched between a spindle and a base plate, as seen in Fig. 1. The testing plate geometry, characterized by the spindle diameter and testing gap, depends on the bitumen's stiffness. In general, a geometry consisting of a spindle diameter of 25 mm and a gap of 1 mm is used at intermediate to high temperatures (40 °C $\leq$ T $\leq$ 80 °C), where the stiffness of the bitumen is relatively low ($|G^*| \leq 500$ kPa). At low temperatures (T $\leq$ 30 °C), a geometry of 8 mm diameter and 2 mm height is used.

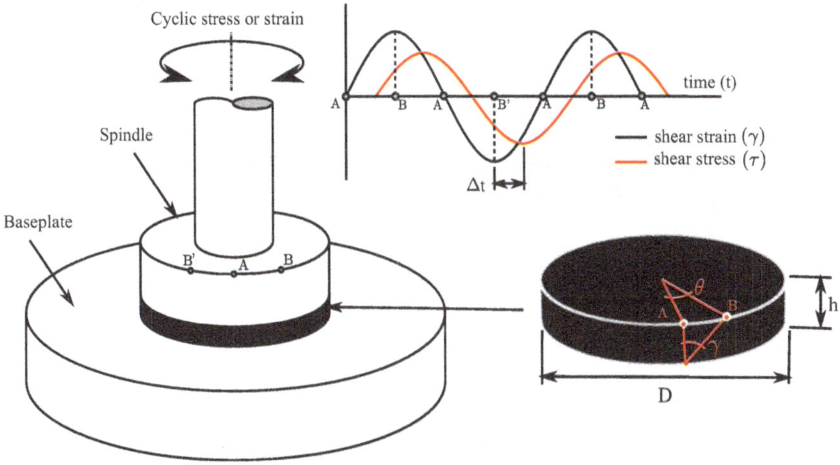

Fig. 1 Principle of operation of DSR [4]

DSR tests can be performed in either stress-controlled or strain-controlled mode. In stress-controlled mode, a fixed shear stress, τ, is applied to the bitumen, and the response shear strain, γ, is measured. In strain-controlled mode, the strain is fixed, and the response stress is measured. The main difference between both testing modes is that at strain-controlled conditions, the bitumen tends to store energy because of the constrained deformation. In contrast, at stress-controlled conditions, the material freely dissipates energy into permanent deformation.

2.3 *Materials*

The materials are three popular bitumen of 50/70 penetration grade. They are labelled as B1, B2 and B3 and were acquired from three separate sources/provenances [21, 23]. The bitumen samples were studied in three aging conditions that included unaged, short term aged using the rolling thin film oven test (RTFOT) and long term aged with the pressure aging vessel (PAV).

2.4 *Experimental Methods*

DSR was the chosen equipment for comprehensive testing of the rheology of bitumen and mastic to obtain a detailed performance-based understanding of binder and mastic ranking and behaviour. The Anton Paar Modular Compact Rheometer 502 was the DSR used for evaluation of rheology, permanent deformation, low-temperature

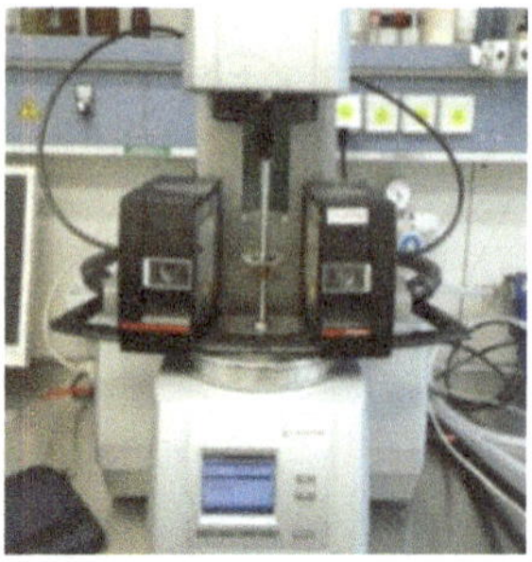

a DSR test device

b Specimen for strain, frequency sweep and SSCR test

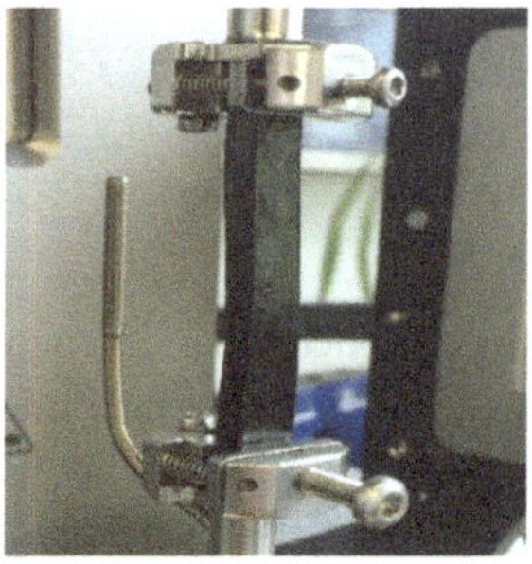

c Specimen for Dresden cryogenic stress tests

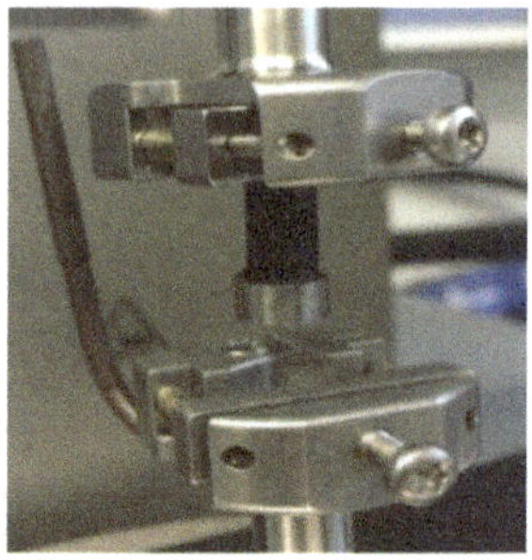

d Fatigue test specimen

Fig. 2 DSR tests on bitumen

resistance and fatigue performance of the bitumen. The tests conducted include, the strain and frequency sweep tests for rheological study, single stress creep recovery (SSCR) test for permanent deformation analysis, both of which were done using the parallel plates. The low-temperature resistance was analyzed using the Dresden cryogenic test (DCT) with prismatic specimen and finally, fatigue resistance was tested using stress-controlled tests on a cylindrical column specimen (Fig. 2).

2.5 *Viscoelastic Performance*

The strain and frequency sweeps were conducted on unaged, RTFOT and PAV aged bitumen and mastic to obtain the dynamic shear modulus and the phase angle values that are important to understand the rheology of the materials. These results are presented in this section. Firstly, the linear-viscoelastic (LVE) limit was determined for every material at hand by conducting strain-sweep tests at temperatures of -10,

Fig. 3 2S2P1D rheological model

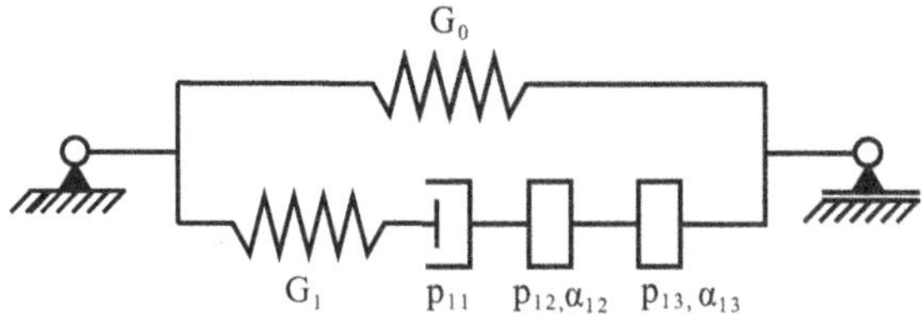

10, 30, 50 and 70 °C and a frequency of 50 Hz. The LVE limit is generally defined by the strain value where the dynamic shear modulus is equal to 95% of its initial modulus value. These LVE limit strains were further reduced by 20% to make certain that the frequency sweeps are conducted within the linear viscoelastic region. To study rheology, strain and frequency sweeps were done in the temperature range of −10 to 70 °C.

Two frequency sweeps were performed for each of the unaged and aged bitumen. The first frequency sweeps were carried out at lower temperatures in the range of −10 to 30 °C using the 8 mm parallel plate and the typical 2 mm specimen thickness that is standard for the 8 mm plate. The second frequency sweeps were conducted at higher temperatures in the range from 30 ° C to 70 °C using the 25 mm diameter parallel plate and the standard 1 mm specimen thickness. The frequency was increased from 0.0159 to 76 Hz and the results obtained were used to calculate the master curves by use of the Williams-Landel-Ferry equation for the time-temperature superposition and were further approximated using the 2 spring, 2 parabolic, 1 dashpot (2S2P1D) rheological model [5]. This model is used for rheological-simple bituminous materials in terms of temperature and frequency, and also explain the material performance using seven parameters, as seen in Fig. 3 [5, 22]. The master curves of the bitumen are shown in Fig. 4. These graphs were constructed using the frequency sweep test data modelled with the 2S2P1D model for interpretation and comparison of results.

The frequency sweep test data modelled, using the time-temperature superposition principle and the 2S2P1D rheology were model is presented in terms of the master curves for dynamic shear modulus and phase angle in Fig. 4.

By observing Fig. 4, the dynamic shear modulus, as well as the phase angle for different bitumen and the change amongst the different sources and aging conditions can be visually seen. From Fig. 4a, which presents the dynamic shear modulus master curve for unaged and aged bitumen, B1 and B3 tend to overlap, while B2 is seen to have slightly lower modulus values than the others at lower reduced frequencies (i.e. higher temperatures). Regarding the master curves of RTFOT aged bitumen, B1 has the greatest shift. It is followed by bitumen B2 with 134% increase at higher temperatures and as low as 2% at lower temperatures while B3 shows almost overlapping results for unaged and RTFOT aged, with a numerical increment ranging from 20 to 6.5% as temperatures lower. For PAV aged materials, all three bitumen show a relatively high increase, as expected. On the other hand, if the master curves for phase angle are observed, with RTFOT aging the phase angles reduce slightly, however a much higher decrease is visible for PAV aged materials. As aging increases, the

Fig. 4 Master curves using 2S2P1D model at 20 °C [23]

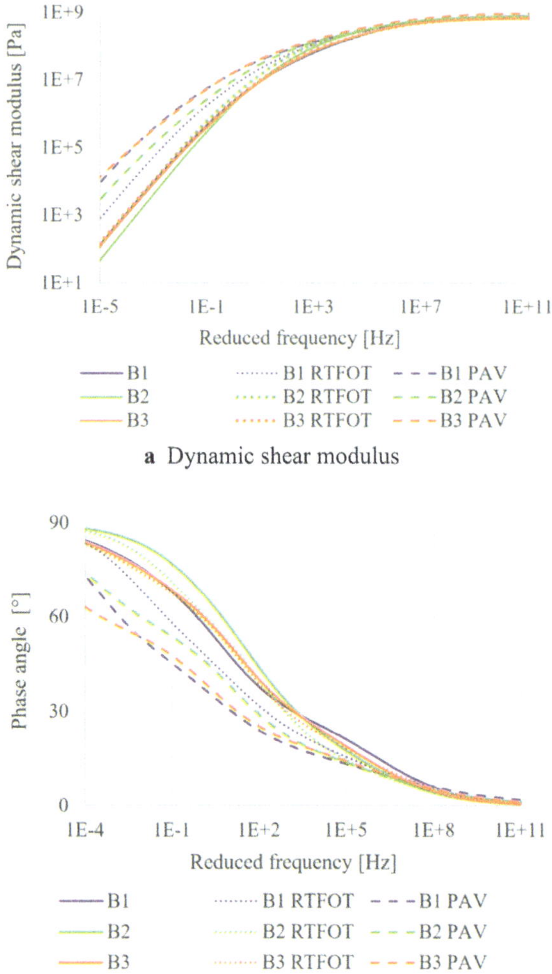

a Dynamic shear modulus

b Phase angle

stiffness increases along with phase angle reductions, that signify a more elastic, rather than a viscous behavior. Overall, B1 and B3 with overlapping curves for PAV aging would be expected to perform similar and better than B2 due to their increased stiffness and reduced phase angles, signifying an increased elasticity at higher temperatures. This would enable these materials to resist the stresses, and recover faster. A better numerical evaluation of the effect of aging could be observed using the aging index. From this test, the parameters of shear modulus at 20 and 60 °C at 10 Hz in RTFOT aged conditions were chosen in order to compare and rank the materials in the later section. These conditions were chosen as these are typical conditions for permanent deformation and fatigue criteria of asphalt [22].

2.6 Aging Index

The aging of bitumen can be expressed numerically in terms of the aging index (A.I.). This index is calculated in terms of ratios corresponding to several physical grading tests and is given by the following equation

$$A.I. = \frac{P_{aged}}{P_{unaged}}, \tag{1}$$

where P_{unaged} is any physical parameter (e.g. viscosity, softening point, stiffness, etc.) of the unaged bituminous materials and P_{aged} is the same physical parameter as for an aged bituminous material.

The A.I. was calculated for dynamic shear moduli (bars) and phase angles (lines) at 20 and 60 °C at a frequency of 10 Hz. The lower the aging index value, the lower is the effect of the aging sensitivity on the material.

From Fig. 5a which represents the A.I.$_{stiffness}$ (with full colors bars for RTFOT and dashed for PAV), it can be observed, that at 20 °C for RTFOT, all materials have almost the same A.I., in the range of 1.35–1.5, while at 60 °C, B1 has relatively high A.I. with 2.8, than the 2.0 and 2.1 of B2 and B3. For PAV aged materials, at 20 °C, B3 shows a much higher susceptibility to aging with the highest value of 3.9 while B1 and B2 are both around 2.5. At 60 °C. However, B3 has the lowest A.I. closely followed by B1, with 60% higher than B3. From the results, it can be concluded that B3 was most affected by RTFOT aging, with B2 being the most susceptible and B3 the least susceptible to PAV aging at higher temperatures. For phase angles (Fig. 5b), the changes at each temperature and aging suggest very similar values, with almost horizontal lines for every temperature and aging type. The phase angle A.I. for PAV is almost double the RTFOT values for the same temperatures. The main parameters for this section include the A.I. (stiffness) at 60 °C for RTFOT aged materials, as rutting occurs at earlier stages of the lifetime and at higher temperatures, while the A.I. (stiffness) at 20 °C for PAV materials with respect to fatigue. For comparison and ranking, the A.I. at 60 °C for RTFOT and 20 °C for PAV were taken into consideration for purpose of rutting and fatigue evaluations.

2.7 Plastic Deformation Performance

For evaluating the rutting performance, single stress creep recovery (SSCR) tests were undertaken. This test follows the same procedure as Multiple Stress Creep Recovery (MSCR) test where a sample is subjected to stress for one second, followed by nine seconds without stress, allowing the material to relax at a temperature of 60 °C with the 25 mm plate. This step is repeated consecutively 10 times. The only difference between MSCR and SSCR is that in MSCR tests multiple stresses are used, while in SSCR tests only a single stress of 3.2 kPa is used. From each cycle

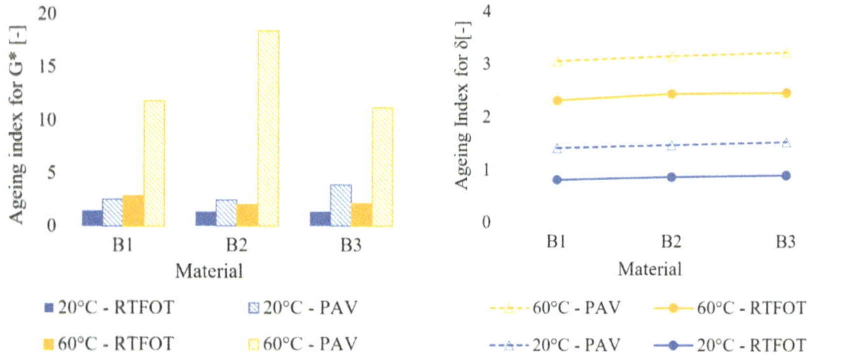

a Dynamic shear modulus **b** Phase angle

Fig. 5 Aging index at 10 Hz [23]

Fig. 6 Results of MSCR tests [23]

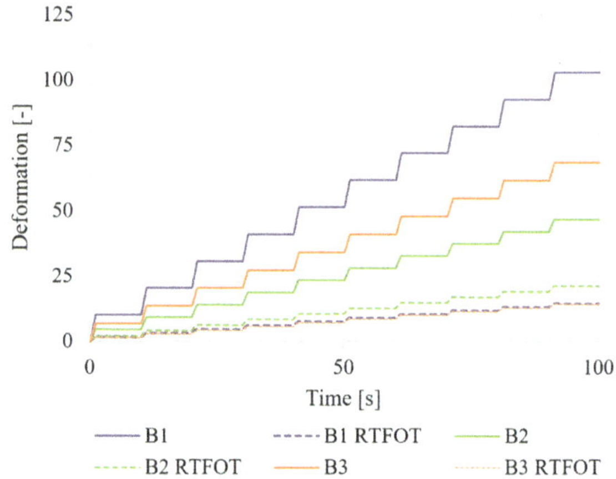

two variables, non-recoverable compliance, J_{nr}, and percentage recovery, %R, were calculated and averaged.

SSCR tests were undertaken to evaluate the rutting performance. These tests were carried out on unaged and RTFOT aged materials. PAV aged materials were not tested, as rutting is crucial in the initial lifespan of a pavement and fatigue is more critical in the long term performance. The test results are presented in Fig. 6 showing the deformation with time for ten cycles of loading and unloading.

From Fig. 6, it can be seen, that the unaged materials have much higher deformation than the RTFOT aged materials, with values for unaged being in the order of 105 to 45 to being less than 25 for aged. There are vast differences between the unaged materials, with B1 having a significantly higher cumulative deformation.

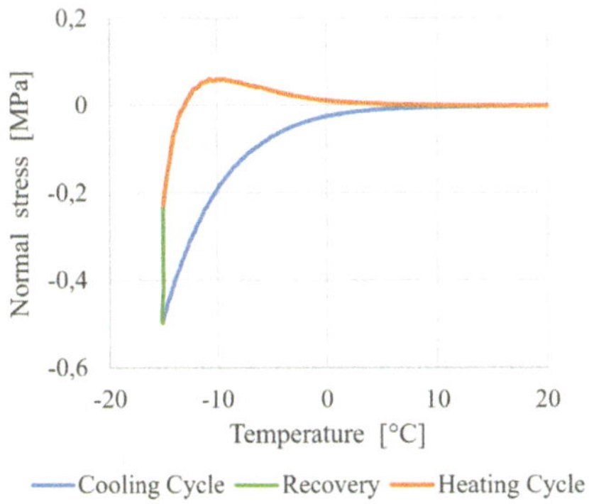

a Full cryogenic cycle for a bitumen

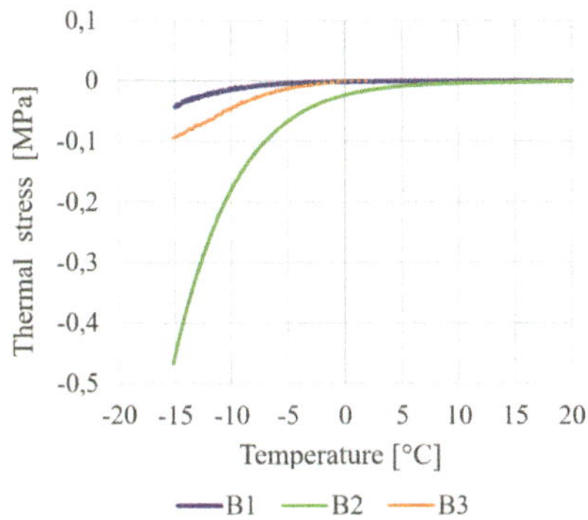

b Cryogenic curves for all bitumen

Fig. 7 Results of cryogenic test [21]

2.8 Dresden Cryogenic (DDC) Test for Low Temperature Performance

Cryogenic tests were performed in order to determine the thermal stresses induced in the bitumen when subjected to low temperatures. This test procedure was developed at Technische Universität Dresden to be conducted using the DSR. The idea was derived from the European standard EN 12697-46, which entails testing of asphalt prismatic specimens subjected to low temperatures without being subjected to any forces. The sample was a rectangular prismatic specimen with 50 mm in length, 4 mm in thickness and 9 mm in width that was fixed between the upper spindle and lower base plate, respectively, and subjected to a constant reduction of temperature of 10 °C per hour. The temperature range of the test was from 20 to −15 °C. There was no loading applied to the materials, instead the axial force exerted on the specimen by the changing temperature conditions was measured during the experiment. It is possible to convert the axial force to thermal stresses by dividing it by the area of the specimen [22].

Figure 7a presents the results of the cryogenic stress tests in terms of the cooling and heating cycles. The heating cycles display a full recovery of the initial stresses when the testing temperature returns to its initial value of 20 °C. The slope of the heating cycle is seen to be different than the slope of the cooling cycle. At temperatures around −10 °C, a change of curvature was observed in the heating cycle curves. The reason for this change in curvature could be the expansion of the material due to the increasing temperatures. Figure 7b shows the cryogenic curves for all bitumen till −15 °C. It can be seen, that the bitumen exerts different levels of cryogenic stress, with B1 having the lowest stress exerted, while B3 has almost two times that of B1, and B2 is three times that of B1. The differences in the low temperature performances could be attributed to the stiffness of the materials. The higher the stiffness of a material, the higher is the expected cryogenic stress regardless of bitumen as shown by Rochlani et al. [21, 22].

2.9 Fatigue Resistance

Using a DSR, shear stress-controlled fatigue tests were undertaken on cylindrical column specimens at 20 °C and frequencies of 10 Hz. Seven to eight stress-controlled tests were conducted for each temperature-frequency combination. Test conditions of 20 °C and 10 Hz are the standard conditions for fatigue testing in Germany for bituminous materials [21, 22, 28].

In these tests, the initial stiffness was taken as the stiffness value at 100 load cycles. The number of load cycles to failure was determined using the transition point approach. The results were further analyzed with the dissipated energy calculation at various load cycles.

The specimens had a height of 20 mm, a top and bottom diameter with rings of 8 mm and without rings of 7 mm, while the actual dimension of the tested part had a height of 11mm and a 6 mm diameter (Fig. 8a). Fig. 8b shows the silicone mould used to make the samples. The sample preparation procedure starts with heating the bitumen/mastic to a temperature of 160 °C. The mixture is then stirred for several minutes to ensure a homogeneous condition. The material is poured into a specially designed mould made of teflon or silicone. Finally, the mould is cooled down to room temperature and stored in a refrigerator at a temperature of 2 °C. Metal rings were used on top and bottom to clamp the specimen in the machine to avoid direct bitumen contact with the fixtures. This was done as bituminous materials are viscoelastic and this material property results in relaxation effects, which could cause a loss of contact.

The bottom end was fixed and the top end was subjected to torque in the DSR. In a straight sample, the strain is expected to be distributed evenly throughout the specimen. In order to avoid adhesive failure between the ring and bitumen, the total contact area of the ring and bitumen was kept at least three times the testing cross-section of the column [17]. A common problem observed with these specimens are that they tend to break near the metal rings due to the stress concentration in this region. To reduce the stress concentration at the bitumen-ring interface, the testing column was given a lower diameter (6 mm) than the inner diameter of the ring (7 mm). There have been no adhesion problems between the ring and bitumen and the use of this specimen shape and size has shown results of good accuracy (R^2 value greater than 0.9) for bitumen and mastic [17, 22, 24].

The fatigue test results were analyzed using the dissipated energy ratio approach. Using this criterion, the energy dissipated by a material can be expressed using the product of the number of load cycles and the complex shear modulus. The number of cycles at failure is denoted by the value of N corresponding to the highest value of $N|G^*|$. For these stress controlled tests, the stress levels were chosen in a way to maintain an initial strain level at 100 load cycles within the range of 0.5–2.2% as visible in Fig. 9. Due to this, the unaged bitumen were tested at a stress range of 100 to 200 kPa, RTFOT at 150 to 250 kPa and PAV at 250 to 400 kPa. This is expected due to an increase in stiffness with an increase of aging, therefore, causing a lower strain for the same stress. From Fig. 9, it can be seen that with aging, the slope of the initial strain-Nmax functions tends to increase.

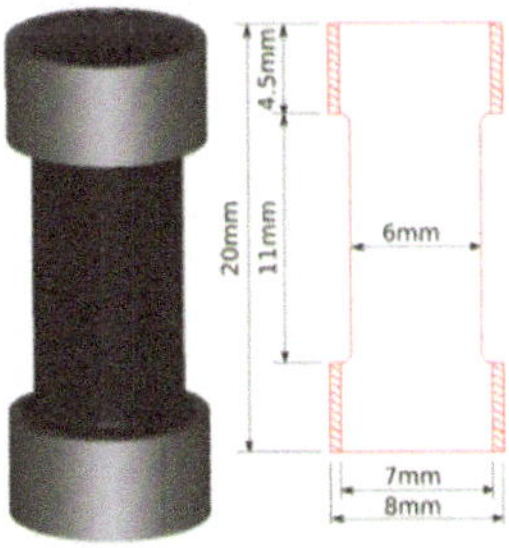

a Fatigue sample dimensions

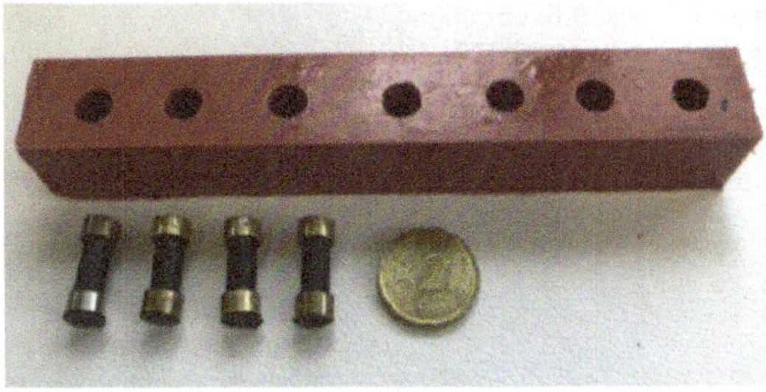

b Silicon mould

Fig. 8 Fatigue sample [21]

Fig. 9 Strain fatigue curves for the bitumen tested

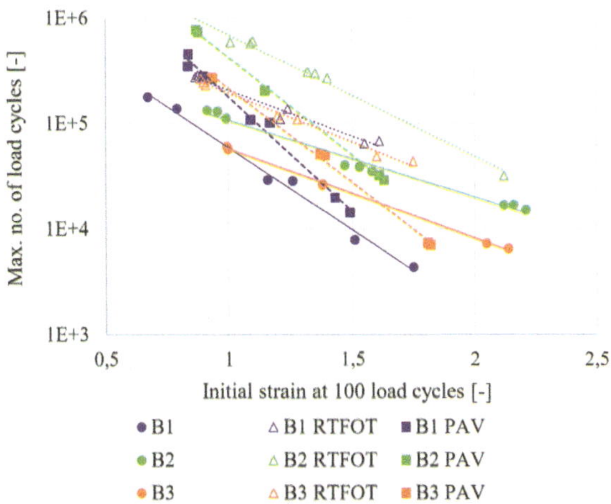

The fatigue curves for applied stress versus the number of load cycles at failure are presented in Fig. 10. From this figure it can be observed that there is a clear distinction between the unaged, RTFOT and PAV aged bitumen curves, with the slope for all the materials other than B1 being relatively similar. The slope could be an indicator of the type of failure, the testing temperature and frequency. Also, all materials showed high R^2 values of at least 0.96, indicating a high accuracy of the test.

If the ranking is considered, B2 seems to have the highest fatigue curves, with RTFOT being significantly higher than the other two materials. B3 fails the fastest for unaged and aged materials. For comparison purposes, PAV aged material can be considered the main parameter of this test for ranking purposes in terms of the stress required by the bitumen to achieve 100,000 load cycles. All the values are given in Table 2.

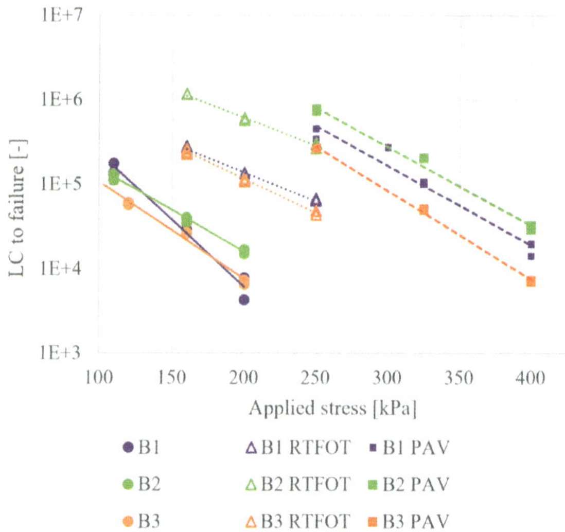

Fig. 10 Stress fatigue curves for the bitumen tested [22]

Table 2 Fatigue performance parameter

Material	Stress (kPa) @ 1E+5 LC	Material	Stress (kPa) @ 1E+5 LC	Material	Stress (kPa) @ 1E+5 LC
B1	123	B1 RTFOT	222	B1 PAV	325
B2	118	B2 RTFOT	320	B2 PAV	350
B3	105	B3 RTFOT	212	B3 PAV	290

2.10 *Performance Diagram*

For a summarizing and comparative evaluation of the behavior of bitumen and mastic mixtures, a network diagram was developed based on the approach developed by Rochlani [21, 22] for the ranking of bitumen and mastic properties. The following six performance-oriented material criteria were considered:

- shear stiffness (complex shear modulus) at 20 and 60 °C and 10 Hz at unaged condition,
- fatigue behavior (stress at a fatigue load change rate of 100,000) at PAV aged condition,
- low temperature behavior (cryogenic stress reached at −15 °C) at unaged condition,
- aging sensitivity with regard to stiffening (aging index at 60 °C, 10 Hz) and
- resistance to plastic deformation (Jnr value).

Figure 11 shows the performance diagrams in which the bitumen can be directly compared, evaluated and ranked according to their performance-relevant properties. The axes are presented in such a way that, the larger the area inside, the better is

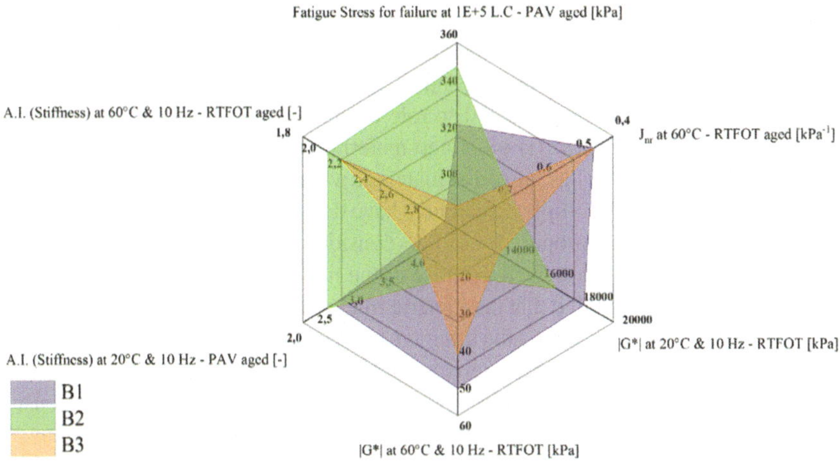

Fig. 11 Performance diagram for the evaluation of the behavior of the examined bitumen 50/70 [21]

the overall performance. Therefore, from Fig. 11, bitumen B1 is the best overall performing material. Furthermore, on the basis of this ranking, it is possible to select an optimum bitumen for a particular filler or vice versa. For example, if permanent deformation is more critical for the pavement one constructs / considered, then if the Jnr axis is observed, instantly, B1 or B3 could be chosen. However, if fatigue is more critical, the axis for fatigue can be seen and instantly, choice of B2 would be optimum. If low temperature behaviour is crucial, then bitumen B1 would be a better alternative [21].

3 Mechanical Characterization of Mastic

3.1 Background

A filler can be defined as any granular material used in asphalt with a grain size less than 75 μm. When a filler is mixed with bitumen, a mixture called asphalt mastic is formed. The bitumen to filler ratio greatly influences the performance of the mastic and consequently that of the asphalt. Once the filler exceeds 50% by mass or 30% by volume of the mastic, its effect becomes more prominent [7]. The filler has been seen to influence the mastic performance mainly for high temperatures and low frequencies. It was observed that a smaller grain size of the filler positively affects the performance as the contact area of the filler particles and bitumen is increased [22]. Additionally, the type of the filler has shown to greatly influence the stiffness properties of the mastic.

Filler was initially considered as a part of the aggregate system, where its main purpose was to fill the voids in between the coarse aggregates. Further studies, however, indicate that due to its fineness and surface characteristics, it performs much more than just filling voids [9, 26]. The role of the filler within an asphalt mixture can be divided into two parts: (i) as an inert material, acting as a void 'filler' between coarse aggregates, and (ii) as an active material, when interacting with the bitumen at the interface [2, 13]. Multiple researchers have demonstrated that the geometrical, chemical and mechanical properties of fillers considerably influence the response and performance of the mastic as well as those of the final asphalt mixtures [2, 20–22].

It has also been concluded that the mineral filler may affect asphalt paving materials in multiple ways by stiffening the bitumen, and by altering the moisture resistance, workability and compaction characteristics of asphalt mixtures [10, 19, 22]. Further studies have also confirmed that fillers with a high specific surface area and density would lead to better mastic performance, as the filler would have superior bitumen adsorption properties (i.e. denoting it as a 'strong filler') [10]. A filler is said to affect the mastic performance based on the type of filler used, its nature (i.e. acidic or basic, physio-chemical properties), and its concentration in the mixture [9, 11]. In another study conducted on fillers in terms of their surface free energy measurements, it was concluded that optimisation of a mixture is possible based on the amount of filler added [1].

This section aims to identify the modifying effect of mineral fillers on the overall behavior of bitumen. For this purpose, the shear viscoelastic behaviour and the performance of one reference bitumen and four mastic were compared.

3.2 Materials and Methods

The materials include one base bitumen of penetration grade 50/70 and four mastic made with Dolomite, Limestone, Granodiorite and Rhyolite fillers. The bitumen and the filler were mixed at 160 °C, at a mixing speed of 60 rpm and produced mastic of a binder-filler ratio of 1:1.6 by mass [22]. This ratio was chosen in order to simulate the bitumen-filler ratio of a stone mastic asphalt, SMA 11S, which is the most popular used asphalt in Germany for high volume roads.

In order to investigate the filler effect on the long-term performance of the mastic, the materials were artificially aged in the laboratory. Typically, a combination of the RTFOT– at 163 °C for 75 min and the PAV at 100 °C for 20 h are adopted to simulate the effect of short term and long-term aging for bitumen, respectively. Owing to the fact that RTFOT is a time-consuming and impractical aging procedure for mastic due to the cleaning and handling issues, the method of RTFOT+PAV was replaced by simply aging the mastic in the PAV for 25 h at 100 °C and 2.07 MPa. This choice was based on conclusions of Migliori and Corte which stated that 5 h of PAV is identical to the standard RTFOT aging procedures [16, 22].

The following abbreviations are used throughout the following sections to refer to the different mastic: mastic with Dolomite filler: 'Dolomite', mastic with Granodiorite filler: 'Granodiorite', mastic with Limestone filler: 'Limestone', mastic with Rhyolite filler: 'Rhyolite' [22].

To gain a better understanding of the properties of the different fillers, a detailed study of the physical and chemical properties of the four fillers was conducted. The results include measurements from specific gravity tests, particle size tests, BET-specific surface area (SSA) tests, scanning electron microscopy (SEM) imaging and X-ray fluorescence spectrometry tests [22].

SEM analysis (Fig. 12) of the fillers was undertaken in order to gain a closer look into the microstructure/micromorphology of the filler materials. The scale of observation extended from 8 to 300 μm. Figure 12 shows a scale of observation of 30 μm. From these microscopic images, it was observed that Dolomite and Limestone filler showed a majority of the particles having a finer grain size, with Dolomite having most particles with similar diameter. Rhyolite showed an overall well graded distribution of particle sizes with a significant portion of particles being larger than 20 μm. This filler also had a rougher texture than the rest of the fillers. Granodiorite had a huge variation between the particle sizes, ranging from very small (order of less than 5 μm) to larger particles (more than 20 μm). Their geometry showed very angular shapes, while those of the other fillers are more rounded.

Fig. 12 SEM images [22]

Table 3 Physical properties of the fillers [22]

Material	Specific surface area (m^2/g)	Pore volume (ml/g)	Average pore size (nanometer)	Density (g/cm^3)
Rhyolite	6.6294	0.0603	36.38	2.62
Limestone	4.1904	0.0168	16.04	2.72
Dolomite	6.4282	0.0202	12.57	2.85
Granodiorite	18.4665	0.0411	8.90	2.74

Table 4 Oxide composition of the fillers investigated in % [22]

Material	Na_2O	MgO	Al_2O_3	SiO_2	K_2O	CaO	TiO_2	Fe_2O_3	Others
Dolom.	0.27	26.02	2.26	5.23	0.61	61.97	0.07	1.91	1.60
Granod.	3.22	2.44	18.09	61.70	3.42	2.73	1.04	6.35	0.58
Limest.	0.18	2.02	1.29	1.94	0.19	92.97	0.12	0.82	0.46
Rhyol.	1.42	0.41	19.65	65.69	8.98	0.26	0.26	3.04	0.24

Furthermore, for a deeper look into the filler's physical properties, studies on the density, specific surface area and pore volume of the fillers were done (Table 3).

Results from these tests permit to conclude that the Rhyolite filler has the lowest density, while the Dolomite has the highest density (Table 3), which is 8.7% larger than that of the Rhyolite. In terms of the specific surface area, it is observed that the Granodiorite filler has the greatest area, which is between 4.4 and 2.8 times larger than all other fillers, whilst the Limestone filler presents the smallest specific surface area value. Rhyolite had the highest pore volume and highest pore sizes. Granodiorite had the smallest pore sizes, however the pore volume was the second highest, whilst Limestone had the lowest pore volume, but the second highest average pore size [22].

In terms of the chemical analysis, Table 4 presents the results of the various oxide profiles acquired from XRF tests. It can be observed that the oxide compositions of Granodiorite filler are similar to those of the Rhyolite filler, with the highest percentage of SiO_2 (more than 60%), followed by Al_2O_3 (18–19%). On the contrary, CaO dominates the dolomite and limestone fillers with 61.97% and 92.97% respectively. Additionally, the dolomite filler also contains a significant portion of MgO of the order of 26%, which is almost negligible in other fillers [22].

Based on the procedure described for bitumen (Sect. 2), similar procedures were used for mastic and the results are presented in the following sections.

3.3 Viscoelastic Performance

Similar frequency sweeps were done on the base bitumen and the four mastic. The 2S2P1D model was used to further approximate the rheological data acquired. The

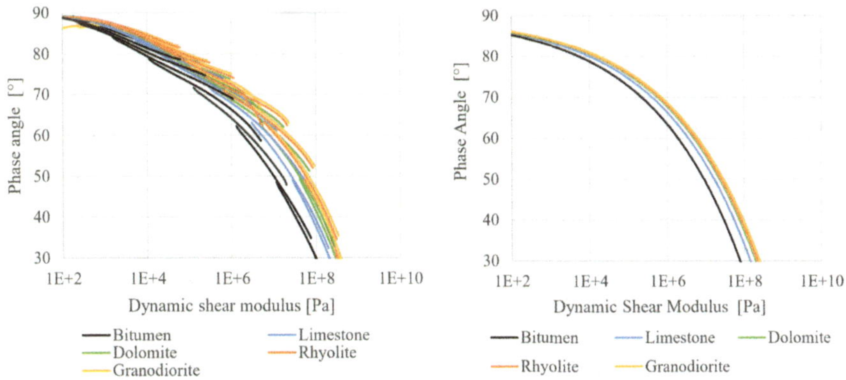

a Test results **b** 2S2P1D model

Fig. 13 Black diagrams for the unaged bitumen and mastic tested [22]

results are presented in Fig. 13 in the form of black diagrams. As can be observed, the addition of mineral fillers shifts the curves slightly to the up-right zone of the diagram, thus leading to a stiffer response. The low dynamic modulus region in the black diagram represents the high temperature rheological properties of bitumen and mastic while the higher modulus denotes the region of lower temperatures. It could be understood from Fig. 13 that within the low dynamic modulus region, the phase angle of both, the plain bitumen and mastic, gradually tends to converge to around 88°. As the phase angle approaches closer to 90°, it indicates that the tested materials are approaching the full viscous state. However, even though the same order of materials is observed throughout with bitumen having the lowest phase angles and Granodiorite having the highest, it can be observed that the three mastics-Granodiorite, Dolomite and Rhyolite are very close to one another, almost overlapping. However, Limestone tracks a slightly lower curve, with 3° to 5° lower phase angle values for any given dynamic modulus than the other mastic. Overall, the results are relatively close (i.e., differences of less than 10% in all cases) and show that the phase angle has not been significantly influenced by the type of filler.

Figure 14 shows that the dynamic shear modulus master curves of the mastic were above the master curve of the base bitumen throughout the full frequency range, which is an expected result that could be attributed to the stiffening effect of fillers. The Granodiorite and the Rhyolite mastic presented the highest moduli. A reason explaining the lowest shear stiffness of the Limestone and the highest stiffness of the Granodiorite amongst the mastic is the SSA of the fillers, since the Limestone filler had the lowest SSA while the Granodiorite the highest. A lower SSA would lead to less adsorption of the bitumen film layer, thereby forming a poorer bond that results in a lower stiffness, and vice versa [22]. At low temperatures, the Granodiorite showed the highest shear stiffness and the Limestone the lowest stiffness. At higher temperatures, all mastic were close to one another [22]. The phase angle master curve, presented in Fig. 15, confirms that the filler does not significantly influence

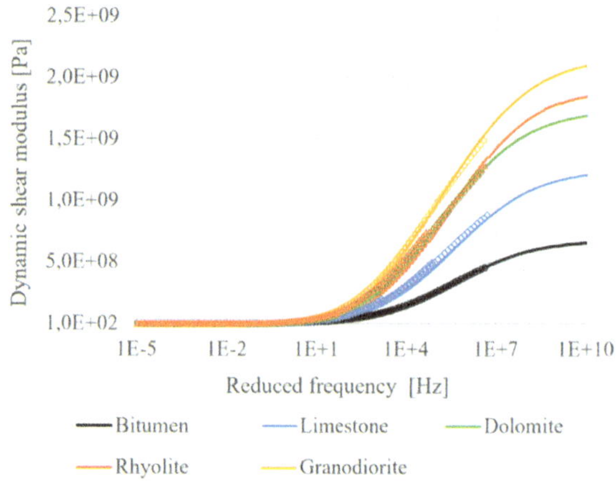

Fig. 14 Master curves of dynamic shear modulus of the unaged bitumen and mastic (lines—model, points—measured data) [22]

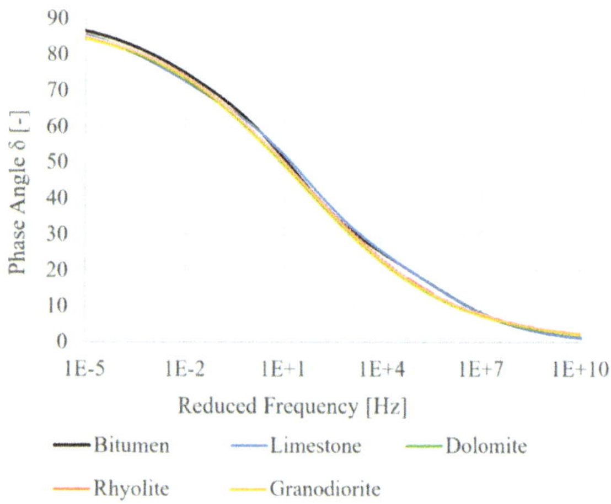

Fig. 15 Master curves of phase angle of the unaged bitumen and mastic (lines—model, points—measured data) [22]

the phase angle, since all master curves overlapped and were relatively close to one another.

3.4 *Ageing Index*

The dynamic shear modulus at 20 °C and 10 Hz was used for analysing aging effects, as it represents common design considerations. Moreover, the A.I. was also analysed using the dynamic shear modulus of the materials at 60 °C and 10 Hz, as the high temperatures is relevant when assessing permanent deformation susceptibility. The results are presented below.

It can be observed from Fig. 16, all mastic were more susceptible to aging than the bitumen. At both temperatures, the A.I. stiffness suggests that the Granodiorite mastic

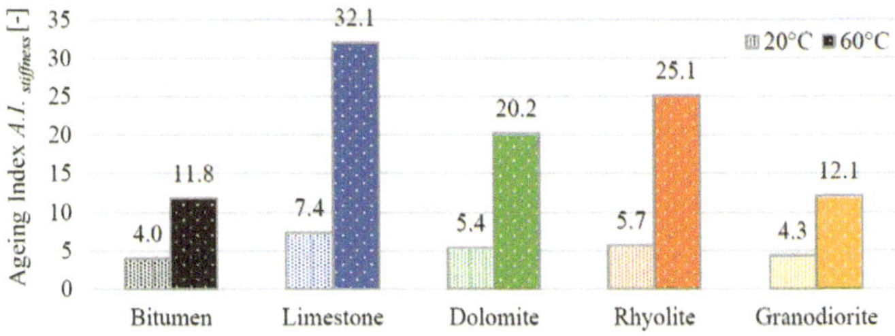

Fig. 16 Stiffness aging index at 20 and 60 °C at a frequency of 10 Hz [22]

Table 5 Results of the SSCR tests [22]

Material	Jnr (kPa^{-1})	R (%)	Cumulative Damage after 10 cycles (%)
Bitumen	0.33	10.02	94.40
Granodiorite	0.10	15.85	3.06
Dolomite	0.33	10.05	10.60
Rhyolite	0.29	2.53	9.16
Limestone	0.30	2.22	9.72

is the least susceptible to aging while the Limestone mastic is the most susceptible (the A.I. value of the Limestone mastic is 1.72 and 2.65 larger than for the Granodiorite mastic at 20 °C and 60 °C, respectively). In fact, the oxidation of the Granodiorite is almost the same as in the base bitumen at both temperatures (i.e., A.I. differences of 7.5 and 2.5% at 20 and 60 °C). The results also show that the impact of aging on stiffness is highly dependent on temperature and that the differences are larger at higher temperatures; this is observed, for example, on the fact that the values of A.I. for all mastic are between 2.8 and 4.3 times larger at 60 °C than at 20 °C.

3.5 *Plastic Deformation Peformance*

The test results of the bitumen can be compared numerically using the values of the parameters Jnr and R % obtained based on the SSCR test explained in the previous section. The results are given in Table 5, while the SSCR test curves for 60 °C aged and unaged are depicted in Fig. 16. The rutting parameters were determined for unaged mastic as rutting usually appears in early years of life time. Based on guidelines from the German and Superpave standards, the lower the Jnr value, the better is the rutting performance [22].

From Table 5 it can be observed, that the Granodiorite mastic displays the highest recovery percent of 15.85%, which is between 1.6 and 7.1 times larger than for the other mastic. Based on the guidelines, Granodiorite is expected to have the best

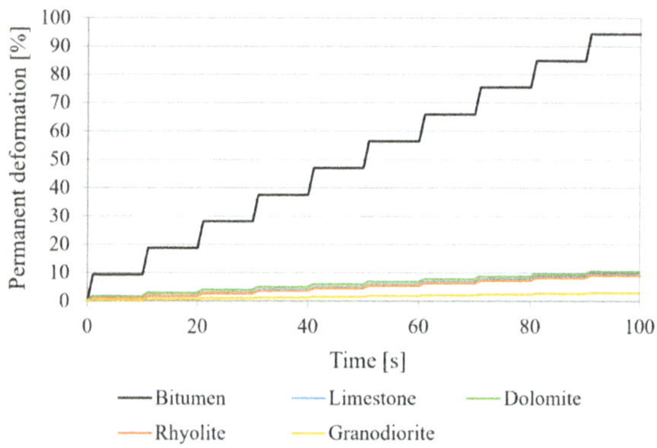

Fig. 17 Creep curves at 3.2 kPa at 60 °C for unaged mastic [22]

rutting performance as it has a significantly lower J_{nr} value of 0.10 while all the other materials are three times this value and significantly similar, including bitumen.

Figure 17 shows the creep curves of the unaged materials. It can be seen that bitumen had the highest deformation among all the materials. The mastic with Granodiorite filler was observed to have the lowest overall deformation, while the rest of the mastic are relatively close to one another. Granodiorite mastic had the highest dynamic shear modulus and hence the stiffness can be an indicator of the rutting performance.

3.6 Low Temperature Performance

Similar cryogenic tests were done on mastic as that for bitumen (Sect. 2.8). The results for these are presented here. From the results, it can be observed that all the materials exhibited a slight relaxation of thermal stresses by the end of the two hours recovery period. It was observed that while the bitumen exerted the lowest initial thermal stress, it also showed the least relaxation of approximately 0.25 N. Based on the axial force after the relaxation, the parameter for the cryogenic performance of the bitumen and mastic tested was chosen (Fig. 18) [22].

In order to determine a cryogenic stress parameter for the mastic characterisation, the normal force after stress relaxation at −15 °C (percentage of recovery in the thermal stress exerted) was taken (Table 6). It can be observed from Table 5 that plain bitumen showed the highest recovery of 64.64% and exerted a much lower thermal stress than the mastic (i.e. approximately just 20% of the stress exerted by the mastic). While among the mastic, Rhyolite exerted the highest stress relaxation (57.23%), closely followed by Limestone (55.23%). However, Limestone had the lowest thermal stress among the mastic (−0.0634 MPa), followed by Rhyolite, which expended a 23% higher stress than Limestone. On the other hand, Granodiorite showed the least recovery of 36.65% and exerted the highest thermal stress of all the mastic. From the results, it can be said that Limestone mastic had the high-

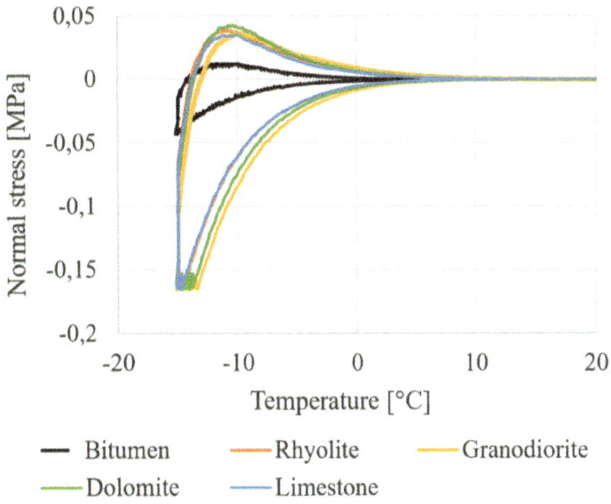

Fig. 18 Results of the cryogenic stress tests for all materials tested [22]

Table 6 Cryogenic stress parameter [22]

Material	Normal stress after 2 h at −15 °C (MPa)	Recovery (%)
Bitumen	−0.0171	64.64
Limestone	−0.0634	55.23
Dolomite	−0.0819	47.11
Rhyolite	−0.0795	57.23
Granodiorite	−0.1040	36.65

est low temperature performance and Granodiorite mastic has the least favourable performance among the materials tested.

3.7 *Fatigue Resistance*

Fatigue tests on mastic were carried out at 20 °C using cylindrical samples of plain bitumen and all five fillers at a frequency of 10 Hz. Each sample was tested nine times, with three samples subjected to the same stress so as to undergo the same strain. The results were analysed using the Dissipated Energy Ratio as previously explained.

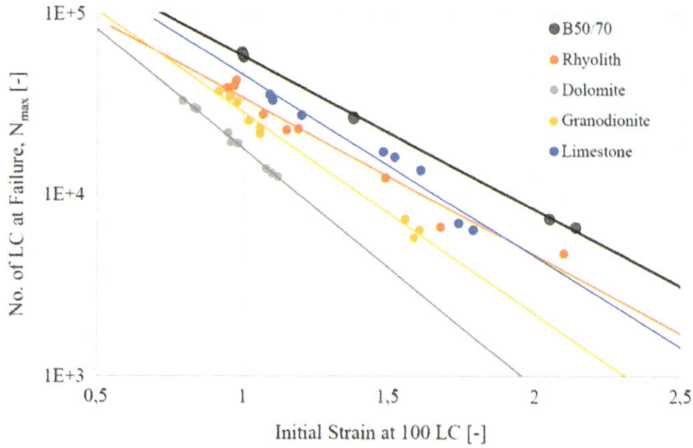

Fig. 19 Fatigue curves for the unaged bitumen and mastic tested

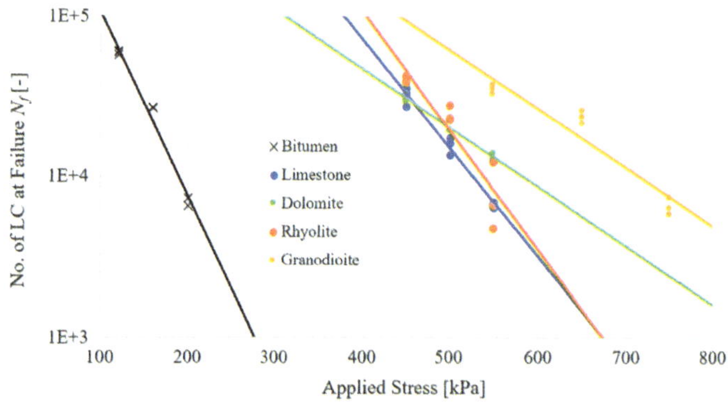

Fig. 20 Fatigue curves for the unaged bitumen and mastic tested [22]

Figures 19 and 20 present data from the fatigue tests. It can be observed that the mastic had a greater fatigue life than the control binder 50/70. From Fig. 19, which shows the graph of the initial strain at 100 load cycles versus the load cycles at failure, it can be seen that at high load cycles of value close to 100,000, Dolomite had the lowest initial strain for a given LC at failure value, followed by Rhyolite, Granodiorite, and lastly, by Limestone. From Fig. 20, it can be also observed, that at higher number of loading cycles at failure at around 100,000 cycles (i.e., right hand side of the figure), the Granodiorite mastic presents the best performance throughout, followed by the Rhyolite, the Limestone and then the Dolomite mastic. However, for failure at lower loading cycles at approximately around 10,000 cycles (i.e., left hand side of the Figure), the Dolomite performs as the second best, followed by the Rhyolite and the Limestone, respectively. Rhyolite and Limestone are very close to each other [22].

Table 7 Fatigue performance parameter [22]

Material	Initial shear modulus (kPa)	Strain @ 100000 LC (%)	Load cycles to failure @ 450 kPa (–)
Bitumen	11,174	0.72	10
Granodiorite	55,567	0.51	95,000
Dolomite	54,011	0.43	31,000
Rhyolite	41,191	0.46	46,000
Limestone	33,212	0.66	32,500

Table 7 lists the number of average load cycles to failure at a stress level of 450 kPa, which was the selected fatigue performance parameter. These data corroborate previous observations, showing that the fatigue lives of all mastic are between 31.5 and 95.0 times larger than that of the control bitumen, and that the fatigue resistance of the Granodiorite is between 2.0 and 3.0 times larger than the other mastic [22].

3.8 Performance Diagram

In order to synthesise and summarise the results of the mastic behaviour, a performance graph was developed as previously explained for the evaluation of bitumen/mastic properties. This evaluation presents the mastic stiffness (dynamic shear modulus at 20 °C and 10 Hz), fatigue behaviour (number of load cycles until the macro crack criterion is reached at a stress level of 450 kPa, or N_f), low temperature behavior (cryogenic stress remaining after the relaxation phase, calculated from the axial force produced by the material to prevent expansion), aging sensitivity to stiffening (aging index or $A.I._{stiffness}$ at 60 °C, 1.59 Hz) and, finally, the resistance to plastic deformation (Jnr value of the unaged mastic) [22].

Figure 21 shows the performance diagram for all mastic tested. The performance diagrams were structured in such a way that an improvement in a material property can be expected with an increasing distance from the origin of the spider diagram. Thus, mastic that covered a large area in these diagrams would be expected to have a superior performance. As a result, performance diagrams permit to easily assess, compare, and rank the mastic based on their performance-relevant properties. Based on this ranking, it is possible to select an optimal filler for a specific bitumen. For the four mastic evaluated, Granodiorite mastic had the largest area within the performance diagram and, hence, that it seems to be the most suitable filler to provide a high-performance behavior in asphalt mixtures, followed by the Rhyolite, the Dolomite and the Limestone, for the specific tested base bitumen [22].

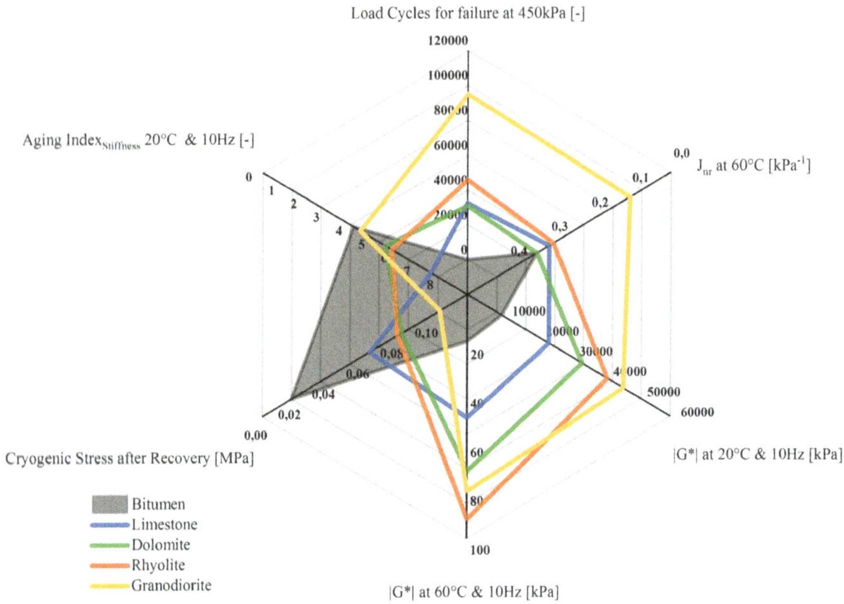

Fig. 21 Performance diagrams for the bitumen and the mastic evaluated [22]

4 Mechanical Characterization of Mortar

4.1 Background

Several studies [25, 27] have concluded that the service life of flexible pavements depends significantly on the nature, quality and composition of the bituminous matrix that binds the coarse aggregates. This matrix, referred herein as mortar, is a mixture of bituminous binder, filler (aggregates with average grain size of less than 0.075 mm) and sand (aggregate particles with an average grain size of less than 2 mm).

The viscoelastic nature of the mortar dictates the macroscopic temperature-frequency dependency of the asphalt mixture. Furthermore, changes in the mortar due to aging will have profound effects on the fatigue resistance of the mix, hence affecting its long-term durability. Therefore, a comprehensive laboratory characterization of mortar is a necessary prerequisite for mesoscale/multiscale numerical modeling to match the holistic behavior of the mixture.

Previous experience of the authors has revealed that common equipment used for bitumen testing, such as the DSR or viscometers, are not suitable for mortar testing because of settling and workability problems. Analogous, traditional tests of asphalt concrete, such as indirect tensile test, bending beam test or uniaxial test, show a limitation because of distortion of the specimen. In this context, it can be said that there is a gap in terms of a laboratory testing equipment capable of characterizing the behavior of mortar.

After taking into account the need of a testing equipment for determining the viscoelastic properties of mortar, the first part of this section presents a novel shear tester, known as DDST, aimed to characterize the rheological behavior of asphalt concrete at different length scales with special focus on the mortar scale. The second part of this section compares the rheological properties of bitumen, mortar and asphalt determined with the results of the DDST.

4.2 *Dresden Dynamic Shear Tester (DDST)*

4.2.1 Experimental Rig

The DDST is a novel testing equipment that can be used to determine the coefficients of viscoelastic models of asphalt concrete, mortar and bitumen. The DDST is a direct dynamic shear box with normal stress applied. It consists of two stacked rings separated by a gap of 1 mm thickness to allow free relative movement between them. An asphalt concrete/mortar/bitumen specimen is installed inside the DDST using a special adaptor that fits into the hubs of the rings. The DDST was designed to be driven by a universal testing machine (UTM) with temperature and relative humidity control (Fig. 22). The load cell based piston of the UTM provides movement in the shear direction, and an in-built pneumatic pressure actuator applies load in axial direction. Shear and axial displacements are measured by four external LVDT.

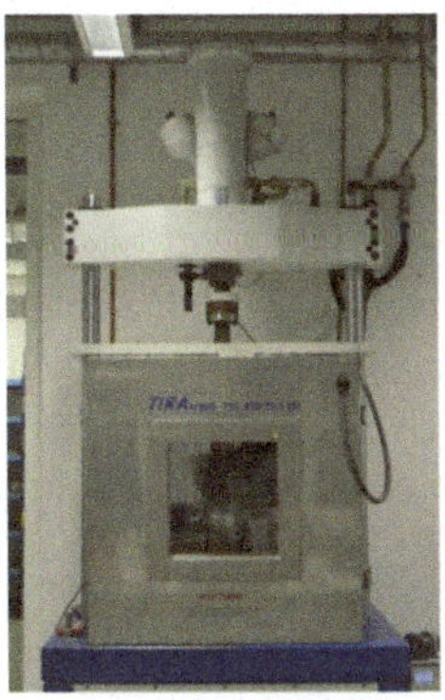

a Overall assembly

b Test device inside the temperature chamber

c Test device

Fig. 22 DDST device [4]

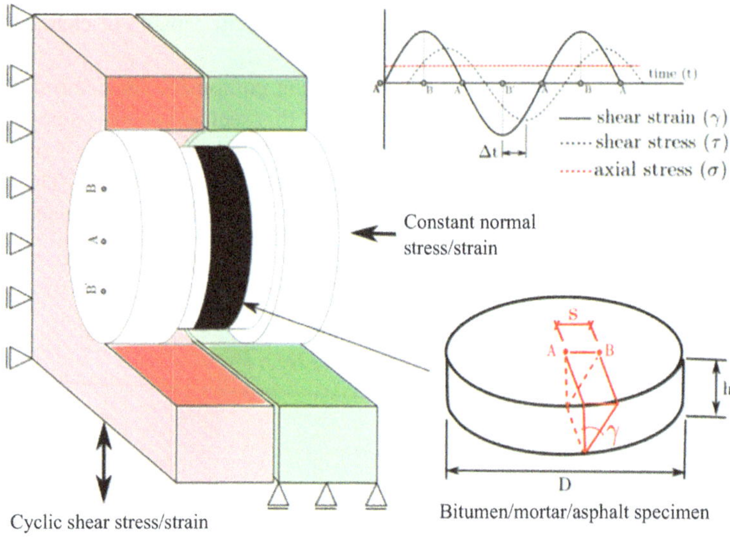

Fig. 23 DDST principle of operation [4]

4.2.2 Principle of Operation

The principle of operation of the DDST is as follows: the specimen is sandwiched between two disc-shaped platens, one of them (movable platen) is allowed to move in the shear direction while the other one (fixed platen) remains fixed during testing. The test is performed by oscillating the movable platen about its vertical axis at different frequencies and temperatures, as seen in Fig. 23.

4.2.3 Specimens

The specimens are cylindrical in shape with the dimension given in Fig. 24. Asphalt concrete specimens are fixed to the plates of the DDST by using a two-component epoxy adhesive. Bitumen and mortar specimens are prepared by pouring the material into a mold made of teflon. The binding property of bitumen ensures fixed bonding between the bitumen/mortar and the steel plates of the DDST.

4.3 Rheological Characterization of Bitumen, Mortar and Asphalt Concrete in the DDST

The time-frequency dependence of asphalt concrete properties is linked to the rheological behavior of the bituminous matrix that binds the coarse aggregates. In order

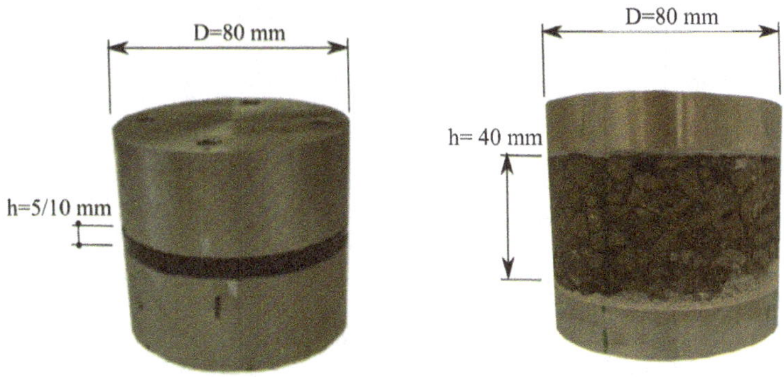

a Bitumen and mortar specimen **b** Asphalt concrete specimen

Fig. 24 DDST specimens [4]

to understand the macroscopic behavior of asphalt concrete, it is necessary to understand the behavior of its constituents at different length scales. This section aims to investigate and compare the rheological properties of asphalt concrete at three different scales of observation:

- At the bitumen scale, with a characteristic length in the range of μm.
- At the mortar scale, with a characteristic length in the range of mm.
- At the asphalt scale, with a characteristic length in the range of cm.

4.3.1 Materials

A Stone Mastic Asphalt (SMA) with a maximum grain size of 11 mm (SMA 11 D S) was selected as reference material Fig. 25. The SMA was characterized by a strong coarse aggregate skeleton with a maximum grain size of 11 mm, a bitumen content of 6.9% by weight, a bulk density of 2436 kg/m^3 and a void ratio of 2.1% by volume.

The binder of the mix was an unmodified penetration graded bitumen of the type 50/70. The mortar of the asphalt mixture was produced by not adding the mineral aggregate fractions with grain size larger than 2 mm. Thus, the resulted mortar was a mixture of bitumen, filler and sand with a volumetric composition of 38.2% bitumen, 28.5% filler and 33.3% sand.

The mineral filler was limestone with the grain size distribution shown in Fig. 26a. Limestone is a sedimentary rock composed of chemically active materials mainly made of calcite (93% CaO, 2% SiO_2, 2% MgO, 3% others). Limestone fillers are found abundantly in aggregates used in hot mix asphalt mixtures in Germany. The other mineral in the mortar was a Diabase sand with the grain size distribution illustrated in Fig. 26b. Diabase is an igneous rock composed mainly of chemically inactive particles of quartz.

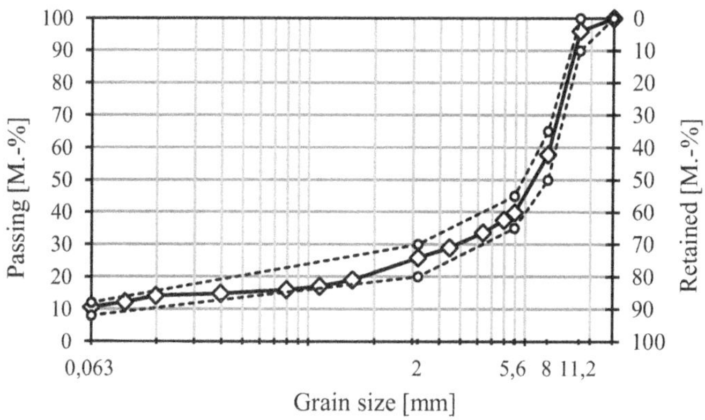

Fig. 25 Grain size distribution of asphalt mixture [4]

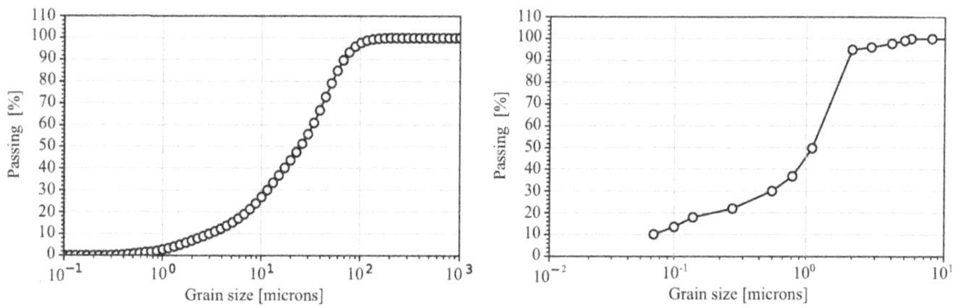

Fig. 26 Grain size distribution of mortar's minerals [4]

4.3.2 Experimental Procedure

Frequency sweep tests were performed in the DDST at three different temperatures in order to collect data to describe the viscoelastic behavior of the materials. Small strain levels were targeted to ensure that the materials were tested within their linear viscoelastic region. The temperature frequency combinations at which the tests were carried-out are shown in Table 8.

4.3.3 Results

The test results, in the form of master curves of dynamic shear modulus are presented in Fig. 27. The master curves show a stiffening effect that shifts the curve towards higher values as the observation scale increases. The shape of the mortar and bitumen curves is similar. The main difference between bitumen/mortar and asphalt concrete

Table 8 Temperatures and frequencies for frequency sweep test in the DDST

Material	Temperature (°C)	Frequency (Hz)
Bitumen	0; 10; 20	0.1; 0.4; 0.8; 1; 1.5; 2; 4; 5; 8; 10
Mortar	0; 10; 20	0.1; 0.4; 0.8; 1; 1.5; 2; 4; 5; 8; 10
Asphalt concrete	10; 20; 30	0.1; 0.4; 0.8; 1; 1.5; 2; 4; 5; 8; 10

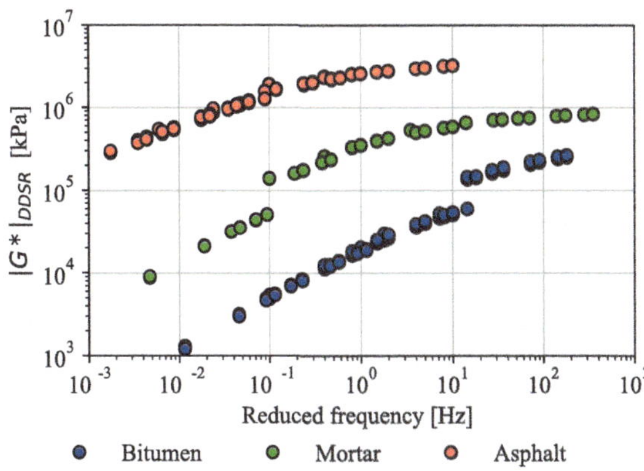

Fig. 27 Master curve of dynamic shear modulus [4]

curves is appreciated at high temperatures (low frequencies). The reason for the difference in shapes can be explained because at high temperatures the aggregate contribution to the shear modulus of asphalt mixture is more evident.

5 Mechanical Characterization of Asphalt Concrete

In order to characterize the behavior of asphalt mixtures used in road construction the following aspects have to be considered:

- Stiffness and stiffness time-temperature dependency.
- Permanent deformation and rut development.
- Fatigue and degradation by fatigue.
- Low temperature stress development and low temperature cracking.

The first aspect corresponds to the elastic-viscoelastic material behavior and the other three are related to pavement distresses. Four different asphalt materials were characterized in this section in terms of stiffness, permanent deformation and fatigue behavior to identify the differences in performance due to the variation of the bitumen type and aggregate gradation.

Table 9 Description of the materials

Mix type	Bitumen type	Bitumen content (wt %)	Air voids (%)	Bulk density (g/cm^3)	Aggregate mineralogy
SMA 11 D S	Pen 50/70	6.72	1.8	2.557	Basalt
AC 11 D S	Pen 50/70	6.04	0.6	2.601	Basalt
SMA 11 D S	PmB 25/55-55	6.70	2.8	2.553	Basalt
PA 11 D S	PmB 25/55-55	6.00	25.9	2.019	Basalt

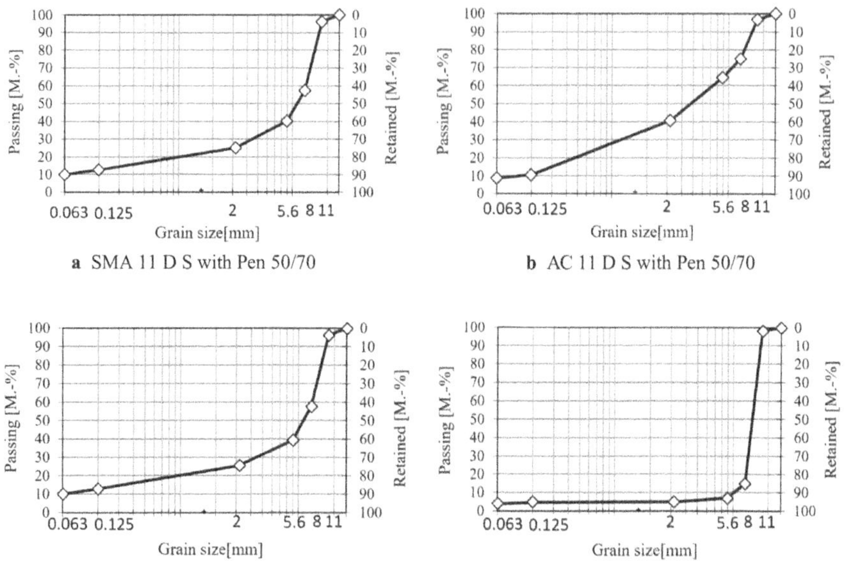

a SMA 11 D S with Pen 50/70 **b** AC 11 D S with Pen 50/70

c SMA 11 D S with PmB 25/55-55 **d** PA 11 D S with PmB 25/55-55

Fig. 28 Grain size distribution of asphalt mixtures [4]

5.1 Materials

Four different asphalt wearing mix types were investigated: two SMA mixes with different bitumen type, one asphalt concrete mixture and one open porous mixture. Table 9 presents the description of all mixes and Fig. 28 shows the grain size distribution.

The materials were produced in a batch type plant in Geilenkirchen (Germany) at a mixing temperature of 165 °C. Afterwards, the mixtures were transported to the test track of the institute of highway engineering at RWTH Aachen University (25 km distance) where they were laid down and compacted. Several cylindrical cores (150 mm diameter and 300 mm height) were drilled out from the track and delivered to the Technische Universität Dresden for testing.

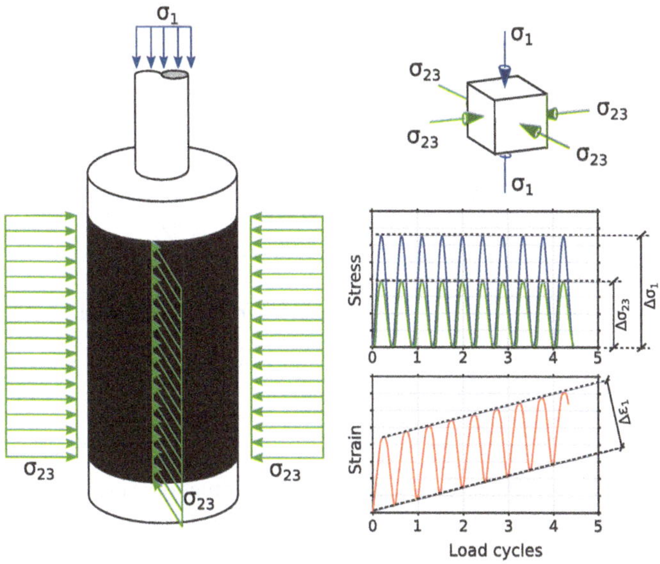

Fig. 29 RLTT. Principle of operation [4]

5.2 *Stiffness and Stiffness Time-Temperature Dependency*

The stiffness time-temperature dependency of all materials was determined with the results of RLTT. The RLTT evaluates the viscoelastic response of a cylindrical asphalt specimen by three-axle-compression. The specimen is subjected to a cyclic confining pressure, σ_{23}, and to a cyclic vertical load, σ_1, as seen in Fig. 29.

The triaxial equipment used to perform the tests consisted of a test frame, a hydraulic unit, a test cylinder and a climatic chamber. Axial deformations were measured by two external inductive displacement measuring systems and an internal magnetic measuring system. The external systems measured displacements within the whole length of the specimen. The internal system measured displacement within the center one-half of the specimen. The samples were cylindrical in shape with size of 150 mm diameter and 300 mm long.

In a RLTT, the material is considered transverse isotropic (i.e., with one axis of symmetry). For transverse isotropic materials, the stress-strain relationship is given by the following equation

$$\Delta\epsilon_1 = \frac{\Delta\sigma_1}{|E^*|} - 2\nu\frac{\Delta\sigma_{23}}{|E^*|}. \tag{2}$$

Zeissler [30] showed that if the amplitude of one of the stress components is constant, then the trend of the strain function in dependence of vertical and horizontal stresses is linear. This linear trend was used to obtain the following functional relationship between vertical strain and stresses

$$\Delta\epsilon_1 = a_1\Delta\sigma_1 + a_2\Delta\sigma_{23} + a_3\Delta\sigma_1\Delta\sigma_{23}. \tag{3}$$

where a_1, a_2 and a_3 are material parameters dependent of temperature and frequency.

The last term of Eq. (3) represents the dependence of the vertical strain on the cell pressure. For stress independent materials, the surface function is reduced to

$$\Delta\epsilon_1 = a_1\Delta\sigma_1 + a_2\Delta\sigma_{23}. \tag{4}$$

The dynamic elastic modulus in the RLTT can be determined by equating the derivatives of Eqs. (2) and (3) with respect to σ_1, as follows

$$|E^*| = \frac{1}{a_1 + a_3\Delta\sigma_{23}}. \tag{5}$$

The master curves of absolute modulus in axial direction were constructed based on the time-temperature superposition principle. Figure 30 shows the master curves of the four materials for different confining pressures. It can be observed that the elasticity modulus of the open graded material (PA 11) shows high stress dependency at high temperatures. This is explained by the stone-to-stone contact of the aggregate skeleton. Open and gap graded asphalt mixes, such as open porous asphalt and stone mastic asphalt, show high non linear behavior due to the contact and interlocking of stones. This stress dependent behavior is not evidenced on the dense graded asphalt mixture AC 11S.

Analogous, the absolute modulus was calculated for the stress independent approach using

$$|E^*| = \frac{1}{a_1}. \tag{6}$$

Figure 31 shows the comparison of the master curves of all materials. It is clearly observed, that the stiffness of the open porous asphalt is much lower than the stiffness of the stone mastic asphalt and the asphalt concrete. On the other hand, no major differences were found on the stiffness of the other three materials. One can highlight the fact that, at high temperatures (lower part of the curve), the SMA 11S with PmB showed the highest elasticity modulus. Instead, at intermediate and low temperatures the AC 11S was the stiffest material.

5.3 Permanent Deformation

The increase in traffic volume and environmental temperature, evidenced in recent years, has triggered road safety problems associated with rutting. Rutting is a major form of pavement distress characterized by a surface depression in the wheel path.

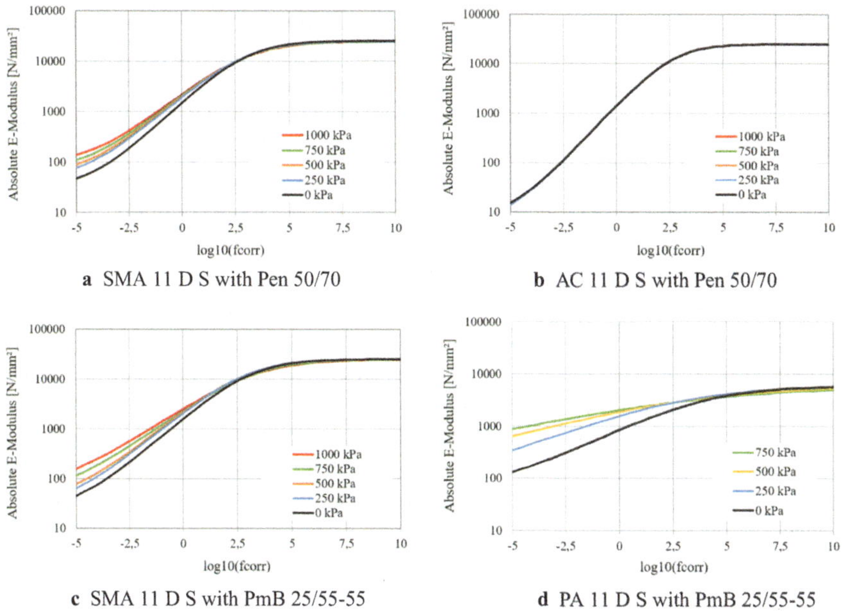

Fig. 30 Master curves, effect of confining pressure

Fig. 31 Master curves of the asphalt mixes tested

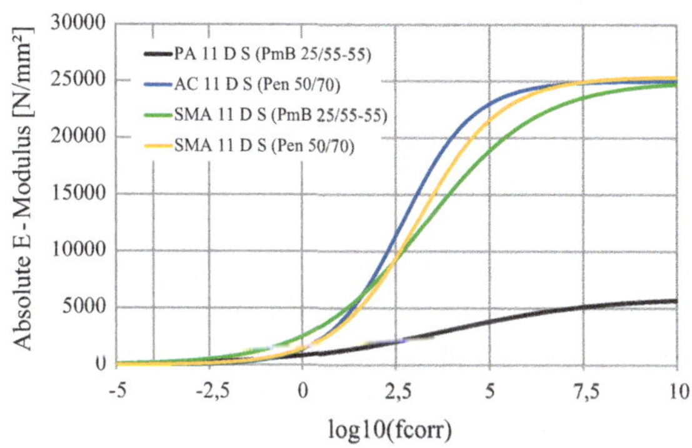

For thick asphalt pavements, rutting occurs mainly due to inelastic deformation of the asphalt wearing course, specially at high temperatures. This is a major concern nowadays because of the inclement temperature rise due to global warming.

The repeated load uniaxial test is usually used to characterize the rutting propensity of asphalt materials [29]. This test is essentially a time sweep in which a sinusoidal axial load is applied to a cylindrical specimen at a constant frequency and temperature for many repetitions. The permanent deformation of the specimen is measured with displacement transducers. Alternatively, a state of the art technique, known as digital image correlation (DIC) can be used. This technique is well suited for the characterization of materials properties in the plastic ranges.

The permanent deformation of all materials was studied using the repeated load uniaxial test. Vertical deformations were measured over the whole length of the specimen by two linear vertical displacement transducers (LVDT) as well as with a DIC system installed outside of the temperature chamber. The DIC system was used to measure strains over the whole specimen's area. The DIC system includes two high speed cameras with 12 mm lenses that were located in front of the specimen on a tripod. The cameras were set with a speed of 200 fps with a full resolution of 1280 × 1024 px. DIC analysis software-VIC 3D-was then used to apply the correlation algorithm between images taken before and after deformation.

The permanent deformation of the asphalt mixtures was studied considering large strain theory. Figure 32 shows exemplary digital images obtained from the DIC system at different load cycles in a plastic test on the AC 11S. It can be observed that the vertical strain distribution is not homogeneous throughout the whole surface of the specimen. High strains are developed within the center one-half of the specimen. At both specimen's ends the vertical strains are much lower due to boundary effects associated with the interaction of the specimen and the loading plates of the testing device.

Figure 33 and Fig. 34 show the results of the permanent deformation tests at 50 °C and 20 °C, respectively. The tests were performed at 8 Hz with a resting time of 2 seconds every 6 load cycles. From the results it is clearly observed, that the permanent deformation resistance of the SMA is much higher than the plastic resistance of the AC. SMA mixtures are characterized by a good resistance to rutting due to their coarse aggregate skeleton.

5.4 *Fatigue*

The fatigue resistance of all asphalt mixes was measured with the indirect tensile test (ITT). The ITT is a practical and effective test that can be used to determine the elastic tensile properties and fatigue resistance of asphalt mixtures. In the ITT, a cylindrical-shape specimen is loaded by vertical radial compression and the response horizontal deformation is measured.

Several time sweeps were performed on all materials at different stress levels until specimen failure. The criterion used to characterize fatigue failure was based on the dissipated energy ratio (DER). Failure was determined at the point in which the DER reaches a maximum throughout the test.

Figure 35 shows the results of the fatigue tests in the form of Wöhler curves. It is observed that the curves of both asphalt materials with polymer modified bitumen are shifted upwards in almost a parallel manner with respect to the asphalts with plain bitumen, which corresponds to a higher fatigue life for the same elastic strain.

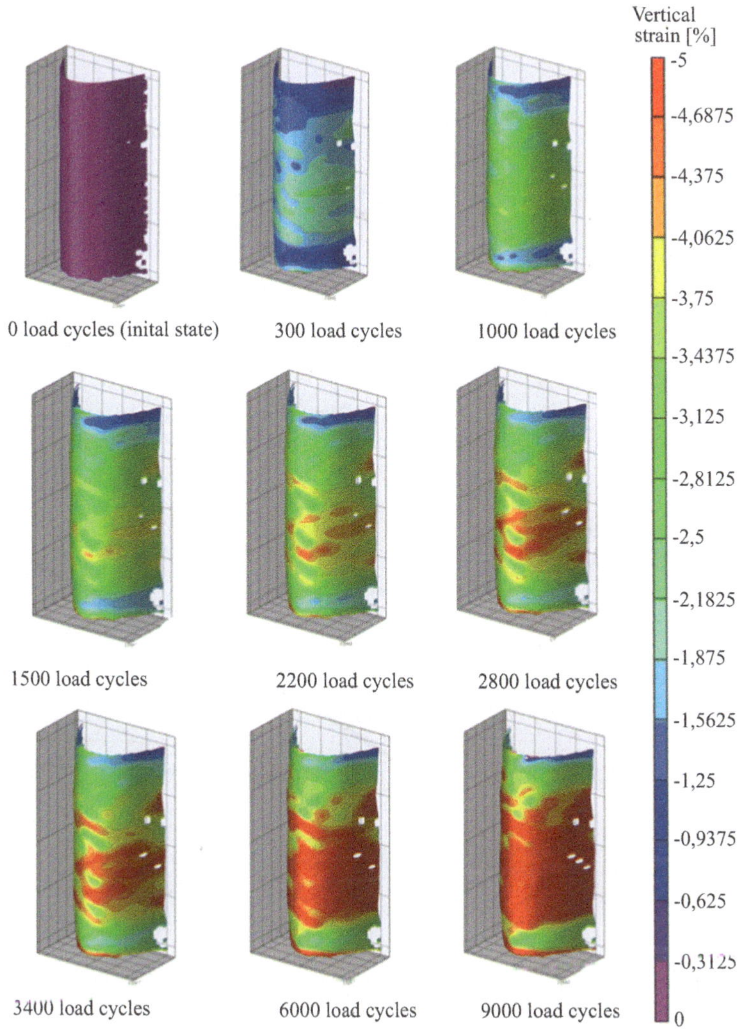

Fig. 32 Accumulation of vertical strains (AC 11S) [3]

6 Conclusions and Outlook

A clear tendency has been observed in the asphalt pavement industry towards characterizing and modeling of road construction materials on a nano (nm...μm), micro (μm...mm), meso (mm...dm), and mega (dam..km) scale [6, 12, 18]. The nano scale of asphalt concrete is still a widely unknown scientific territory. Some advances in this area has been done by investigating the colloidal structure of bitumen [13–15]. The knowledge is broader on the other side of the scale, at mega level of observa-

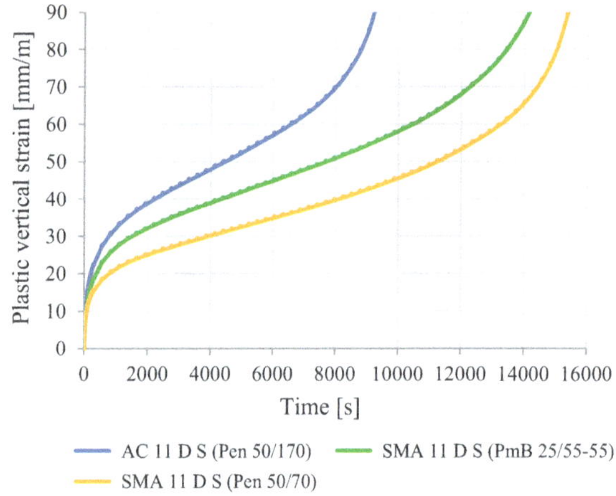

Fig. 33 Permanent deformation curves at 50 °C with a vertical load of 100 kPa

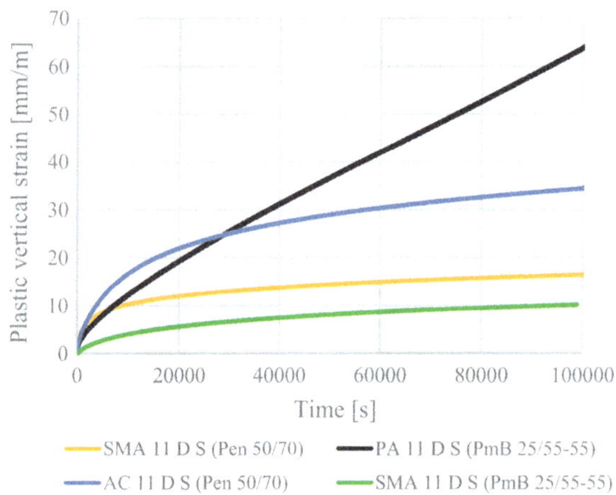

Fig. 34 Permanent deformation curves at 20 °C with a vertical load of 250 kPa

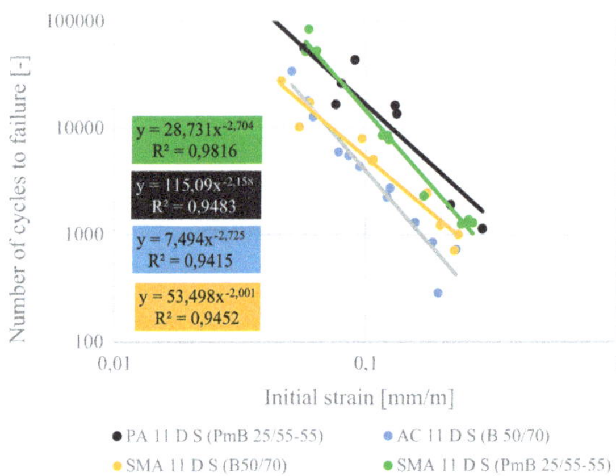

Fig. 35 Fatigue curves

tion. However, the costs associated with scientific testing on a 1:1 scale are very high. Furthermore, the parameters of large scale tests, such as climate, traffic loads, and construction geometries, limit the general value of the results. Development of advanced testing at micro and meso scales has made much progress in the last decades, specially for bitumen and asphalt.

In this context, this chapter presented state-of-the-art characterization techniques of asphalt materials at micro and meso scales. The results of the laboratory procedures were envisaged to be used within a holistic experimental-numerical framework aiming to simulate the vehicle-tire-pavement interaction.

All experimental tests on bitumen and mastic (Sects. 2 and 3) were foreseen to determine material properties that can be related to major pavement distresses, such as rutting at high temperatures, fatigue at intermediate temperatures and cracking at low temperatures. Some key material indicators were selected for the construction of performance diagrams. These diagrams are useful tools that can be used not only for quality control but also as material fingerprints.

An important contribution to the state of the knowledge was presented in Sect. 4. In this section, a novel shear tester, known as DDST, was described. The DDST is an equipment that can be used to characterize the shear response of mortar under different temperature and frequency conditions. The DDST is a versatile research tool that was conceptualized to determine the viscoelastic properties of mortar as input to multiscale modeling of asphalt concrete. The ability to predict asphalt performance using multiscale modeling techniques, would enhance the design of flexible pavements. It would be advantageous if road engineers could account for the effect of the asphalt constituents and bring out tailored mechanical responses by optimizing specific parameters at the microscopic level, such as the bitumen in the mortar or the shape/gradation of the aggregates.

Finally, in Sect. 5, four different asphalt mixes were industrially produced and tested in the laboratory. The results showed that the bitumen type and the aggregate gradation are two of the most important factors that affect the performance of the mixture.

References

1. Alfaqawi, R., Airey, G., Grenfell, J.: Effects of mineral fillers and bitumen on ageing of asphalt mastics properties. In: Proceedings of the 10th International Conference on the Bearing Capacity of Roads, Railways and Airfields (BCRRA), Athens (2017)
2. Antunes, V., Freire, A., Quaresma, L., Micaelo, R.: Influence of the geometrical and physical properties of filler in the filler-bitumen interaction. Constr. Build. Mater. **76**, 322–329 (2015)
3. Behnke, R., Canon Falla, G., Leischner, S., Händel, T., Wellner, F., Kaliske, M.: A continuum mechanical model for asphalt based on the particle size distribution: numerical formulation for large deformations and experimental validation. Mech. Mater. **153**, 103703 (2021)
4. Canon Falla, G.: Characterization and modeling of asphalt concrete from micro-to-macro scale. Ph.D. thesis, Institute of Pavement and Urban Engineering, Technische Universität Dresden (2021)

5. Di Benedetto, H., Delaporte, B., Sauzéat, C.: Three-dimensional linear behavior of bituminous materials: experiments and modeling. Int. J. Geomech. **7**, 149–157 (2007)
6. Eberhardsteiner, L., Hofko, B., Blab, R.: Multiscale modeling to predict hot mix asphalt stiffness behavior. In: Proceedings of the 4th Chinese-European Workshop on Functional Pavement Design (CEW 2016), Delft (2016)
7. Elnasri, M., Airey, G., Thom, N.: Experimental investigation of bitumen and mastics under shear creep and creep-recovery testing. In: Proceedings of the 2013 Airfield & Highway Pavement Conference, Los Angeles, California (2013)
8. Grover, A., Little, D., Bhasin, A.: Structural characterization of micromechanical properties in asphalt using atomic force microscopy. J. Mater. Civil Eng. **24**, 1317–1327 (2012)
9. Guo, M., Tan, Y., Hou, Y., Wang, L., Wang, Y.: Improvement of evaluation indicator of interfacial interaction between asphalt binder and mineral fillers. Constr. Build. Mater. **151**, 236–245 (2017)
10. Kandhal, P.S., Chakraborty, S.: Effect of asphalt film thickness on short- and long-term aging of asphalt paving mixtures. Transp. Res. Record **1535**, 83–90 (1996)
11. Kavussi, A., Hicks, R.G.: Properties of bituminous mixtures containing different fillers. J. Assoc. Asphalt Paving Technol. **66**, 153–186 (1997)
12. Lackner, R., Blab, R., Eberhardsteiner, J., Mang, H.: Characterization and multiscale modeling of asphalt - recent developments in upscaling of viscous and strength properties. In: Proceedings of the III European Conference on Computational Mechanics, Lisbon (2008)
13. Lesueur, D.: Evidence of the colloidal structure of bitumen. In: Proceedings of the ISAP International Workshop on the Chemo-Mechanics of Bituminous Materials, Delft (2009)
14. Lesueur, D., Lázaro Blázquez, M., Andaluz Garcia, D., Ruiz Rubio, A.: On the impact of the filler on the complex modulus of asphalt mixtures. Road Mater. Pavement Des. **19**, 1–15 (2018)
15. Loeber, L., Muller, G., Morel, J., Sutton, O.: Bitumen in colloid science: a chemical, structural and rheological approach. Fuel **77**, 1443–1450 (1998)
16. Migliori, F., Corté, J.F.: Comparative study of rtfot and pav aging simulation laboratory tests. Transp. Res. Record **1638**, 56–63 (1998)
17. Mitchell, M., Link, R., Mo, L., Huurman, M., Wu, S., Molenaar, A.: Research of bituminous mortarfatigue test method based on dynamic shear rheometer. J. Test. Eval. **40**, 103738 (2012)
18. Partl, M., Bahia, H., Canestrari, F., de la Roche, C., Di Benedetto, H., Piber, H., Sybilski, D.: Advances in Interlaboratory Testing and Evaluation of Bituminous Materials: State-of-the-Art Report of the RILEM Technical Committee 206-ATB. Springer, Netherlands (2012)
19. Prowell, B., Zhang, J., Brown, E.: Aggregate Properties and the Performance of Superpave-Designed Hot Mix Asphalt. Transportation Research Board, Washington D.C. (2005)
20. Rieksts, K., Pettinari, M., Haritonovs, V.: The influence of filler type and gradation on the rheological performance of mastics. Road Mater. Pavement Des. **20**, 1–15 (2018)
21. Rochlani, M.: Performance-based characterisation of Bitumen and Mastic using the DSR. Ph.D. thesis, Institute of Pavement and Urban Engineering, Technische Universität Dresden (2021)
22. Rochlani, M., Leischner, S., Canon Falla, G., Wang, D., Caro, S., Wellner, F.: Influence of filler properties on the rheological, cryogenic, fatigue and rutting performance of mastics. Constr. Build. Mater. **227**, 116974 (2019)
23. Rochlani, M., Leischner, S., Canon Falla, G., Goudar, P., Wellner, F.: Influence of source and ageing on the rheological properties and fatigue and rutting resistance of bitumen using a dsr. In: Proceedings of the 9th International Conference on Maintenance and Rehabilitation of Pavements (Mairepav9). Zürich (2020)
24. Rochlani, M., Leischner, S., Wareham, D., Caro, S., Falla, G.C., Wellner, F.: Investigating the performance-related properties of crumb rubber modified bitumen using rheology-based tests. Int. J. Pavement Eng. 1–11 (2020)
25. Schüler, T., Jänicke, R., Steeb, H.: Nonlinear modeling and computational homogenization of asphalt concrete on the basis of xrct scans. Constr. Build. Mater. **109**, 96–108 (2016)
26. Wang, H., Al-Qadi, I., Faheem, A., Bahia, H., Yang, S.H., Reinke, G.: Effect of mineral filler characteristics on asphalt mastic and mixture rutting potential. Transp. Res. Rec.: J. Transp. Res. Board **2208**, 33–39 (2011)

27. Wimmer, J., Stier, B., Simon, J.W., Reese, S.: Computational homogenisation from a 3d finite element model of asphalt concrete linear elastic computations. Finite Elem. Anal. Des. **110**, 43–57 (2016)
28. Wörner, T.: TP Asphalt-StB. Teil 24 Spaltzug-Schwellversuch -Beständigkeit gegen Ermüdung. Forschungsgesellschaft für Straßen- und Verkehrswesen (FGSV). Köln (2020)
29. Wörner, T.: TP Asphalt-StB. Teil 25 B 1 Einaxialer Druck-Schwellversuch - Bestimmung des Verformungs - ver hal tens von Walzasphalt bei Wärme (2020). Forschungsgesellschaft für Straßen- und Verkehrswesen (FGSV). Köln (2020)
30. Zeissler, A.: Investigation of the stress-dependent material behavior of asphalt. Ph.D. thesis, Institute of Pavement and Urban Engineering, Technische Universität Dresden (2015)

Experimental and Simulative Methods for the Analysis of Vehicle-Tire-Pavement Interaction

Jan Friederichs, Guru Khandavalli, and Lutz Eckstein

Abstract This subproject of the research group FOR2089 focuses on the vehicle-induced road load and the interaction between vehicle, tire and road. The here presented measurement methods are not only used to validate the different simulation models established by the research group members, but also to parametrize and optimize physical tire models for the application to real road topology as well as asphalt texture models. In comparison to models with a single-point road contact, a discretized tire footprint interacts locally with the road, which allows the investigation of ground pressure and shear stress distribution on varying surfaces. In previous studies, this local road load has not been validated at this level of detail. By a holistic analysis of the tire's influence on the vehicle and the road at the same time, a more realistic vehicle-tire-pavement behavior can be predicted by the simulation models. This chapter is separated into two parts. The first part mainly focusses on methods regarding vertical forces. As heavy-duty vehicles cause the highest loads on the main traffic routes, the methods for vertical dynamics are applied for heavy-duty purposes. The influence of different component model approaches on the road load are presented and validated using a hydraulic axle test rig. The second part presents methods for analyzing horizontal forces in the tire-road interface on a passenger car level to take advantage of specialized measurement systems. The influence of asphalt modulation on the tire force transmission mechanisms and the vehicle dynamics are presented. Furthermore, the friction characteristics on asphalt is investigated with a special regard to future tire measurement on artificial surfaces with asphalt texture.

Keywords Tire · Pavement · Interaction · Vehicle dynamics · Friction · Simulation · Validation

Funded by the German Research Foundation (DFG) under grant EC 412/1.

J. Friederichs (✉) · G. Khandavalli · L. Eckstein
Institute for Automotive Engineering, RWTH Aachen University, Aachen, Germany
e-mail: jan.friederichs@ika.rwth-aachen.de

M. Kaliske et al. (eds.), *Coupled System Pavement—Tire—Vehicle*,
Lecture Notes in Applied and Computational Mechanics 96,
https://doi.org/10.1007/978-3-030-75486-0_5

1 Methods for Analyzing the Vehicle-Tire-Pavement Interaction Focused on Vertical Forces

The prediction of the future vehicle population is a necessary investigation for a more realistic derivation of robust requirements for durable pavements. While the individual traffic will rise only slightly, studies forecast an increase of freight traffic on German roads of around 81% in 2050 compared to 2007 [20, 23]. Semi-trailer trucks transport the majority of the goods and an increase of the vehicle mileage of around 100% is expected over the next 25 years [23]. Due to the high wheel loads, the increasing freight traffic has a major impact on road damage. As the dynamic wheel loads can increase the pavement contact stresses in comparison to the static wheel loads, simulation and vibration models are used to investigate the dynamic wheel loads on road surfaces in the following sections.

1.1 Modeling of Tire Characteristics

The tire is the force-transmitting component between road and vehicle and plays the key role in all further investigations. According to Fig. 1a, there are three main tire model types available, which differ in the model approach and the use case.

Empirical tire models, such as Magic Formula (MF) [19], use mathematical formulas to calculate the tire characteristics. While their small calculation effort allows real-time applications, the tire-road-interaction is determined in only one tire-road point contact element. Physical tire models, such as FTire [9] or CDTire [7], are based on natural scientific descriptions and build up a discrete contact patch, which enables a more realistic investigation of dynamic wheel loads on road topography, such as Belgian blocks (cf. Fig. 1b). In finite element models, material and kinematic descriptions are combined in a precise calculation of tire characteristics.

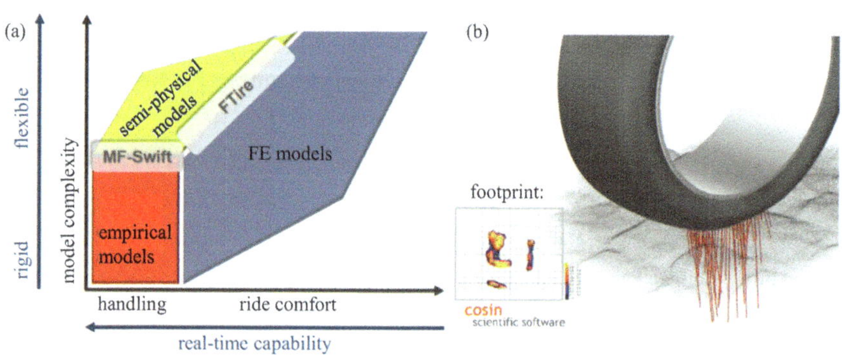

Fig. 1 **a** Classification of tire model types, **b** FTire's discrete contact patch on Belgian blocks

Due to a high parameterization, implementation and computational effort, this model type is not commonly used for multi-body simulation (MBS) of full vehicles, which need to be performed for a holistic analysis of the vehicle-tire-road interaction.

As a standard parametrization process, the FTire and MF models require a certain set of measurements, which has been performed for two different commercial vehicle (CV) tires on suitable tire test benches. The handling characteristics, such as brake slip or side slip behavior, are measured on drum tire test rigs suitable for a high demand of wheel loads and the large tire dimension. The measurements of passenger car tires, which are used in the second part, are carried out on flat-trac test benches. The measurement on a flat surface has the advantage that the influence of the drum curvature does not need to be compensated for the analysis on flat pavements. Both test rigs are coated with corundum paper P120 to approximate the friction characteristic of asphalt. The tire stiffness in all three dimensions as well as the torsion stiffness are measured on a specialized tire stiffness test rig.

For the physical tire model FTire, further tire characteristics are measured. Stiffness measurements on obstacles help to parametrize local bending and deflection properties. Cleat measurements on a rolling drum with varying velocities and loads characterize the tire's modal and damping behavior as well as the reactive forces to external excitations at rolling conditions. Using a pressure imaging mat, the contact patch boundaries and the non-rolling pressure distribution are measured, so that the contact patch reaction to deflection with and without camber is characterized.

In the actual parametrization process, the model's parameters are iteratively optimized until the simulation approximates the overall tire behavior sufficiently.

1.2 Modeling of Vehicle Dynamics of Heavy-Duty Trucks

Due to their size, technologies and number of parts, heavy-duty vehicle simulation models require a high complexity and computational effort. In some cases, complex submodels disproportionately increase the simulation time with minimal impact on the simulation accuracy. In this investigation, the degree of detailing of the simulation model is methodically increased in multiple steps by comparing it to a fully flexible model at equal load conditions to analyze the model efficiency. Figure 2 gives an overview of the investigated component complexities with regard to vertical dynamics analysis. The real-time factor (RTF–ratio of computational time to actual maneuver time) varies from 18 for the simplest model to 240 for the fully detailed model. Different model approaches for the tire, the vehicle body and the suspension and damping system are available. The submodels of a reference vehicle model are highlighted.

Non-linear steel springs are compared to an air suspension system with activated and deactivated load balancing. If one axle deflects during an obstacle crossing and the air springs are linked, the resulting air volume change will be transferred to another air spring, which rebounds to balance the load [27].

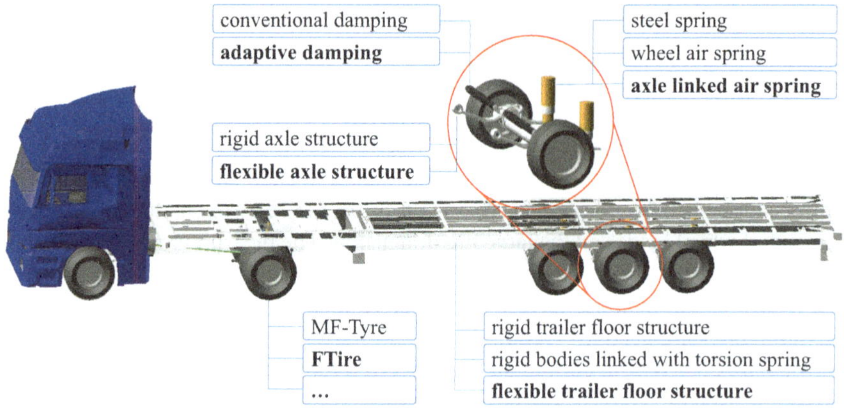

Fig. 2 Overview of vehicle component complexities [27]

The damping behavior is firstly described by conventional dampers, which are defined through a non-linear function between the damper force and velocity. Secondly, a more complex pneumatic damping control (PDC) system is investigated. The system provides an almost constant damping ratio for any load condition by linking load dependent dampers and load dependent air spring systems [27].

Finally, the body structure is considered in three levels of complexity, i.e. rigid body, rigid bodies connected with torsion springs and a fully flexible finite element structure. The axle tubes are considered as rigid and flexible structures. The modes of vibration of the trailer floor have been determined in previous investigations and were limited to a maximum to reduce the model data size and the simulation time [27].

With the help of a sensitivity analysis, the influence of the different levels of details on the vehicle dynamics of a semi-trailer-truck can be investigated and compared to the full-detailed reference model (cf. Fig. 2) on different synthetically generated road excitations (cf. Fig. 3). Only one submodel is altered from the reference model at a time. A single-sided obstacle excitation identifies the influence of flexible torsion structures, such as the trailer floor model and the effects on the resulting wheel loads. Realistic, both sided excitations in form of a bridge connecting element are used to investigate the component characteristics for a mainly lifting movement of the axles.

Figure 3 presents the resulting normalized wheel loads for the two investigated synthetic road excitations. Especially, the choice of the tire models has a major impact on the maximum dynamic wheel load. The single-point contact of the MF-Tire model does not envelope an obstacle like a real tire, which leads to increased deflections and higher wheel loads. Therefore, a pronounced contact patch and its local deformability is essential for the accuracy of dynamic wheel loads on uneven roads.

Furthermore, the influence of flexible bodies is clearly noticeable for single-sided excitations in higher dynamic wheel loads. The changed tactile reaction of the wheel load in the time domain is shown in Fig. 4. A rigid axle leads to the largest alteration

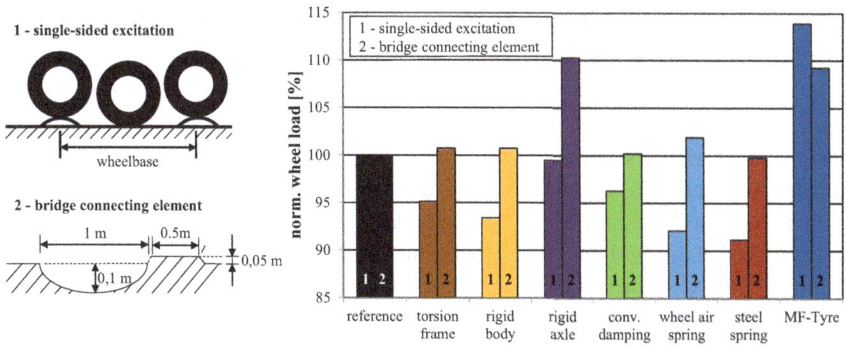

Fig. 3 Comparison of normalized wheel load to different road excitations [27]

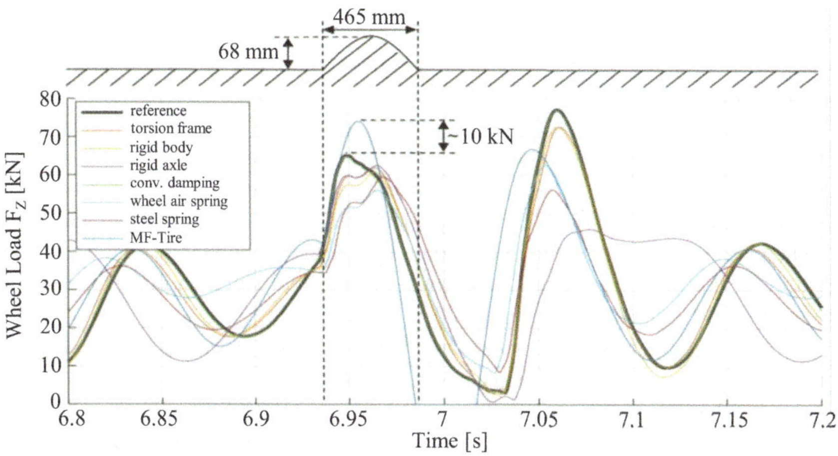

Fig. 4 Effect of different component model complexities on wheel load variation [27]

of the aftershake characteristics considering the reduced oscillation frequency. As the flexible structures strongly affect the wheel load variation, just as an axle-linked load-balancing system, it is not recommended to simplify these submodels for the investigation of non-parallel wheel excitations. Compared to the impact of tire models and flexible structures, the effect of the damping system on the wheel load variation is marginal.

In conclusion, the vertical dynamic simulations have shown that finite element descriptions for body structures and tire models with a discrete contact patch are necessary to achieve high accuracy for load analysis as simplifications of the component models lead varying wheel loads. Therefore, only small simplifications of the component models are permissible. While the investigated obstacles represent excitations on a high macro scale, the excitations of asphalt texture on a micro scale are analyzed in a next discretization step.

1.3 Influence of Asphalt Texture on Wheel Load Variation

Due to the unevenness, the road is one of the largest sources of excitation for the vehicle. This excitation causes vertical movements in the chassis, which results in wheel load variations that stress the pavement. Surface unevennesses are distinguished with regard to micro (<0.5 mm) and macro texture (0.5–50 mm). A further influencing factor is the topology of a road (wavelengths > 50 mm), which causes low frequency excitations of the vehicle. MBS in driving dynamics are usually performed on an even road surface (micro and macro flat) and are used in combination with a mathematical point-contact-tire-model. By using a physical tire model instead, it is possible to use a high-resolution road model. For a realistic modeling of the imprinted excitations, three-dimensional road models are necessary. The level of detail of these models is decisive for the pressure distribution in the tire contact patch (also cf. Sect. 2.1).

With the help of the methodology shown in Fig. 5, real road irregularities with macro and micro texture components are transferred to the computer-aided simulation environment.

The surface of a road section is digitized by means of a non-contact three-dimensional stripe light projection with a maximum vertical resolution of 4 μm. The results are digitally assembled into a road surface using a newly developed Matlab® algorithm and stored in a road modeling format suitable for simulative applications.

The processing of large amounts of data is a particular challenge due to the high resolution. For this purpose, a format must be selected which can be processed in the multi-body simulation environment.

By using the open source format OpenCRG, the data processing of the created three-dimensional surface geometry is simplified. The Curved Regular Grid (CRG) format enables a wide range of applications due to the variable resolution of the data points. For example, long and curved road courses can be implemented for lateral dynamics simulations or road courses with very high texture resolution for driving comfort simulations. The compact storage of the data is of great advantage. The digitally processed road profiles are transferred into a regular profile network, see Fig. 6 [28].

Figure 6 shows that the trajectory of the middle of the road is substituted by a reference line. The lateral topology of the roads can be mapped by cuts running

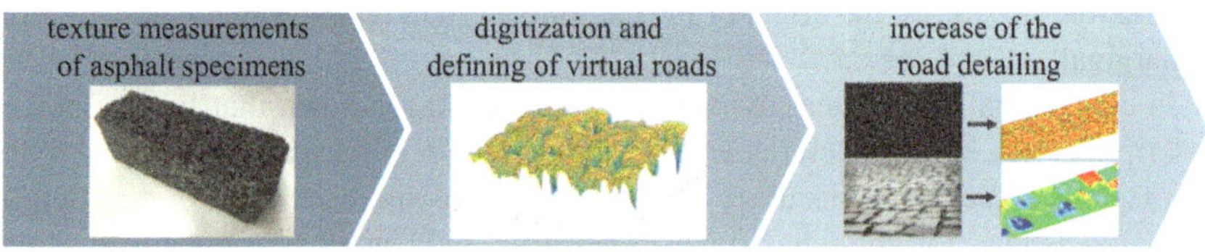

Fig. 5 Procedure for increasing the road detailing in the simulation [28]

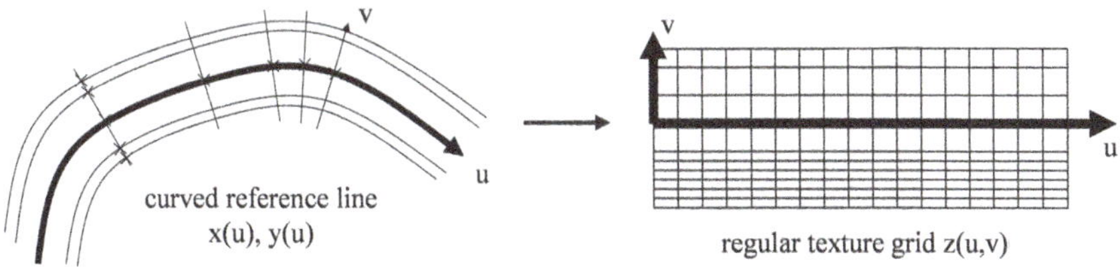

Fig. 6 Principle of the CRG data format [21]

orthogonal to the reference line. With the help of the curved reference line, a real road course or customized trajectories, which are needed in Sect. 2.2, can be constructed.

In further investigations, the influence of micro-scale road definition on the vertical vehicle dynamics is simulated on a typical German Bundesautobahn (BAB) with the representative stone mastic asphalt SMA11S. The road model has been built up according to Fig. 5 without overlaid topology. The FTire model is able to output the pressure distribution of the contact patch, which is shown in Fig. 7 for slipless straight line rolling with constant velocity and wheel load on standard flat road for MBS and on the CRG surface BAB.

The new approach of road detailing causes an inhomogeneous pressure distribution with local pressure peaks. In comparison to the maximum pressure on a flat surface, the peaks on CRG are up to twice as high. Furthermore, areas without road contact at the location of asphalt texture minima are visible. Consequently, the inhomogeneous pressure leads to local micro excitations. Using the semitrailer model of Sect. 1.2, straight-line maneuvers are simulated on the BAB surface to analyze the effect of micro excitations on full vehicle level (cf. Fig. 8).

While the wheel load of the trailer axle is nearly constant on the flat surface, the CRG surface leads to high frequency excitations in the wheel center. The peaks of the wheel load increase up to 0.8%. Under the assumption of the fourth power rule, this results in a higher stress level for the road of approximately 3% [26].

The high frequency excitation does not only affect the dynamic wheel loads, but also leads to varying horizontal tire forces, which is analyzed in detail in Sect. 2. As

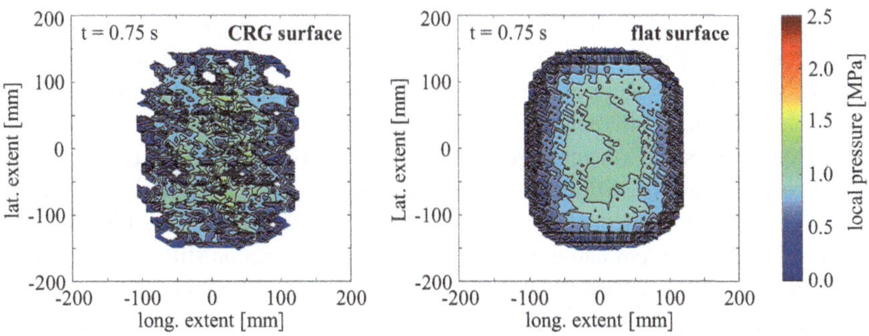

Fig. 7 Pressure distribution in the contact patch on rough CRG (left) and a flat surface (right) [27]

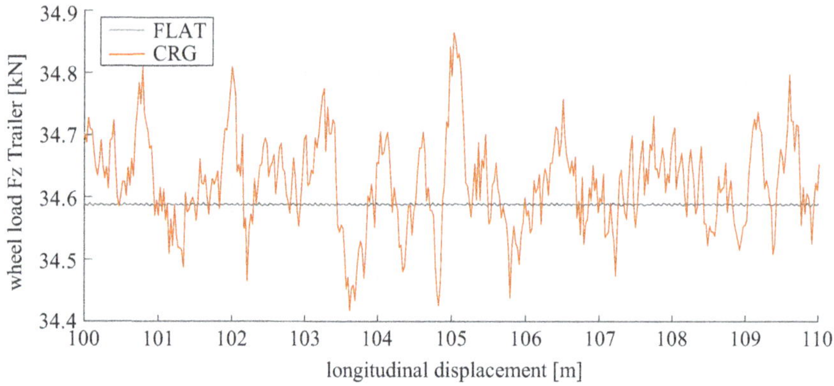

Fig. 8 Straight line rolling on flat and rough CRG surface [27]

the presented full vehicle investigations have been performed only on a simulative basis, a new method of experimental validation has been developed in the next section.

1.4 Generation of Quasi-Rolling Test Rig Excitation Signals for Vehicle Axles

Driving tests on real roads require expensive full vehicle prototypes. To reduce the effort of the extensive determination of load collectives, a new approach combining experimental and simulative methods shall transfer the characteristics of a rolling tire on a road surface to a non-rolling tire used in a heavy-duty axle test rig. In that way, the vehicle-tire-road interaction can be investigated at controllable influencing factors.

The properties of vehicle axle systems are conventionally investigated using servo hydraulic axle test rigs. During these investigations, the tire is standing still. However, the dynamic properties of standing and rolling tires show substantial differences. This results in a decreasing accuracy of the interpreted measurement results, that are exemplary used for the validation of multi-body simulation models. Thus, the aim of generating quasi-rolling test rig excitation signals opposed to the static excitation signals is the improvement of the measurement results and the simulation results.

The key metric to characterize the vertical transfer behavior of tires on a macro excitation level is the vertical tire stiffness. There are different influencing main factors on vertical tire stiffness, such as inflation pressure, construction of the tire and the excitation frequency. Other significant influencing factors are the alteration of the state of movement and the rolling velocity: The dynamic stiffness of a rolling tire is up to 20% lower than the dynamic stiffness of a standing tire [3].

To develop a method for generating quasi-rolling excitation signals, several specifications have to be taken into account. Firstly, the method has to be implementable on simple test rig concepts to ensure independence of the test rig itself. Secondly, the characteristics of rolling tires must be depicted accurately as this is the key aim. Furthermore, different surface unevenness scales and three-dimensional unevenness excitations must be considered to accurately reproduce the situation of the rolling tire on different surfaces. The independence of the method on the axle oscillation system has to be ensured to enable application on different types of suspensions. The applicability to heavy-duty vehicles enables a broader field of application of the method compared to the singular focus on passenger cars. Lastly, the method has to comply with physical test rig limitations.

The objective can be addressed by two approaches. The first approach consists in expanding a stationary test rig used for vertical dynamic testing and modifying the excitation signal as well as single components. The first sub task includes the extension of the test rig concept by implementation of a driven flat belt unit. This extension enables combined vertical and rotational excitation. Due to limitations of the flat belt unit, this approach is only feasible for passenger car tires and thus does not comply with the previously mentioned requirements. For the second approach, the excitation signal can be altered using a suitable method and single components of the test rig can be modified. The modification of the stiffness and damping conditions of the test rig is possible, however challenging to implement. A key aspect of this approach is the adaption of the excitation signal of the test rig's excitation cylinder by modifying the input road profile. The key aspects of the presented approach are shown in Fig. 9.

No existing test rig concepts for velocity dependent, highly dynamic investigations of the vertical dynamic behavior of heavy-duty tires were found in the literature. Thus, the implementation of this method has to start on a theoretical, virtual level. An FTire model is used on this level because of its suitability for heavy-duty tires [9]. This tire model allows calculation of the pressure distribution in the tire tread and yields the opportunity to analyze the contact area between tire and different road surfaces and, therefore, the particle stresses of the road surface. Another significant part of this method is the inclusion of road surface unevenness. Three-dimensional road surface

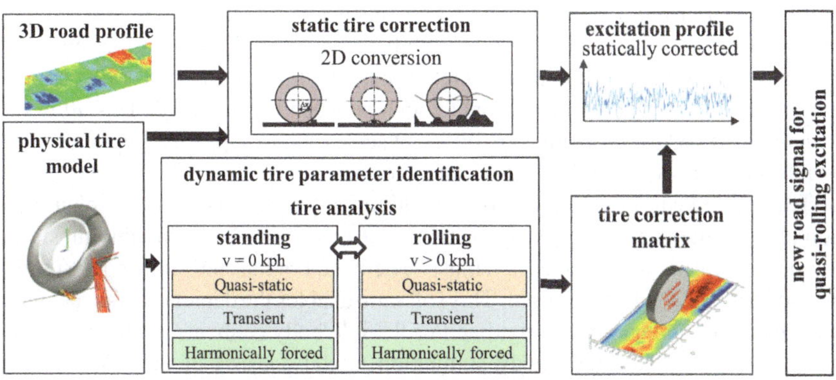

Fig. 9 Approach for generating quasi-rolling excitation signals [28]

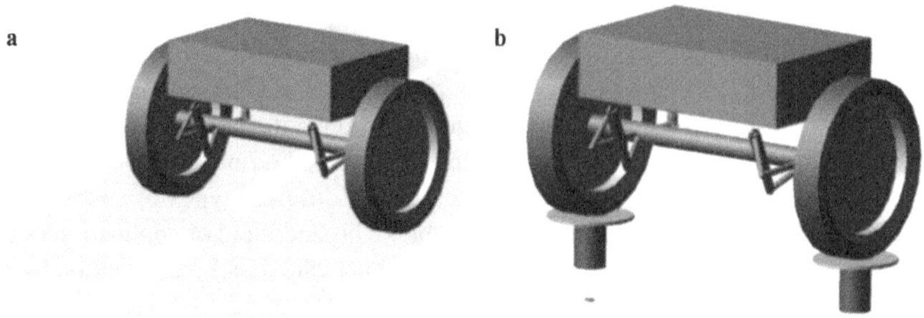

Fig. 10 Virtual test rig setup: **a** free rolling on a road model and **b** non-rolling on actuators [28]

unevennesses have to be converted into two-dimensional road surface excitations, which are suitable for the vertical dynamic test rig.

Development Tools. To realize the method for quasi-rolling test rig excitation signals for vehicle axles, several development tools are necessary. One of these tools is virtual road modeling, which is discussed in Sect. 1.3. Further sophisticated tools to support the method are so-called virtual test rigs. The purpose of these tools, which are implemented in Matlab/Simulink and MBS software, is to determine the different system characteristics for rolling and standing tires. These tools enable the enforcement of a dynamic excitation onto a rolling tire at either the wheel center or the tire/road contact surface. The virtual test rig is set up in two configurations, namely the free rolling configuration and the hydraulic test rig configuration as depicted in Fig. 10.

The free rolling configuration contains free-rolling boundary conditions on three-dimensional road surfaces, whereas the hydraulic test rig configuration depicts a virtual, servo-hydraulic test rig with disc connections to the tires that prevent the tire's rotation. These configurations allow the comparison of the system behavior between rolling and standing tires in the same oscillating system [28].

To use the virtual test rig in combination with complex, three-dimensional road profiles, the transmission of forces and moments in all three spatial directions as well as vibrations from the road surface have to be considered. Therefore, the minimum requirement for the selection of a suitable tire model for this virtual test rig configuration is the discretization of the contact area between tire and road. This allows to scan the road surface unevenness in different scales and to depict the implicit forces, which occur due to tire deformations and shortwave excitations [8, 18].

The hydraulic test rig configuration is additionally necessary for the implementation of the method. To analyze the reaction forces and moments, a suitable substitute oscillation system of the suspension must be defined. For this, the multi-body simulation software ADAMS/car is used. This software enables the depiction of mechanical systems by a finite number of rigid and/or flexible body elements in all six degrees of freedom [12]. The multi-body simulation model consists of three levels: The template as a basis of single mechanical components, subsystems as combinations of templates and the assembly as combination of subsystems.

Based on these virtual test rigs, a signal processing method can be defined for converting the three-dimensional road profile into a two-dimensional excitation signal for an axle test rig. The key aspects of these methods are described in the following paragraphs.

Geometric Correction of Excitation Profiles. Before the three-dimensional unevenness excitation signal (x, y, z) can be used for vertical excitation, it has to be reduced to two dimensions (z, time). In case of transient as well as stochastic unevenness excitations, the method of quasi-static filtering provides plausible system responses [28, 29]. A physical tire model such as FTire is used to realize the quasi-static geometrical filtering method. In the virtual test rig environment in free rolling configuration, the tire is rolled over the three-dimensional road profile that is to be reduced slowly and with constant wheel load. By rolling the tire over the road profile, the trajectory of the wheel center can be extracted. This trajectory can be defined as the result of the quasi-stationary tire reaction to the unevenness excitation and can be transformed into a time-dependent excitation signal. With this approach, the basic tire characteristics such as tire dimension, horizontal and vertical stiffness, inflation pressure and wheel load are already taken into account when generating the corrected excitation profile.

Tire Correction Matrix. To include the velocity-dependent behavior of rolling tires into a two-dimensional, geometrically corrected excitation signal, additional adaptions to the unevenness profile are necessary. These adaptions can be realized using a tire correction matrix.

The road excitation signal can be divided into three components [16]: The harmonic component, the stochastic component and the transient component. All three components are investigated separately on the virtual test rig in both configurations. In the free rolling configuration, different velocities are applied. The excitation can either be conducted at the wheel center or in the tire/road contact patch. In case the wheel center is excited, inertia effects of the rim or proportional tire masses have to be considered additionally. Figure 11 shows the structure of the tire correction matrix.

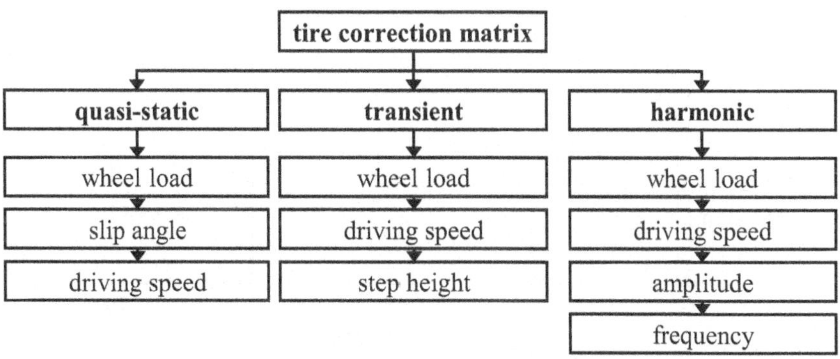

Fig. 11 Structure of tire correction matrix [28]

The tire correction matrix is created individually for each tire-inflation-pressure combination. Depending on the excitation signal, the individual parameter sets are defined using suitable data from the virtual test rigs [28].

Quasi-rolling Excitation Profiles. For the application on a stationary, vertical dynamic test rig, the two-dimensional unevenness signal is corrected using the tire correction matrix. The quasi-static, filtered original signal has to be analyzed regarding the amplitude characteristics in the frequency domain to enable the application of the tire correction matrix with a defined velocity and wheel load. This is achieved by applying a Fast Fourier Transformation (FFT) to the signal, which results in transforming the time domain signal into the frequency domain. The frequency spectrum of the signal consists of amplitude and phase data that is adapted by using the correction matrix. To enable the adaption, a frequency-based weighting of the spectrum is performed according to VDI 2057-1 [24] for comfort rating of vehicles. The weighting function is used to depict the importance of the three different excitation components that are included in the tire correction matrix. Hence, the different excitation forms can be considered systematically: The weighting for the following investigations is calculated with 40% quasi-static, 20% transient and 40% harmonic signal ratio. An exemplary weighting function is shown in Fig. 12.

This weighting ratio is adjustable depending on the converted excitation signal. The weights are adjusted depending on the share between micro texture (e.g. German Bundesautobahn) and macro texture (e.g. Belgian blocks) in the original signal. Using the frequency dependent weighting of the tire evaluation factors from the tire correction matrix, the necessary weighting function for each boundary condition can be developed. A corrected excitation signal spectrum by application of an exemplary tire correction matrix is depicted in Fig. 13. The weighting function is used for the amplitude modification of quasi-static, filtered road profiles to generate quasi-rolling excitation signals for a stationary, quasi-static test rig.

Validation and Application of the Developed Method. The validation of the presented method is carried out in two steps. The first step is the comparison of simulation results. Here, the virtual test rigs are used firstly separately with standing and rolling excitation signals and successively with quasi-rolling excitation profiles. The second step of the validation is the verification of the generated excitation signals

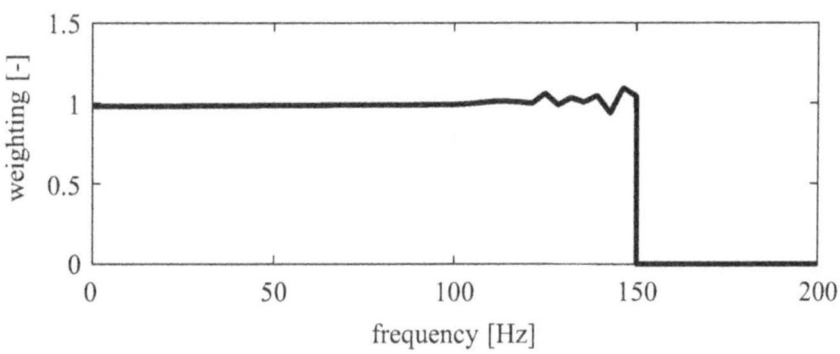

Fig. 12 Exemplary weighting function [28]

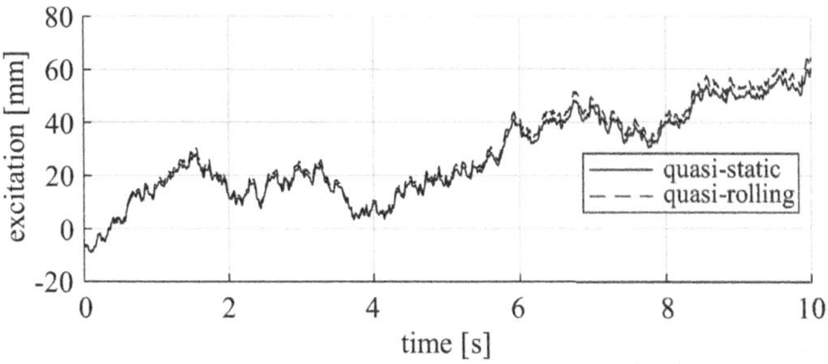

Fig. 13 Resulting excitation signal spectrum after application of the tire correction matrix [28]

on a real servo-hydraulic test rig. In both steps, the robustness of the method is investigated by varying parameters such as tire types, inflation pressures and vertical wheel loads.

The virtual validation consists of two different simulation setups. On the one hand, the free-rolling tire is investigated on a three-dimensional unevenness profile. On the other hand, the tire with constrained rotational degree of freedom is investigated using the quasi-rolling excitation signals generated by the presented method. The free rolling dataset is used as a reference. This approach is valid because a validated tire model is used for this simulation approach in combination with a validated multi-body model of a commercial vehicle suspension system.

Furthermore, a large variety of different parameter sets is used to investigate the robustness of the presented method. Figure 14 gives an overview of the investigated parameter combinations.

The investigation of different road profiles ensures the robustness of the previously discussed weighting function to consider different distributions of micro and macro textures. Exemplary, the comparison of wheel travel, wheel load, axle acceleration

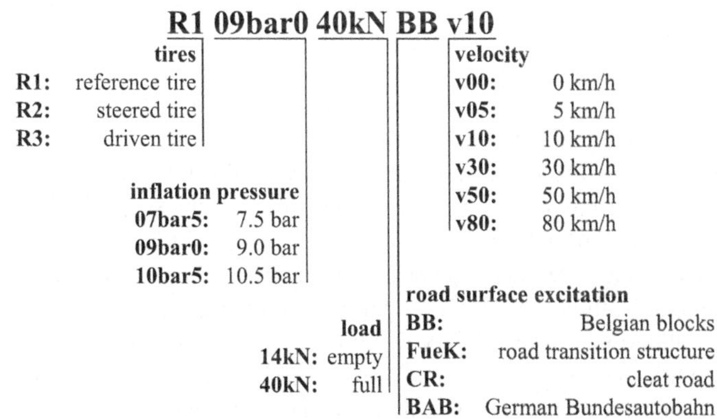

Fig. 14 Overview of different virtual validation variants [28]

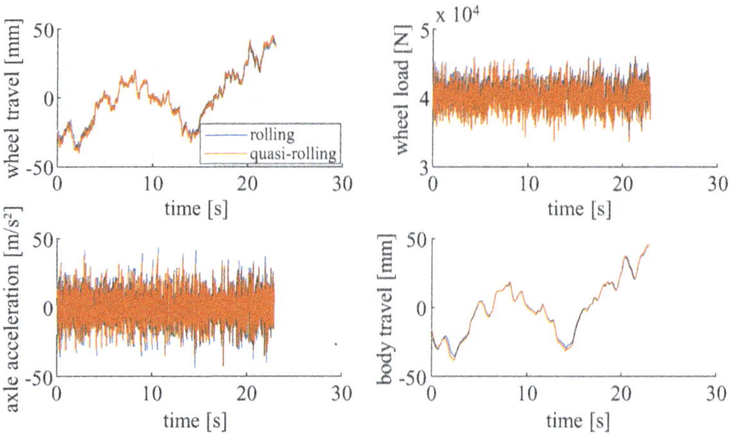

Fig. 15 Validation results for German Bundesautobahn excitation, 80 km/h [28]

and body travel for the German Bundesautobahn road surface between free rolling excitation with the virtual test rig in free-rolling configuration and constrained, quasi-rolling excitation in the hydraulic test rig configuration is shown in Fig. 15.

In this exemplary result, a high correlation between real rolling and quasi-rolling excitation can be achieved. Comparable results can be reproduced for the other parameter combinations depicted in Fig. 14. For detailed insights, the reader is referred to [28].

Experimental Validation. The final validation step includes the usage of a real, servo-hydraulic test rig for commercial vehicle axles. The test rig consists of two hydraulic cylinders that are connected to a substitute oscillation system of a commercial vehicle axle. The substitute system is built up using commercial vehicle tires, air suspension and dampers, as well as an axle frame tuned to have elastic properties comparable to a semi-trailer frame (see Fig. 16). For the axle frame, two rotational degrees of freedom are constrained to secure the setup. As measurement signals, forces in the tire contact discs as well as accelerations on wheel hub and axle frame are detected. Furthermore, relative displacements between wheel hub and frame and the air spring pressure are acquired during the measurements.

For the excitation, the same signals as for the virtual validation are used. Because in case of the experimental validation, the properties of the suspension's substitute oscillation systems differ from the system used for the virtual validation, absolute results are expected to deviate. However, tendencies shall be comparable for virtual and experimental validation. Different parameters are varied in an extent similar to the virtual validation. An exemplary result of the reactive vertical loads in the tire contact area at different speeds is shown for German Bundesautobahn and Belgian blocks as a comparison between simulation and experiment in Fig. 17.

Fig. 16 Servo-hydraulic test rig for commercial vehicle axles [28]

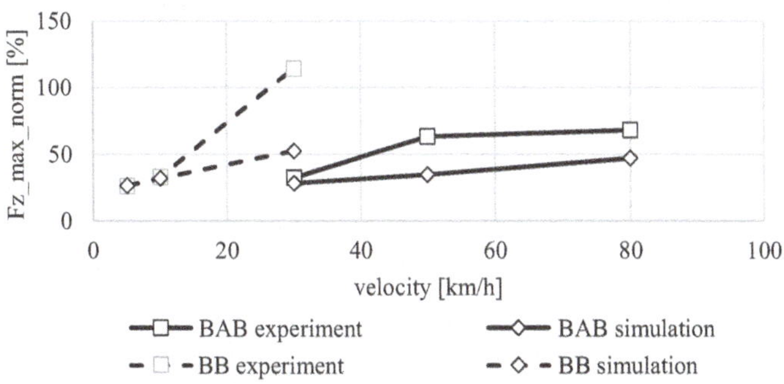

Fig. 17 Comparison between virtual and experimental investigation [28]

As mentioned before, the quantitative difference between simulation and experimental results is justified by different suspension configurations in simulation and experiment. Qualitatively, the velocity dependence of the reactive vertical loads is comparable in simulation and experiment. The deviation for Belgian blocks at 30 km/h results from a rolling movement of the axle in the experiment, whereas the rolling degree of freedom is constrained for the simulation model. Overall, the qualitative correlation between the different validation steps is high. The quantitative, relative difference between simulation and experiment remains below 20% even though the simulation model is built up for a deviating suspension.

In conclusion, the suitability of quasi-rolling excitation signals can be proven. By application of the presented methodology, the effort for vertical dynamic investigations on suspension components can be reduced by a significant amount. This is realized because the extensive determination of load collectives in driving tests on real road surfaces is not necessary anymore [28].

1.5 Suspension Control for Heavy-Duty Axles

The unevenness of the road is the primary source of vertical vibrations experienced by the vehicle and the passengers. These vibrations are unfavorable to both ride comfort and the safety of the vehicle. Further, especially with heavy vehicles, these vibrations cause significant amount of wheel load variations that lead to road wear with time. Suspension systems are employed in vehicles to reduce these vertical movements. Suspension systems, in general, comprise an elastic element such as a coil spring or air spring that carries the static load and a damping element that absorbs the energy from the oscillating element.

Passive suspension systems have fixed spring and damper characteristics whereas controllable suspension systems enable dynamic modification of these characteristics with the help of mechatronic systems. In general, different levels of control actions are possible, such as the modulation of the damper force, modulation of the spring characteristic or replacing both elastic and damping element with a force actuator [22]. The most common type of controllable suspension system is where the damper force is dynamically altered to suit a specific control objective such as ride comfort or ride safety. The methodical procedure for designing a road protecting damper controller is presented in the following.

System Modeling. Using simulation methods, the vertical dynamics corresponding to the truck axle, which has also been used in the previous sections, are analyzed with the help of the virtual axle test rig developed in the ADAMS environment (cf. Fig. 18). The passive suspension of the axle comprises an air spring and a passive damper. The passive damper is replaced with an actuator to transform the model into a controllable suspension system. A controller determines the amount of force exerted by the actuator depending on the feedback from certain vehicular measurements. The controller is developed in Simulink and is subsequently linked to the multibody axle model with the help of a Dynamic Link Library (dll).

Road Excitation. A bump excitation signal and a stochastic excitation signal are considered to compare the performances of the dampers in the time domain and the frequency domain, respectively. The bump road signal comprises a 10 cm high bump whereas the stochastic road signal is an excitation signal with the unevenness similar to that of a German highway "Bundesautobahn". Figure 19 shows the road unevenness patterns where u is the longitudinal travel, v is the width of the road patch and z is the height of unevenness.

Control Strategies. Several control strategies have been proposed over the years to determine the appropriate control action given a state of the vehicle. In order to obtain the best possible performance, the control strategies deal with several challenges such as the conflict between ride comfort and ride safety objectives, limited amount of actuator force or damping coefficient and limited maximum possible suspension deflection. These control strategies are classified into several categories such as classical control, robust control, optimal control and predictive control [22]. However, the ability to implement any of these strategies is highly dependent on

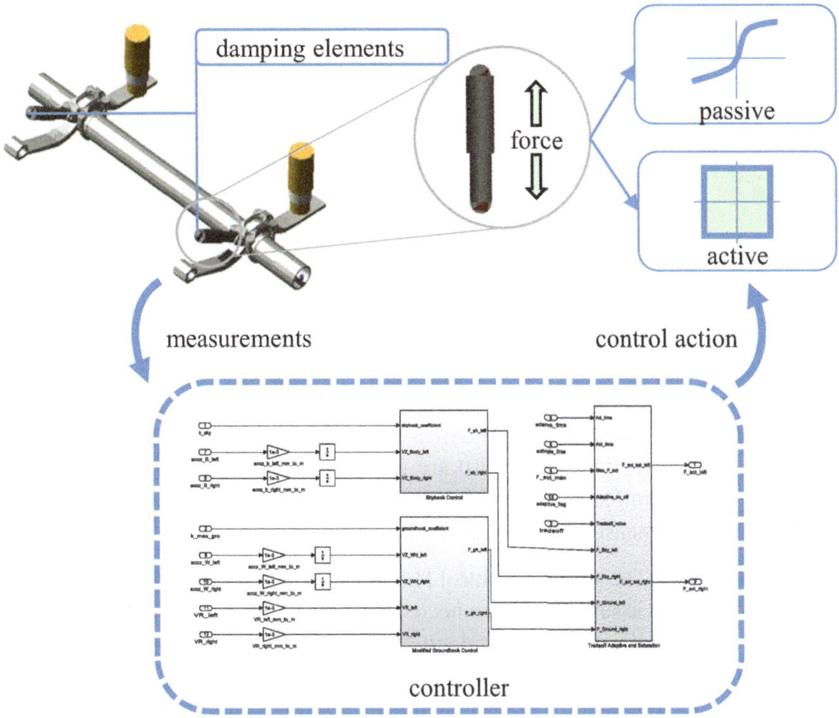

Fig. 18 Truck axle and suspension system—passive and active damping

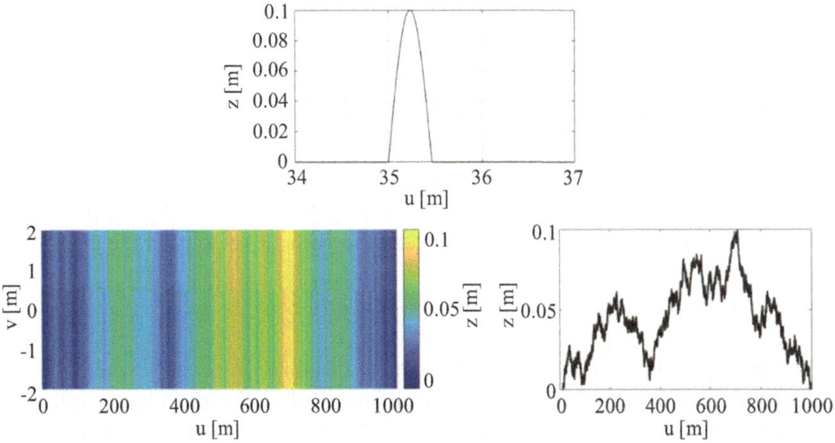

Fig. 19 Bump excitation signal (top) and stochastic road unevenness pattern (bottom)

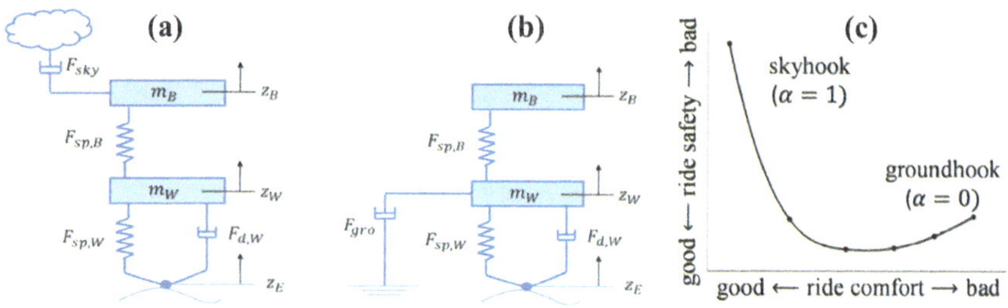

Fig. 20 **a** Ideal skyhook control, **b** ideal groundhook control and **c** pareto front

the on-board computation effort and availability of certain measurements from the vehicle. The accuracy of these measurements highly affects the controller's performance. Considering these factors, classical control schemes such as skyhook control and groundhook control (cf. Fig. 20) are prevalently used in modern vehicles by the virtue of their simple, yet effective, methodologies and at the same time require realistic vehicular measurements.

Skyhook Control. The skyhook control [14] is a classical comfort oriented strategy where the damper is assumed to be connected between the sprung mass (body) and the sky so that, in an ideal case, the vertical movements of the sprung mass are completely isolated from that of the axle. However, in reality, the actuator is mounted between the body and the axle and the skyhook actuator force F_{sky}, given by the product of the skyhook damping coefficient k_{sky} and the body vertical velocity $\dot{z}_B$, affects both the body and axle movements but reduces the body movement significantly,

$$F_{sky} = -k_{sky} \cdot \dot{z}_B. \tag{1}$$

Modified Groundhook Control. The ideal groundhook control [25] is a classical road-holding oriented strategy where a fictitious damper is assumed to be connected between the axle and the road, which helps to reduce wheel loads and subsequently to minimize road wear. The groundhook damping force is given by the product of the groundhook damping coefficient k_{gro} and the vertical velocity of the axle $\dot{z}_W$,

$$F_{gro,ideal} = -k_{gro} \cdot \dot{z}_W. \tag{2}$$

Absence of the passive component results in instability especially when exposed to high frequency excitations. As a result, the groundhook control law is reformed with additional passive force, which is defined by the suspension deflection velocity $\dot{z}_B - \dot{z}_W$,

$$F_{gro} = F_{passive}(\dot{z}_{BW}) - k_{gro} \cdot \dot{z}_W. \tag{3}$$

Hybridhook Control. Hybridhook control is a combination of both skyhook and groundhook strategies to deal with the conflicting objectives by setting a compromise $\alpha \in [0, 1]$ between the skyhook force and groundhook (cf. Fig. 20c),

$$F_{hy} = \alpha \cdot F_{sky} + (1 - \alpha) \cdot F_{gro}. \tag{4}$$

Performance Analysis. The performances of various control strategies can be compared to that of the reference passive system to understand the vibration behavior of the axle. The ride comfort and road-holding (ride safety) offered by different suspension systems are quantified by the resultant vertical body acceleration and the resultant wheel load variations. As a result, these measurements are compared to each other in order to evaluate the controllers' performances in the time domain and the frequency domain. In the subsequent discussion, the performances of the passive system, groundhook control, skyhook control and a hybridhook control with $\alpha = 0.75$ (tradeoff: 75% skyhook and 25% groundhook) are considered.

Figure 21 shows the response of the truck axle when driven on the bump road with a longitudinal velocity of 30 km/h. It can be observed that the skyhook controller results in the minimum peak acceleration value (~23% less than passive) immediately after encountering the bump. However, the skyhook control results in the highest number of oscillations. As can be expected, the groundhook control results in a behavior opposite to that of skyhook with the highest peak acceleration (55% higher than passive). The hybridhook controller with 75% tradeoff between the skyhook and groundhook controllers results in a 34% higher peak acceleration than that of the passive system but returned to the zero acceleration state earlier than the passive system.

Considering the wheel load response of the control strategies, the groundhook control results in the best performance with minimum fluctuations in the wheel load. Both groundhook and hybridhook controllers offer superior performance when

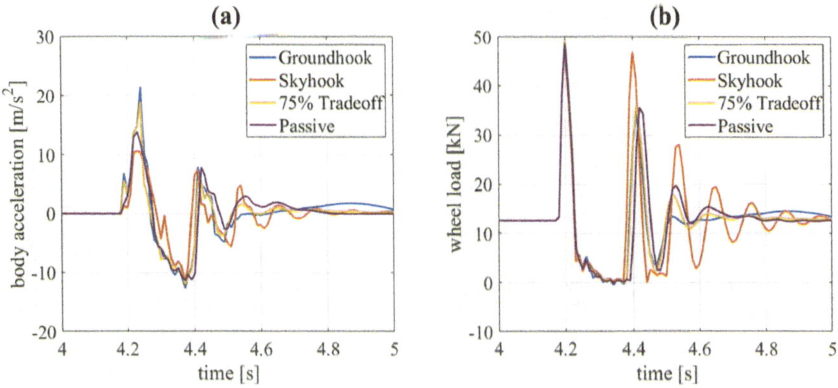

Fig. 21 **a** Body vertical acceleration and **b** wheel load for the bump excitation

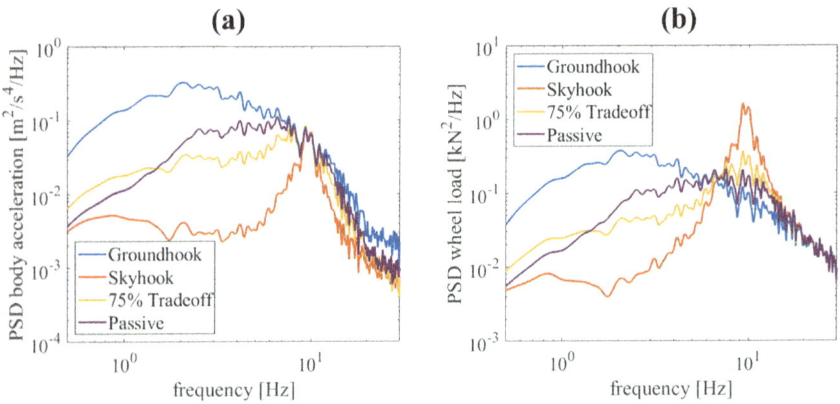

Fig. 22 Power Spectral Densities (PSD) of **a** body vertical acceleration and **b** wheel load for the stochastic excitation

controlling the dynamic wheel load oscillations. Taking both ride comfort and ride safety into account, the hybridhook controller results in improved performance.

The frequency response of the system in terms of Power Spectral Densities (PSD) for the stochastic road excitation signal for the truck driven with a longitudinal velocity of 80 km/h is shown in Fig. 22. The Root Mean Square (RMS) values of the body vertical acceleration and the wheel load over various frequency ranges are shown in Fig. 23. The values confirm the behavior of the control systems as depicted by the PSDs of body vertical acceleration and the wheel load. The contrasting behaviors of the skyhook and groundhook system are clearly visible affecting both ride comfort and ride safety in the opposite manner. The skyhook controller results in diminished vertical acceleration and dynamic wheel load in the body eigenfrequency region (~1.8 Hz). As a result, the skyhook controller offers the best ride comfort when

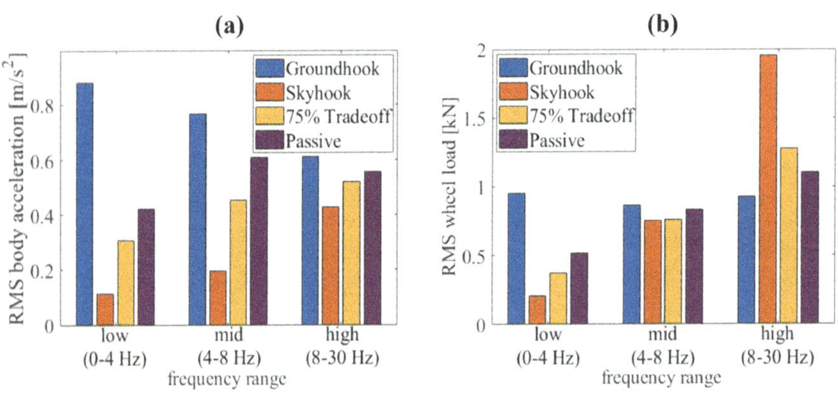

Fig. 23 Root Mean Square (RMS) values of **a** body acceleration and **b** wheel load over various frequency ranges

compared to that of the other systems. However, the skyhook controller delivers poor performance when excited with high frequency signals, especially for the ride safety.

Groundhook control, on the other hand, results in reduced wheel loads in the wheel eigenfrequency (~10 Hz) and higher frequency ranges, which is beneficial for ride safety. The hybridhook controller with a 75% tradeoff between skyhook control and groundhook control delivers balanced performance for both ride comfort and ride safety.

The passive system performs marginally better in reducing the dynamic wheel load in the wheel eigenfrequency region. For overall performance, it can be inferred that the hybridhook controller offers benefit in improving ride quality better than other systems by offering appropriate compromise between ride comfort and ride safety. To further improve the damper's performance with regard to ride comfort, intelligent control systems can be employed as discussed in [15].

2 Methods for Analyzing the Vehicle-Tire-Pavement Interaction Focused on Horizontal Forces

Driving dynamics characteristics are influenced by many factors within the vehicle-tire-pavement interaction. Especially, the forces in the tire contact patch have a major effect on the vehicle behavior and the road load. The following sections present experimental and simulative investigations of the tire behavior with the aim of a more realistic simulation on asphalt focusing on the horizontal forces.

2.1 Measurement and Simulation of Stress Distributions in the Tire Contact Area

The contact patch is the force interface between tire and road. Due to the tire's geometry, the tread profile and the asphalt structure, the ground pressure is unevenly distributed, which has a direct influence on the wheel load and on the local friction characteristic between rubber and road.

According to Sect. 1.1, the ground pressure distribution of non-rolling tires is already considered in the parameterization of physical tire models. However, the shear stresses in the contact patch, which are caused by turning maneuvers of a rolling tire are not validated to date. In previous works of [1] and [4], sensors based on triaxial force transducer pins have already been used for the measurement of dynamic footprints of free rolling truck tires. However, the influence of applied slip angles, which lead to higher shear stresses, is not investigated.

In this work, tire measurements with constant conditions are performed with a force matrix sensor (FMS) by A&D and ika's mobile tire test rig (cf. Fig. 24). Its integrated wheel guidance unit for a test tire is able to decouple the motion of the

Fig. 24 Mobile Tire Test Rig [2] and the Force-Matrix-Sensor

trailer with an active control of the wheel load, the track guidance and the camber angle. An inertial measuring unit, a measuring hub and laser sensors determine the forces, moments and movements of the tire [2]. All of the presented measurements were carried out with different passenger car tires and at a rolling velocity of 30 km/h. The FMS consists of 40 aligned sensor-pins with an area of 8 mm × 8 mm each. Based on strain gauges, a pressure measuring range of 1.77 MPa in all three spatial directions is realizable. Two laser barriers trigger the data acquisition and determine the rolling velocity, so that the time-dependent sensor-pin data arrays can be transferred into a path-dependent map with longitudinal and lateral extend, cf. Fig. 25.

In comparison to measurements with a pressure imaging approach for a non-rolling tire, the lateral tread grooves cannot be displayed for the free rolling tire due to the resolution of the sensor. While the contact patch width remains the same, its length decreases by approximately 10% for different wheel loads due to the increased vertical stiffness for the rolling tire at 30 km/h. The ground pressure peak locations correlate for the non-rolling tire and the rolling tire. Especially, the inward pointing rib edges of the longitudinal tread profile cause the highest ground pressure values. Due to the similarities, the measurement of the static footprint is already helpful for the prediction of the ground pressure distribution of a free rolling tire.

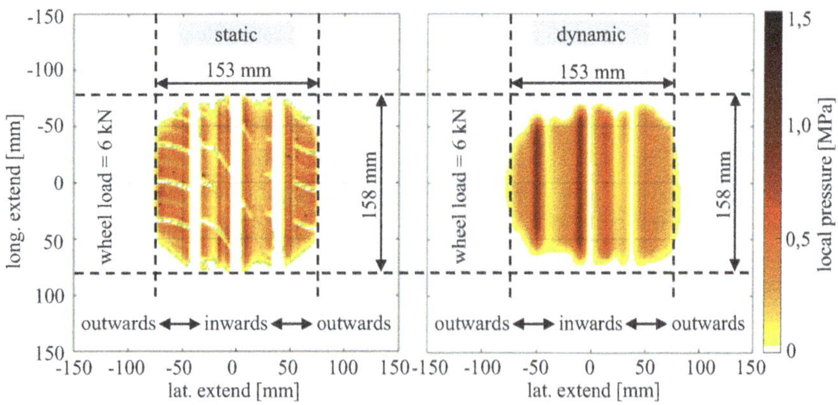

Fig. 25 Comparison of static and dynamic ground pressure distribution measurement at 30 km/h

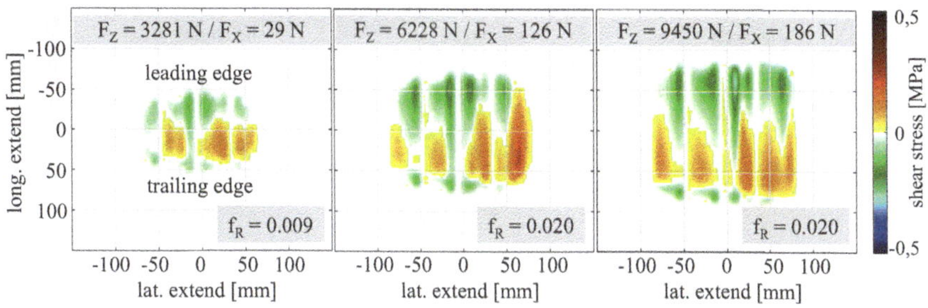

Fig. 26 Longitudinal shear stresses of the free rolling tire at different wheel loads at 30 km/h

The longitudinal shear stresses for a free rolling tire under varying wheel loads are shown in Fig. 26. Due to the deflection kinematics of the rolling tire, the shear forces at the leading edge run into the opposite direction to those at the trailing edge. The force-free lines within the longitudinal tread blocks mainly run diagonally from the outside to the inside slightly shifted in front of the center of the contact patch. In result, the proportion of the forces resisting the tire's forward movement prevail the forces driving the tire forward. This phenomenon is quantified by the rolling resistance f_R, which is calculated by the ratio of resulting longitudinal and vertical forces.

While the longitudinal shear stresses approximately increase linearly with the wheel load, the lateral stresses show a higher fluctuation in the measurement for the free rolling tire as camber and side slip angle fluctuations can only be prevented up to a certain degree. However, also in the lateral shear stress distribution, counteracting sections occur due to a deflection in rolling condition, which is shown in Fig. 27. The resulting lateral force always points in the same direction, which is an indicator for an existing plysteer or conicity effect. In the tear-off area, this dominating shear stress direction prevails.

Keeping the wheel load constant at around 6 kN, the counteracting shear stresses are shifted towards the outer tire side with increasing slip angle, which is shown in Fig. 28. For the tires investigated, the lateral shear stresses are unidirectional for slip angles higher than 2°. Due to the counteracting forces, the highest lateral shear

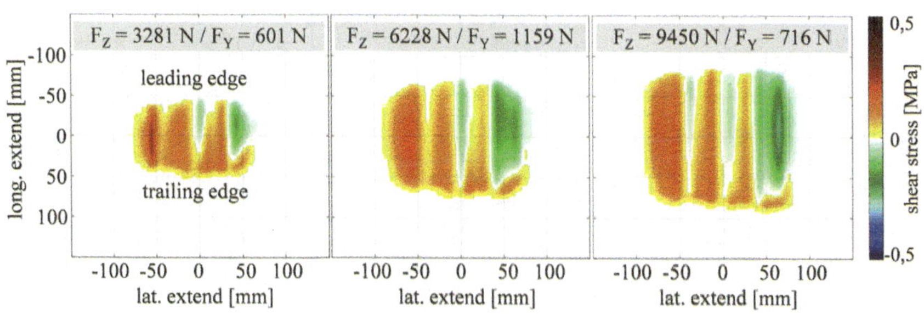

Fig. 27 Lateral shear stresses for the free rolling tire at different wheel loads at 30 km/h

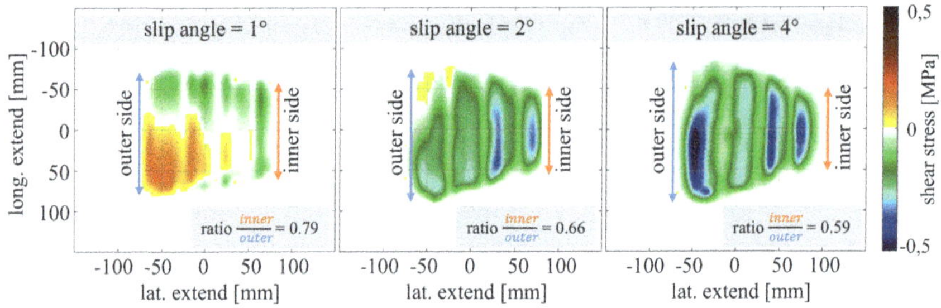

Fig. 28 Lateral shear stresses for tire with different slip angles at 6 kN wheel load at 30 km/h

stresses, which contribute most to the resulting tire side force, are found at the inner side. Further increase of the slip angle results in a shift of the peak ground pressure sections towards the outer side of the tire. Consequently, also the maximum shear load is increasingly shifted to the outer side of the tire. Furthermore, the outer contact patch boundary becomes more trapezoidal with increasing slip angle, shown by the ratio in Fig. 28.

Increasing the slip angle not only affects the lateral forces, but also the longitudinal forces, which is presented in Fig. 29. While the calculated mean longitudinal shear stresses correlate to the accumulated tire force, the minimum and maximum stresses show different behavior. The longitudinal shear force peaks, which are directed against driving direction, only increase for small slip angles up to 2°. Further turning of the tire leads to a reduction of the peak values. As the minimum values as well as the proportion of the shear stresses, which are directed in driving direction, monotonously decrease, an overall increase of the rolling resistance with rising slip angle occurs.

The lateral shear stresses increase in a degressive behavior analogously to laboratory side sweep measurements. While the mean shear forces and accumulated tire

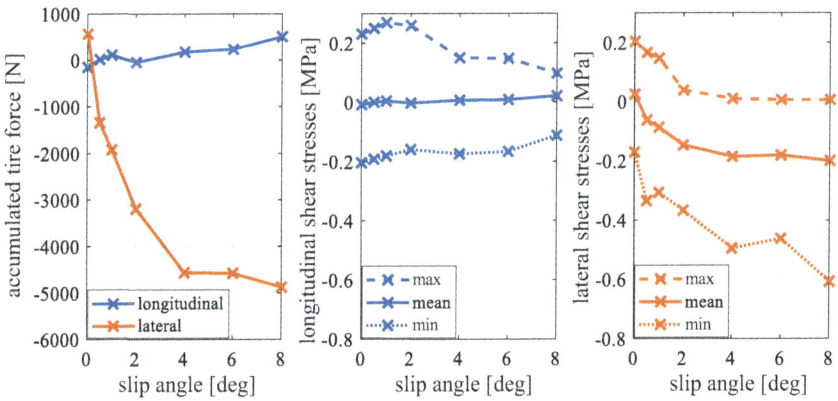

Fig. 29 Overview of the measurement results with increasing slip angle at 6 kN wheel load

side force nearly reach its maximum at 4° slip angle, the peak shear stress tends to keep increasing.

In a next step, the data generated in this experimental study are used to validate the dynamic footprint behavior of the physical tire model FTire.

Optimization and Validation of the Dynamic Footprint Simulation. In the framework of this study, the dynamic footprint behavior of a standard FTire parametrization procedure shall be compared to the results of a more application-oriented parametrization approach. The passenger car tire used in the FMS-study has been parameterized according to Sect. 1.1 and shall be named "standard" FTire-model in the further analysis. The "detailed" approach additionally uses the FMS-data for the iterative parameterization process as a quality characteristic to achieve a higher level of detail for the tread model. In this way, the identification of the tire model's integrated friction characteristics can be validated with not only the resulting forces and moments on the rim, but also locally in the contact patch with the dynamic footprints. For the simulations, the controller functions of the virtual tire test rig from Sect. 1.4 have been extended to include the application of passenger car tires.

Exemplarily, the footprint behavior for measurement and simulation is compared for a high slip angle of eight degrees, so that the divergences between the model approaches are clearly visible. The ground pressure distributions for a wheel load of 6 kN, a velocity of 30 km/h and 0° camber angle are gathered in Fig. 30.

As mentioned before, the measurement shows that peak pressure is found at the outer side of the tread. Further, there is a steep slope at the outer contour boundaries of the leading and trailing edge. The simulated footprint of the standard model shows a different behavior. The ground pressure is homogenously distributed and the outer contact patch boundaries are less trapezoidal. However, the influence of side wall stiffness is slightly visible. The footprint characteristics of the detailed FTire-model resemble the measurement better. The improvements were not only affected by a target oriented identification of the model's different stiffness parameters, but also by using exact geometric data for the outer tread curvatures and the tire profile. These data have been generated using a portable hand scanner (HandySCAN 700TM, AMETEK GmbH). The point cloud of the measured tire section is unwrapped and the outer contour of the tire is derived excluding asperities, such as production residues

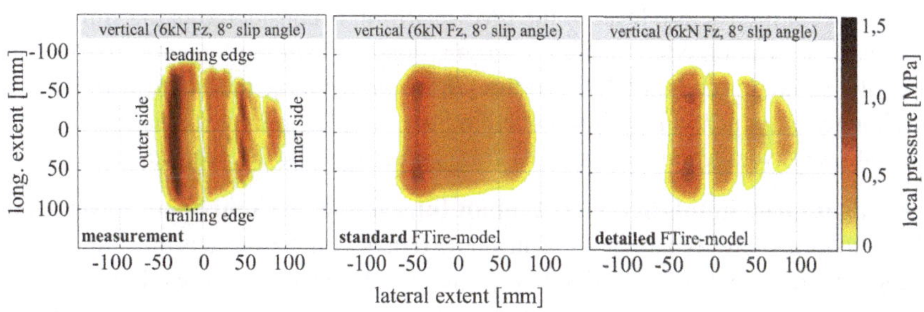

Fig. 30 Measurement and simulation of the ground pressure distribution in the contact patch [6]

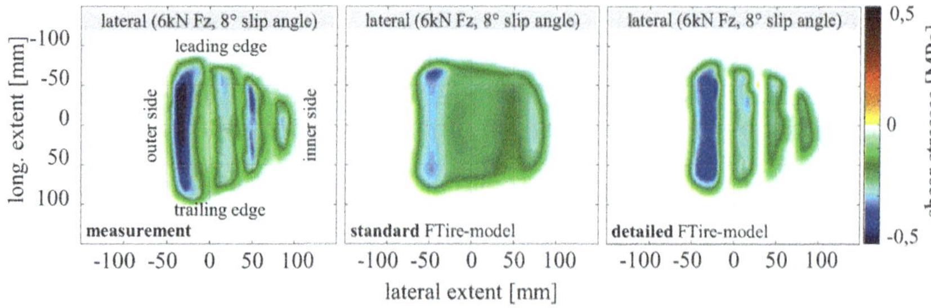

Fig. 31 Lateral shear stress distribution in the contact patch at 30 km/h [6]

from the vulcanization process. Furthermore, a two-dimensional tread pattern is extracted from this measurement. Usually, the longitudinal tread grooves are wider and the lateral tread grooves are often smaller than 2 mm. As the FTire's contact elements are arranged in a fishbone pattern according to a 2D tread pattern input, the required number of contact elements to correctly map the narrow lateral grooves needs to be at least eight times larger than in the standard model. As the computation time increases nearly linear with the amount of contact elements, only models with longitudinal grooves are considered in this study.

Due to the integration of the tread pattern and change of geometry and stiffness parameters, the remaining parameter set including the friction map is no longer valid. Therefore, the friction values need to be changed according to the correlation of the ground pressure and shear stress distribution of the measurements without changing the total horizontal rim forces, which are generated from the laboratory handling tests.

The resulting lateral shear stress distributions for the tire condition of Fig. 30 are shown in Fig. 31. Due to the homogenous ground pressure distribution and the larger contact patch area of the standard model, its shear stresses are lower than in the measurement, while the detailed model shows higher shear stresses especially at the outer tire side. In this case, an increase of the friction coefficients at high ground pressure and a reduction of the values for low ground pressure help to approximate the FMS-results. However, the different tire conditions need to be considered simultaneously to find the best compromise in the dynamic footprint behavior.

The deviating friction level of the FMS-platform is considered in the simulation. For that, the friction coefficients for varying load and sliding velocities on the tire test rig coating corundum P120 and the sensor platform are measured with a linear friction tester, which will be introduced in Sect. 2.3. The relative deviation between the values is averaged and serves as a friction factor of the flat road surface model.

Investigation of Simulative Tire-Asphalt Interaction on Component Level. After the validation of the dynamic footprint characteristics on flat surface, the influence on the shear stress distribution on asphalt texture can be analyzed more reliably. Exemplarily, Fig. 32 shows the transformation of the lateral shear stresses into histograms for rolling tire models at 60 km/h, 8° slip angle and 6 kN wheel load on flat surface

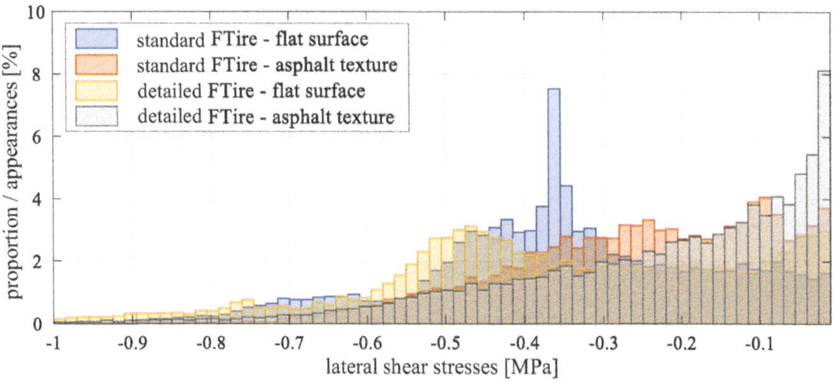

Fig. 32 Histograms of simulated shear stress distributions at 8° slip angle and 6 kN wheel load

and the asphalt road model of Sect. 1.3 (BAB). On a flat surface, the main part of the shear stresses shifts from approx. 0.37 MPa for the standard model to approximately 0.46 MPa for the detailed model due to the increased ground pressures. The simulations on asphalt texture indicate different shear stress characteristics. The detailed tire model has a monotonically decreasing frequency of appearances with increasing lateral shear stress. Due to the high proportion of low shear stresses and the consistently lower proportion of higher shear stresses in contrast to the simulation on flat surface, a reduced tire side force on asphalt texture is deducible from the histograms.

The choice of the surface texture has great influence on the pressure and stress distribution in the tire contact patch and, thus, on the forces that are transferred from the tire to the vehicle and locally to the asphalt layer. Consequently, the modulation depth of the asphalt model has high impact on the force transmission. On the other side, the road model data size has a direct influence on the computational effort, e.g. 1 km^2 of asphalt texture with a grid width of 0.5 mm has a size of approximately 20 Gigabyte. As full vehicle simulations need long roads or large area skid pads, the investigation of reduced grid resolution for larger road segments is necessary.

In the following example, a new asphalt texture is regarded due to the need of real road validation measurements. A CRG road model representing fine-grained asphalt taken from the Aldenhoven Testing Center next to the FMS-platform has been digitalized for a grid width of 0.5 mm and is furthermore step-wise resampled to a maximum grid width of 10 mm using the Matlab cubic griddata. Furthermore, the grid width of the detailed tire model's contact elements is varied between 2 mm and 10 mm.

Figure 33 illustrates the lateral shear stresses in the contact patch of the detailed FTire-model with contact element distance of approximately 2 mm at 50 km/h for different road grid resolutions at the same moment on the identical asphalt segment.

As a consequence of the smoothing effect of grid resolution reduction, the overall contact area increases at the investigated condition by 4.03% for the road model with 4 mm grid resolution and increases by 8.51% for the road model with 10 mm

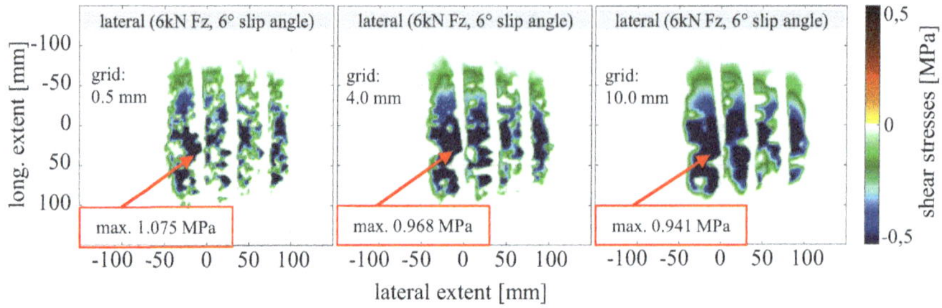

Fig. 33 Lateral shear stresses for different grid resolutions of the asphalt texture [6]

grid resolution. The texture peaks decrease along with the local ground pressure. In result, the maximum shear stress decreases in a more degressive way by 9.95% for the road model with 4 mm grid resolution and 12.47%, which can be lead back to the degressive decrease of the tire model's friction values for increasing local ground pressures. The lateral deflection of the longitudinal tread profile and the overall outer contour boundaries do not change significantly.

These local effects accumulate to a variation of the total transmitted tire side forces, which is shown in Fig. 34 for constant rolling velocity and slip angle. The main differences are the force offset between the standard and the detailed FTire-model as well as the degressive increase of the tire side forces with decreasing grid resolution. The variation of the tire model's grid resolution only has little impact on the tire side forces for both wheel loads, which possibly allows a reduction of the computation effort using a higher tire grid width. The variation of the road grid resolution influences the side forces of all models within a range of c. 700 N for 3 kN and c. 1000 N for 6 kN wheel load.

The grid resolution has a major influence on the simulation with FTire-models on OpenCRG road models. To find the best fitting CRG grid resolution, real road measurements must be performed on the asphalt, which has been digitalized via the hand scanner. Using again the mobile tire test rig, slip angle sweeps at different

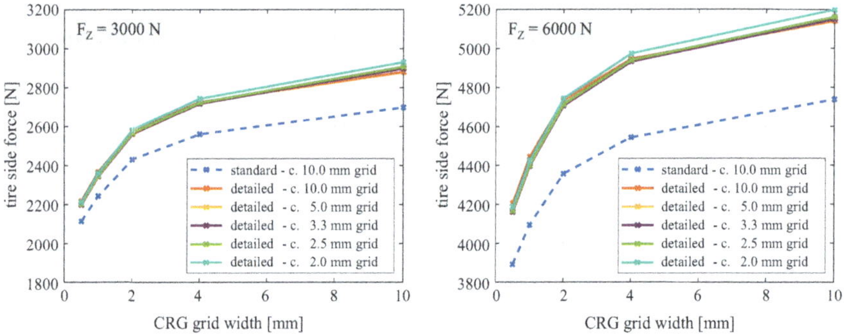

Fig. 34 Tire side forces at 6° slip angle on fine grained asphalt for different model resolutions [6]

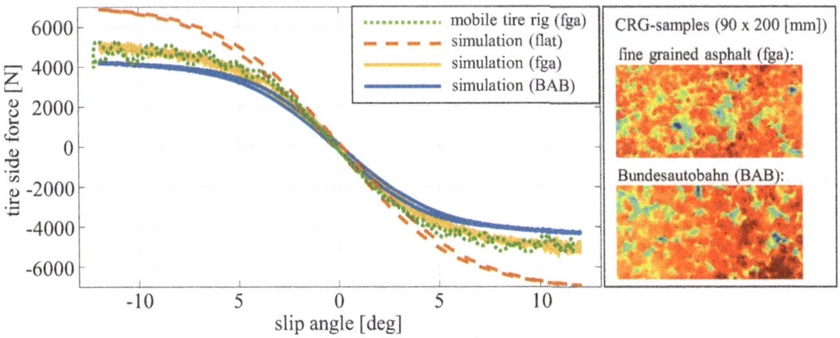

Fig. 35 Slip angle sweeps in simulation and measurement on different surface textures [6]

wheel loads are carried out to determine the actual side force characteristics on the investigated asphalt texture.

Figure 35 compares the measurement at 6 kN wheel load with the simulated equivalents for different surface textures with 0.5 mm grid resolution. There is low frequency fluctuation in the measurement, which can be attributed to road unevenness. Long-wave unevenness is not integrated into the digital roads at this point and can, therefore, not be detected in the simulations. Correlating the real road tests with the simulations of Fig. 34 on fine grained asphalt, a high texture resolution of 0.5 mm grid can be recommended for the simulation on asphalt texture. In this case, the deviation in the side force potential and the cornering stiffness between simulation and measurement is minimal.

However, the big data size accompanied with larger computation time needs to be accepted when simulating the tire-road interaction on realistic asphalt texture.

Having validated the contact patch behavior of the tire model and the necessary grid resolution, further simulations can be performed to predict more reliably wheel forces or ground pressure and shear stress distributions for the evaluation of different asphalt textures. Exemplarily, Fig. 35 also shows the side forces during a slip angle sweep on the BAB-texture. Its cornering stiffness and side force potential is lower than the values for the fine grained asphalt. It is assumed, that the high texture peaks locally lead to high ground pressures, which result in a lower averaged friction value.

2.2 Influence of the Road and Tire Modeling on Driving Dynamics Characteristics

So far, the influence of the asphalt texture is considered on a global friction level of the flat surface in full vehicle simulation. The friction coefficients are measured (cf. Sect. 2.3) or calculated theoretically from the pavement surface morphology and the tire rubber properties [30]. In this new approach, the texture is used directly as a simulation input and no further friction adjustments are implemented.

After the validation of the footprint characteristics of the FTire-model and the determination of the correlating CRG texture resolution on component layer, the influence of the models' level of detail on driving dynamics characteristics shall be determined in further investigations. The full vehicle simulation analysis is carried out with ADAMS/Car, which has already been introduced generally in Sect. 1.2. The integrated demo vehicle of the software represents a typical sports car with rear wheel drive, sporty chassis and a short ratio steering system. Due to the investigated tire dimension of a passenger car, the demo vehicle is used for the simulation instead of the semitrailer model of Sect. 1.2. However, as the investigated tire dimension of 205/55R16 is not the typical choice for a sports car, the driving dynamics limit range may shift significantly, for which this analysis shall not be misinterpreted as a quality analysis of the tire model FTire in the driving dynamics limit range.

Three driving maneuvers are considered: the closed loop maneuver of quasi-static accelerated cornering, the open loop maneuver of step steering and the controlled maneuver of emergency braking. The characteristics of the vehicle model using four detailed FTire models are compared to the one using standard FTire models on the fine grained asphalt texture introduced in Sect. 2.1 and on the regular flat surface.

Quasi-static Accelerated Cornering. The maneuver identifies steering or, respectively, the handling behavior of the vehicle. In this investigation, the vehicle is slowly accelerated on a constant radius of 50 m starting with 5 km/h until an instable driving behavior occurs. The steering controller of the demo car is not changed.

The necessary area of a skidpad with a radius of 51.5 m would result in over 8 km^2 and over 160 Gigabyte data size. To reduce the built up digital asphalt area, a straight road with a length of $2\pi \times 50$ m is generated and curved afterwards by aligning each longitudinal increment to an angle in the global coordinate system according to Sect. 1.3. For a valid simulation, the trajectories of each tire's horizontal movement have to verify all-time road contact after the simulation.

In Fig. 36, the steering wheel angle and the vehicle side slip angle are plotted over the increasing lateral acceleration of the maneuver. Due to the excitations of the asphalt texture, the steering controller constantly performs visible steering adjustments, which does not occur in the simulations with a flat surface. According to

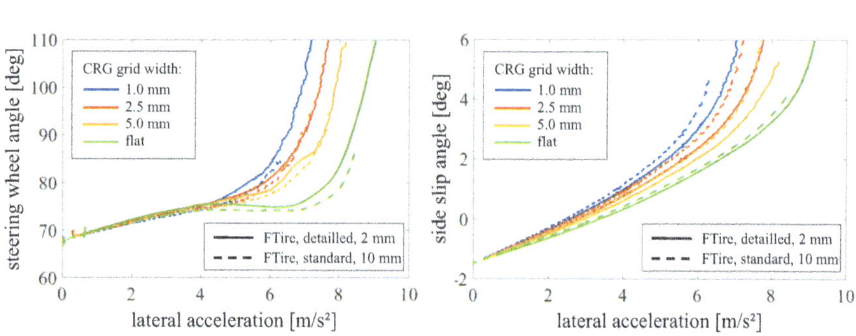

Fig. 36 Analysis of quasi-static accelerated cornering on different texture resolution

Fig. 36, the texture has a significant impact on the self-steering behavior. While the steering wheel angle increases monotonously for the simulations with texture, an oversteering tendency on the flat surface occurs between 5 and 6.5 m/s^2 until the loss of traction at the front axle results in a fast increase of the steering wheel angle.

Up to a lateral acceleration of 4 m/s^2, the steering wheel angle varies under 1° for all grid resolutions and tire model versions. However, the side slip angle increases differently for all simulations from the beginning. To follow the curve, the vehicle needs to apply higher side slip angles at lower lateral accelerations on asphalt texture. This effect amplifies with decreasing grid size of the road texture.

Step Steering. In this open-loop maneuver, the vehicle response behavior to a sudden steering input is investigated. The yaw rate response is one key performance that can be derived from this maneuver. According to ISO 7401:2011 'Road vehicles—Lateral transient response test methods—Open-loop test methods' [11], the vehicle is driven at a constant test speed until a steering angle is applied as rapidly as possible and kept until steady state conditions eventuate. The steering input angle is adjusted step-wise to ascending lateral accelerations that can be derived from steady state cornering (cf. Fig. 36). The yaw rate response is defined as the time between the point when 50% of the steering wheel angle is applied and the target point where 90% of the steady state yaw rate first occurs. For realistic conditions, a steering wheel velocity of 500 deg/s has been chosen in this study, so that the time between 10 and 90% steering input does not outreach 0.15 s.

One main challenge of open-loop maneuvers on texture is the trajectory-based generation of the digital asphalt roads as the cornering of the vehicle is not controlled. To set up maneuver-specific texture lanes, the simulation is firstly carried out on regular flat surface. The incremental angles p within the global coordinate system can be derived from the x-axis of the vehicle $\hat{x}_V$ and the axes of the global origin $(\hat{x}_O, \hat{y}_O)$

$$p = \arctan 2\left(\hat{x}_V \bullet \hat{y}_O, \hat{x}_V \bullet \hat{x}_O\right). \tag{5}$$

The incremental angles for each maneuver are used to bend the same straight asphalt lane to ensure equal texture conditions. As the velocity may decrease due to the maneuver, the corresponding distance travelled for each time step needs to be determined. Afterwards, the driving distance and incremental angle are resampled to the road grid resolution. Due to the varying vehicle behavior on flat surface and texture, the road width should be at least three times the axle track width, so that the tires keep road contact throughout the maneuver. This method is also suitable for other open-loop maneuvers such as sinusoidal steering.

The influence of the grid resolution on the yaw rate response for initial velocities of 60 and 90 km/h is illustrated in Fig. 37. Although the response time varies up to 50%, the offsets between the different model specifications remain rather constant with the exception of the higher lateral accelerations at 90 km/h, which can be traced back to the driving dynamics limit range taking into account Fig. 36.

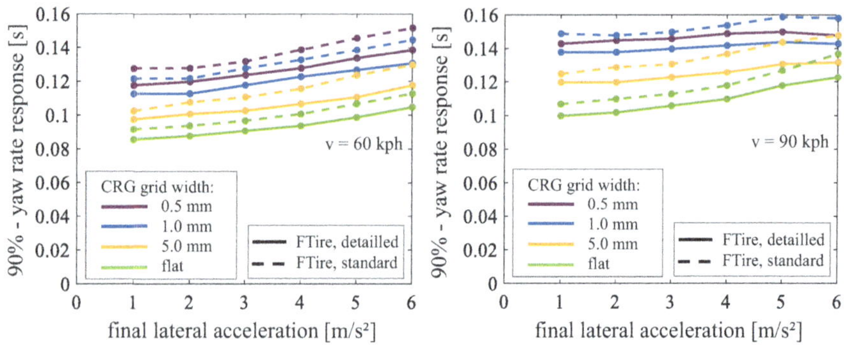

Fig. 37 Analysis of the yaw rate response to sudden steering input for different model resolutions

The higher response time of the standard FTire can be explained taking into account the higher necessary vehicle side slip angle (cf. Fig. 36) as the standard tire model requires higher slip angles to fit the necessary forces at the front and rear axle (cf. Fig. 34). Another possible explanation may be a longer relaxation length due to the difference in the structure stiffness parameters between the two FTire models. The relaxation length is validated only indirectly in the parametrization process.

Emergency Braking. The stopping distance is an important key performance for the safety of a vehicle. It is the sum of reaction distance, which is the time between the hazard detection and the beginning of braking, and the actual braking distance. Due to the importance towards traffic safety, a method to simulate the braking distance on asphalt texture is introduced in the framework of this work.

For current vehicles, the anti-blocking system (ABS) is obligatory. As ADAMS/Car does not provide a standard ABS-system, a custom longitudinal slip or brake force controller has been implemented using Matlab/Simulink and integrated into the multibody simulation software via an dll-file. Usual ABS-systems monitor the longitudinal slip condition and reduce the contact pressure on the brake disc when the tire begins to block. In that way, the tire brakes within the slip range of maximum longitudinal force potential. The reaction frequency and intensity of the controller has a direct influence on the braking performance. As a realistic ABS-function needs a complex controller, this study uses a simplified yet better comparable approach, in which a fixed longitudinal slip value for maximum braking potential is adjustable. To determine the correct value, brake slip component simulations are carried out for a range of wheel loads at the investigated initial rolling velocity on the investigated surfaces. In a next step, the wheel loads in the full vehicle simulation can be derived from the estimated braking acceleration. The final slip values are interpolated from the component simulations. In this example, the longitudinal slip at the front axle is adjusted in a range of 0.14 to 0.16 for all model combinations and investigated velocities. Due to the load change during braking, the wheel loads reduce heavily at the rear axle, which results in adjusted slip values of 0.19 and 0.21 according to the component simulations. The small variation of the values shows that the influence of

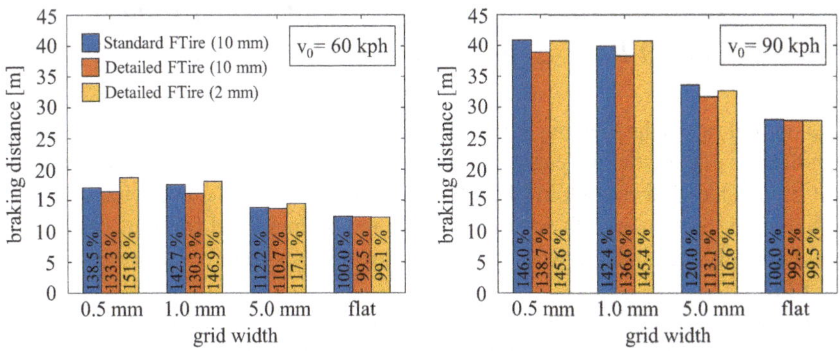

Fig. 38 Analysis of slip controlled braking tests on different texture resolution

the texture on the best braking slip value is small. Within the simulations, the force of the brake pedal is increased rapidly. However, the ABS-controller intervenes directly and keeps the braking slip at a constant value.

The results of the emergency braking simulations are presented in Fig. 38. In contrast to the lateral dynamics simulation, the tire model's grid resolution has a visible impact on the braking distance, for which the results of a detailed FTire with a standard grid width of c. 10 mm are listed in Fig. 38 additionally.

The characteristics between the braking distances, which are depicted as relative values to the results of the standard tire model on flat surface, look very similar for an initial velocity v_0 of 60 and 90 km/h. As expected, the usage of the different tire models results in nearly the same braking distance on a flat surface. With increasing grid resolution of the texture grid however, the braking distance increases up to 150%. Thereby, the detailed FTire with high grid resolution and the standard tire model lead to similar braking distances for the same surface. Only the detailed FTire with reduced grid resolution achieves slightly shorter braking distances. The simulations on texture grid widths of 0.5 and 1.0 mm show very similar braking distances contrasting the simulation results for the lateral dynamics. The values itself correlate with average emergency braking distances (cf. [17]), while the braking distances under 30 m on regular flat surface are very low compared to the state of the art.

In conclusion, the asphalt texture has also a big influence on the driving dynamics behavior. Methods to simulate on actual texture are provided and show that an increasing resolution leads to a reduction of transmittable forces, which leads to an increase of the steering angle and side slip angle requirement. As the offsets appear very linear, the increase of the texture resolution acts like a reduction of the friction coefficient between tire and road. Hence, the friction of the investigated tire and asphalt texture is considered in the next section.

2.3 Experimental Determination and Validation of Rubber Friction on Real Asphalt and Artificial Surfaces

The simulations of the previous sections have been performed with tire models, whose handling characteristics are parametrized on the basis of laboratory measurements on test rigs coated with corundum P120. Based on friction tests, the lower transmissible horizontal forces on asphalt need to be validated by measured friction coefficients. In a first step, the characteristics of linear drawing tests shall be investigated using the FMS.

The analyses of rubber-pavement-friction are performed on a mobile linear friction tester shown in Fig. 39. It can be used for outdoor real road tests or laboratory tests by mounting surface samples directly to its supporting frame. The 60 × 60 mm rubber samples are waterjet-cut from the tire tread and mounted directly to a 3D force transducer, which is connected to an electro-mechanical spindle drive via a sledge construction. For the measurements, the rubber specimen is loaded using weights between 5 and 55 kg and accelerated rapidly to a desired sliding velocity of 0.001 and 0.5 m/s. An additional string potentiometer measures the longitudinal movement of the sledge.

All measurements can be performed in dry condition, but also in wet condition using a water pump with an adjustable volume flow combined with a flat nozzle (cf. Fig. 40b). The required volume flow $q(v_S)$ to build up the desired theoretical water film height h_w of 1 mm depends on the water film width w_w and the sliding velocity v_S in

$$q(v_S) = h_w \cdot w_w \cdot v_S. \tag{6}$$

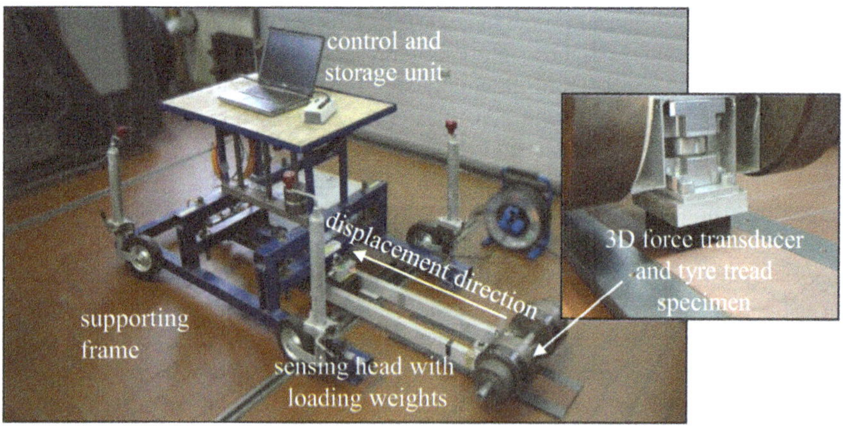

Fig. 39 Mobile linear friction tester in indoor configuration [10]

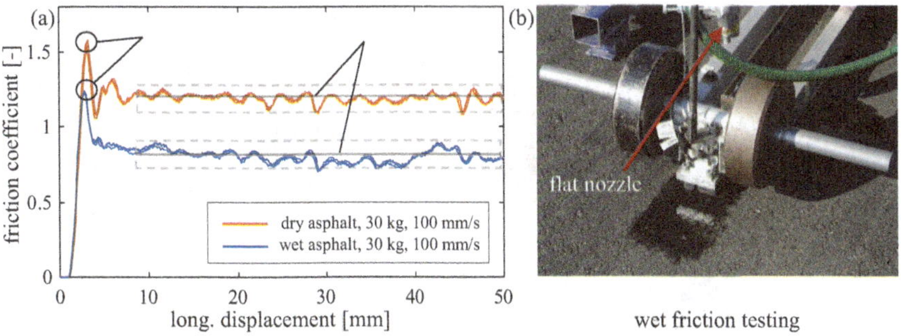

Fig. 40 **a** Investigated friction coefficients and **b** application of wet friction testing

The initial temperature of the rubber core and the surface is assured to stay constant throughout the measurement series as the cooling period between two measurements is monitored with two mounted infrared thermometers. The local temperature change in the contact area during the sliding event is not regulated. Every measurement is repeated three times for each boundary condition in order to check the plausibility and to reduce single event measuring deviation. However, a good repeatability can be achieved, as Fig. 40a overlays all measurement repetitions for each boundary condition.

The ratio of the longitudinal and vertical force at the same time step is defined as the current friction coefficient. According to Fig. 40, two characteristic values are determined for each measurement. In the first phase, the friction increases nearly linearly and reaches the maximum μ_{static}. The main contributor to friction in this phase is the adhesion. The static friction value is only valid, if the adjusted sliding speed has already been reached, for which high acceleration is required. The sliding friction $\mu_{sliding}$ is the average friction coefficient in the steady-state section, where the adhesive forces are reduced and the friction is mainly determined by hysteresis. The example in Fig. 40 shows that due to the water the reduced adhesion results in lower friction coefficients. The coefficients of all boundary conditions are summarized in a friction map.

Local Ground Pressure and Friction Distribution of Sliding Tests. At the beginning of each testing series, the tread samples undergo a flattening treatment, so that the ground pressure distribution is assumed to be homogenous on flat surfaces. For this purpose, a fine-grain belt grinder is used on low speed to carefully oblate the upper tread layer. Due to the reduced temperature input of this treatment, the vitrification effects are avoided. It is assumed that the ground pressure distribution is homogenous.

In reality, the actual ground pressure distribution for tread samples has a decreasing pressure gradient towards the outer edges of the sample and its peak close to the longitudinal tread groove, similar to the static and dynamic measurement of a tire footprint (cf. Fig. 25). Figure 41 compares the ground pressure distribution at static condition using a pressure imaging mat and at sliding conditions using the force

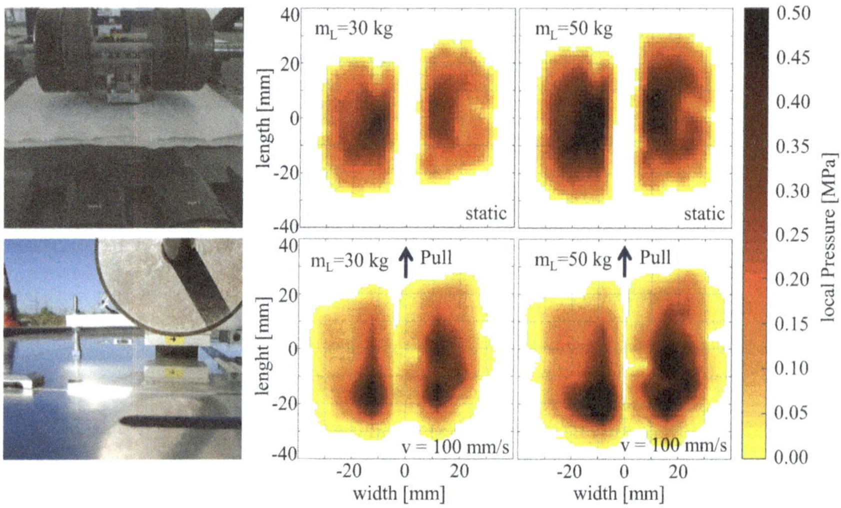

Fig. 41 Comparison of ground pressure distributions of rubber samples in static and sliding condition

matrix sensor of Sect. 2.1 in a special use case. Due to the limited sliding distance of the friction test rig, the FMS laser beams are triggered using extension bars, which is considered in the data evaluation of the internal FMS-velocity.

The difference of the contact area for the static and dynamic measurements is a result of the sensor resolutions and cannot be compared for that reason. However, two main effects influenced by sliding movement can be seen: The shift of the pressure peak against the pull direction and the increased pressure gradient within the tread sample. These effects are quantified in Fig. 42 for all load conditions and sliding velocities.

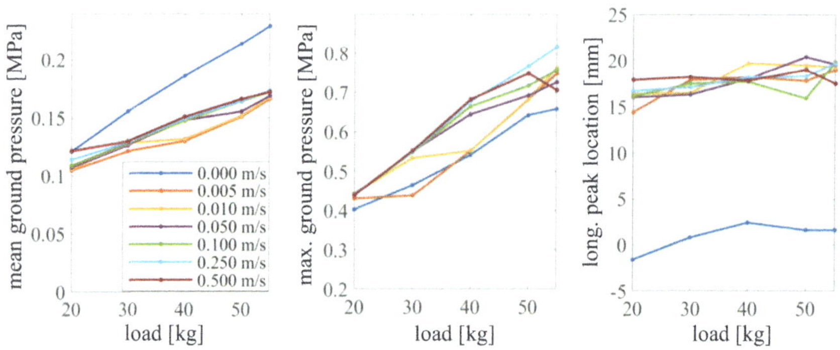

Fig. 42 Pressure distribution characteristics in linear friction tests

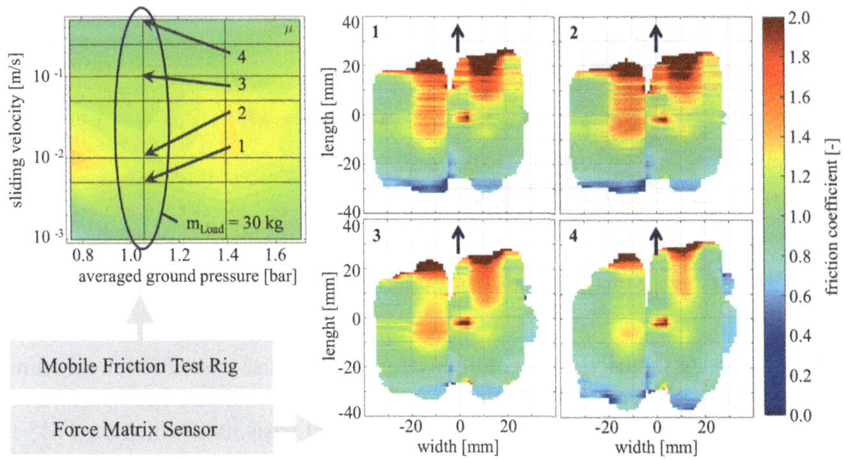

Fig. 43 Comparison of global and local friction coefficients of a sliding rubber sample

On the one hand, the mean ground pressure increases with the load. The difference between the mean ground pressure for low and high sliding velocities is a result of the reduced vertical forces during the sliding movement on the other hand, which is validated by the total vertical forces measured by the linear friction test rig. In contrast, the maximum ground pressure generally rises more quickly with increasing velocity. The longitudinal shift of the pressure peak is less velocity dependent. The shift is located between 15 and 20 mm from the middle of the tread sample against pull direction for all measurement conditions, so that this effect can rather be reasoned with the longitudinal displacement between fixed inner liner and the tread surface.

The resulting sliding friction coefficients of the sensor platform for both measurement systems are shown in Fig. 43. While the values of the friction tester vary between 0.8 and 1.3 for the different measurement conditions, locally the full range of sliding friction coefficients occurs.

Considering Fig. 41, the friction values decrease with rising ground pressures and vice versa. The decline of friction values with rising sliding velocity from the data of the mobile friction test rig is also locally observable within the tread sample in the data of the force matrix sensor. The slightly increased ground pressure for higher sliding velocities may contribute to the reduced friction values. Of course, other effects such as a reduction of adhesion with rising velocities could be overlaid.

Friction Differences between Real Road and Metallic Surfaces. As described in Sect. 2.1, large friction differences can occur for the same tread rubber material on different surfaces. If the road surface changes suddenly, the occurring friction jumps may lead to a brief horizontal oscillation of the tire tread. Especially in the use case of crossing the force matrix sensor with a rolling tire, the surface change from textured asphalt to polished stainless steel has an influence on the measurements.

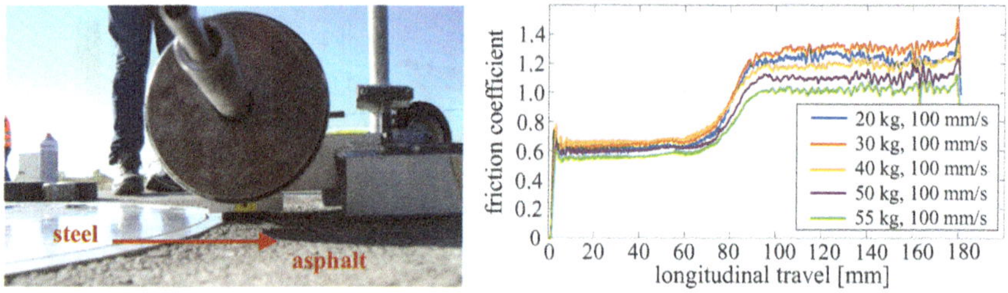

Fig. 44 Friction change from steel surface to asphalt texture

To investigate the friction discontinuity for the FMS use case, friction measurements crossing the surface borders at constant sliding speed have been performed. Figure 44 shows the friction transition from the steel basis, into which the FMS is inserted, to the asphalt. As the steel platform is exposed to changing weather conditions, the ageing process results in a reduced friction level compared to the polished sensor platform. Although the tread length is 60 mm, the relevant friction change only takes place in a longitudinal displacement of c. 25 mm making the friction transition even more intense. Furthermore, the sliding friction coefficients increase up to 200% depending on the measurement condition. Due to this unneglectable influence on the tire-road interaction, new durable materials need to be investigated to minimize the friction discontinuity effects.

A possible solution is the replication of the nearby asphalt texture with an additive manufacturing process. First successes with polymer materials have already been achieved, but ultimately they only reflect the properties of an asphalt road to a limited extent [13]. A limiting factor for the samples was the reproduction of the micro roughness with the used material. Due to the high wear requirements and the scientific progression in metallic 3D-printing, the investigated asphalt could be replicated using stainless steel (1.4404) on the basis of selective laser melting (SLM). Due to a holistic study using a topographic surface to investigate multiple friction laws, the printing process details are described in Chapter "Numerical Friction Models Compared to Experiments on Real and Artificial Surfaces".

Friction Characteristics on Asphalt and 3D-printed Replica. To investigate the friction characteristics of asphalt and its 3D-printed metallic replica, nearby asphalt has been digitalized using a portable 3D handheld laserscanner (HandySCAN 700®) with a height resolution up to 30 μm and a lateral resolution of 50 μm. The post-processing includes tilting, cutting and converting to a volume model using Computer Aided Design methods (CAD). The road surface has been printed without a substrate plate, but directly on a mounting plate of the same material for a positive-locking adaption to the linear friction test rig.

The detailed friction maps for the coefficients μ_{static} and $\mu_{sliding}$ in dry conditions are illustrated in Fig. 45. Each nodal point within the map represents one measuring condition. Both, original asphalt and replica, show an increase of the friction peaks μ_{static} with rising velocity and a slight decrease with rising loads. As the friction

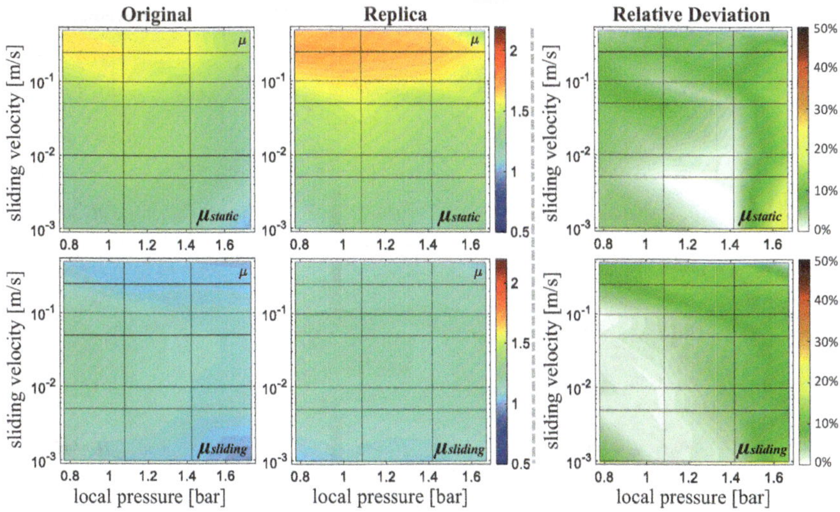

Fig. 45 Friction maps for original asphalt and replica surface in dry condition [5]

characteristics are very similar, the relative deviation outreaches 20% at no point. On average, the static friction only has a relative deviation of 9.0% between the original asphalt and the replica. The sliding friction $\mu_{sliding}$ even undercuts this value with 5.8%. The coefficients show a slight different distribution on both surfaces. The highest values can be found between 10 and 50 mm/s sliding velocity and they decrease with higher and lower velocities. In summary, there is a reasonable correlation of both friction coefficients for the whole map under dry conditions, so that substituting the flat stainless steel surface with artificial asphalt minimizes the friction differences to a sufficient degree (cf. Figs. 44 and 45).

According to Fig. 46, both friction coefficients decrease under wet conditions. Although the friction characteristics correlate to the dry measurements, the reduction of the friction values differs significantly. The averaged relative deviation is 31.9% for μ_{static} and 24.2% for $\mu_{sliding}$. The reasonable correlation of the friction coefficients under dry conditions cannot be achieved at wet conditions, for which the surfaces are investigated with optical methods in a further step.

Due to the complexity of the asphalt surface roughness, basic amplitude parameters, such as the root-mean-square roughness (R_q), the mean profile depth (MPD) or the skewness (R_{sk}) struggle to link roughness and friction. In contrast, the radially averaged power spectral density (PSD) is an effective mathematical function to analyze surface roughness on all scales (cf. [13]). The macro scale down to a wavelength of 200 μm is measured by the portable handscanner. The micro scale down to a wavelength of 20 μm is measured by a high resolution line scanner, which drives unidirectionally on a sledge construction.

Figure 47a shows the PSD for the real asphalt sample, the metallic replica and further investigated material on the basis of polyamide, whose replica surface is

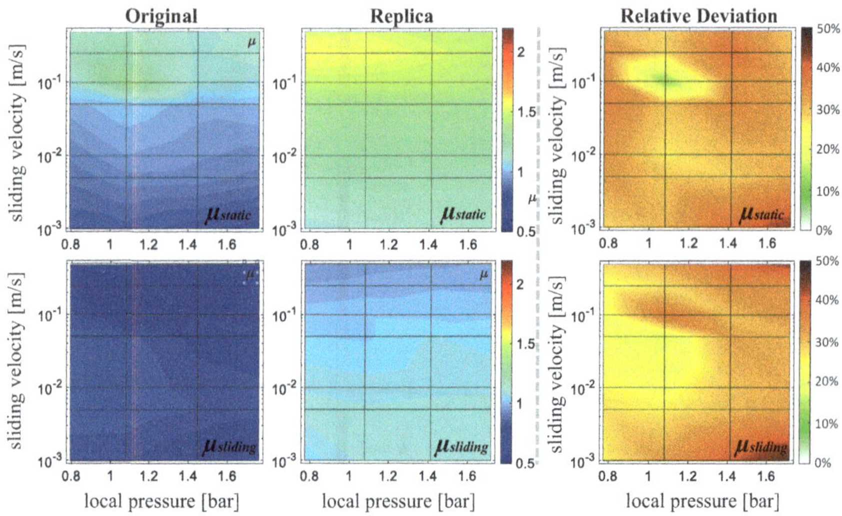

Fig. 46 Friction maps for original asphalt and replica surface at wet condition [5]

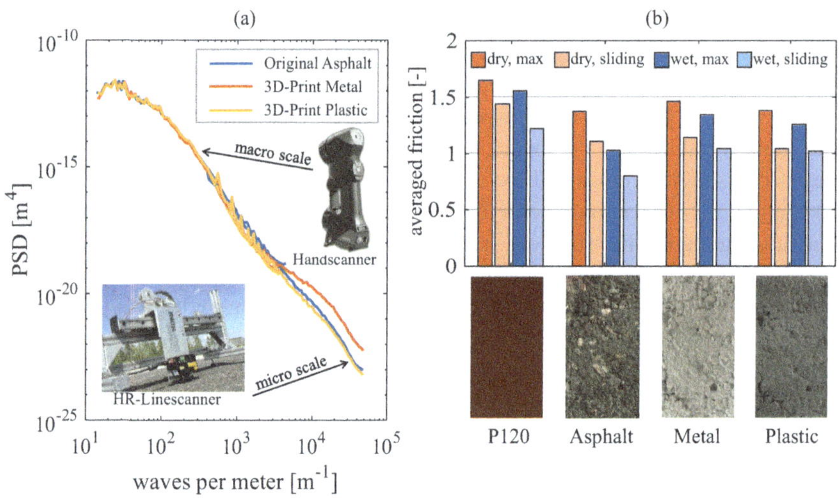

Fig. 47 **a** Radially averaged power spectral density analysis and **b** averaged friction values

build up using a commercial multi jet fusion procedure. Two main differences can be extracted from the diagram. In the micro scale, small printing asperities between 100 and 300 μm lead to an increased PSD for the metallic replica. This more intense micro roughness is a clear contributor to the high wet friction coefficients. At a wavelength of 1 mm the original asphalt specimen has a slightly higher PSD than its replicas, which may also have an impact on the friction. Figure 47b shows the averaged coefficients of the friction maps. Further measurements were also carried out on corundum P120, the standard belt and drum coating for tire test benches. It is shown that the replicas correlate better to the asphalt than P120. The friction level for the plastic replica only has a small offset to the metal replica, although it correlated better in the micro scale. Deviating adhesive effects between the material pairings cannot be investigated in this analysis. However, it is assumed that the adhesion effects decrease more intense on real asphalt under wet conditions than on metal or plastic, as both materials are considered to be more hydrophobic than asphalt.

As most measurement series are carried out at dry conditions, alternative sensor or tire test rig coatings for drum test rigs or flat-trac test benches show high potential for more realistic tire behavior measurements. Especially, the metal replica did not show any measurable signs of wear and, therefore, it shows advantages in long-term durability.

Comparing the averaged friction coefficients of P120 and original asphalt, the deviation of the friction values with c. 25% correlates with the deviation of the emergency brake distances of Fig. 38. The same tire and the same asphalt are considered in the simulations and the experiments. It is a first indicator, that the integration of high resolution asphalt texture into the simulation leads to realistic depiction of the vehicle-tire-road interaction. Further investigations with different asphalt types could validate this hypothesis.

3 Summary and Conclusion

In this chapter, vehicle-tire-road interaction has been investigated using methods for vertical and horizontal force transmission. Physical tire models can be used to simulate three-dimensional excitation profiles on tire and vehicle level. For non-parallel axle excitations, flexible component models must be used to generate realistic load collectives at the full vehicle level. This validated full-vehicle simulation model has been used to generate a realistic road load collective, which is used in Chapter "Simulation Chain: From the Material Behavior to the Thermo-mechanical Long-term Response of Asphalt Pavements and the Alteration of Functional Properties (Surface Drainage)" for a coupled modeling approach of the vehicle-tire-road interaction.

To validate simulative driving maneuvers on a laboratory hydraulic axle test rig, a method that transfers the properties of the rolling tire to a non-rolling axle test rig has been implemented. Building upon this validation, the compromise between load and road protection has been optimized by the development a hybrid damper control.

The validation of the contact patch behavior of a physical tire model allows the investigation of local asphalt load more precisely, e.g. the applied shear stresses on a single grain. It has been shown that the modulation of the texture plays a major role in the force transmission of the tire model on the asphalt and the resulting full vehicle behavior. A detailed analysis with linear friction tests has shown that the decreasing transmittable horizontal tire forces correlate with the friction difference between the investigated asphalt types and test rig coating used for the parametrization of the tire models. Furthermore, the replication of asphalt using additive manufacturing techniques shows a good correlation of the friction coefficients. The materials can be used for future test rig or sensor coatings for a more realistic measurement of tire properties or to validate friction laws using target-aimed artificial surfaces, which is investigated by the research group in the following Chapter "Numerical Friction Models Compared to Experiments on Real and Artificial Surfaces".

References

1. Anghelache, G., Moisescu, R.: Measurement of stress distributions in truck tyre contact patch in real rolling conditions. Vehicle Syst. Dyn. **50**, 1747–1760 (2012)
2. Bachmann, C.: Vergleichende Rollwiderstandsmessungen an Lkw-Reifen im Labor und auf realen Fahrbahnen. Ph.D. thesis, RWTH Aachen University (2018)
3. Bukovics, J., Roca, E., Gerschütz, W., Kolm, H., Scheider, A.: Auslegung von Akustik und Schwingungskomfort. Vieweg-Verlag, Wiesbaden (2008)
4. Fernando, E. G., Musani, D., Park, D-W., Liu, W.: Evaluation of effects of tire size and inflation pressure on tire contact stresses and pavement response. Texas Transportation Institute, http://tti.tamu.edu/documents/0-4361-1.pdf. Last accessed 11 Dec 2020
5. Friederichs, J., Wegener, D., Eckstein, L., Hartung, F., Kaliske, M., Götz, T., Ressel, W.: Using a new 3D-print-method to investigate rubber friction laws on different scales. Tire Sci. Technol. **48**, 250–286 (2020)
6. Friederichs, J., Eckstein, L.: Enhanced prediction of the tire-road-interaction by considering the surface texture, In: Functional Pavements – Proceedings of the 6th Chinese – European Workshop on Functional Pavement Design (CEW 2020). CRC Press, Nanjing (2020)
7. Gallrein, A.: CDTire – State-of-the-art Tire Models for Full Vehicle Simulation. Fraunhofer ITWM, Kaiserslautern (2004)
8. Gipser, M.: Reifenmodelle in der Fahrzeugdynamik: eine einfache Formel genügt nicht mehr, auch wenn sie magisch ist. MKS-Simulation in der Automobilindustrie, Graz (2001)
9. Gipser, M.: The FTire Tire Model. https://www.cosin.eu/wp-content/uploads/ftire_eng_3.pdf. Last accessed 11 Dec 2020
10. Hartung, F., Kienle, R., Götz, T., Winkler, T., Ressel, W., Eckstein, L., Kaliske, M.: Numerical determination of hysteresis friction on different length scales and comparison to experiments. Tribol. Int. **127**, 165–176 (2018)
11. ISO 7401:2011 – Road vehicles – Lateral transient response test methods – Open-loop test methods. International Standard, Geneva (2011)
12. Isermann, R.: Fahrdynamik-Regelung – Modellbildung, Fahrerassistenzsysteme, Mechatronik. Vieweg-Verlag, Wiesbaden, (2006)
13. Kanafi, M.: Rocky road – surface roughness impacts on rubber friction. Ph.D. thesis, Aalto University (2017)
14. Karnopp, D., Crosby, M., Harwood, R.: Vibration control using semi-active force generators. J. Eng. Indus. **96**, 619–626 (1974)
15. Khandavalli, G., Kalabis, M., Wegener, D., Eckstein, L.: Potentials of modern active suspension control strategies – From model predictive control to deep learning approaches. In: 10th International Munich Chassis Symposium 2019, pp. 179–199. Springer Vieweg, Wiesbaden (2020)
16. Knauer, P.: Objektivierung des Schwingungskomforts bei instationärer Fahrbahnanregung. Ph.D. thesis, Technische Universität München (2010)
17. Liu, X., Cao, Q., Wang, H., Chen, J., Huang, H.: Evaluation of vehicle braking performance on wet pavement surface using an integrated tire-vehicle modeling approach. Transp. Res. Rec. **2673**, 295–307 (2019)

18. Oertel, C., Eichler, M., Fandre, A.: RMOD-K – Modellsystem zur Simulation des Reifenverhaltens beim Überrollen kurzwelliger Unebenheiten – Version 6.0. In: International ADAMS User's Conference, Berlin (1999)
19. Pacejka, H.: Tyre and Vehicle Dynamics, 3rd edn. Butterworth-Heinemann - Elsevier, Burlington (2012)
20. ProgTrans, A.G.: Abschätzung der langfristigen Entwicklung des Güterverkehrs in Deutschland bis 2050. Bundesministerium für Verkehr, Bau und Stadtentwicklung, Basel (2007)
21. Rauh, J.: OpenCRG – The open standard to represent high precision 3D road data in vehicle simulation tasks on rough roads for handling, ride comfort, and durability load analyses. http://www.opencrg.org. Last accessed 10 Dec 2020
22. Savaresi, S., Poussot-Vassal, C., Spelta, C., Sename, O., Dugard, L.: Semi-Active Suspension Control Design for Vehicles. Butterworth-Heinemann, Oxford (2010)
23. Shell, A.G.: Shell LKW-Studie – Fakten, Trends und Perspektiven im Straßengüterverkehr bis 2030, Hamburg/Berlin (2010)
24. VDI-Richtlinie 2057-1: Einwirkung mechanischer Schwingungen auf den Menschen Ganzkörper-Schwingungen. Verein Deutscher Ingenieure e. V., Darmstadt (2017)
25. Valášek, M., Kortüm, W.: Semi-active suspension systems II. In: Nwokah, O., Hurmuzlu, Y. (eds.) The Mechanical Systems Handbook. CRC Press LLC, Boca Raton, Florida (2002)
26. Velske, S., Mentlein, H., Eymann, P.: Straßenbau – Straßenbautechnik. Werner Verlag, Düsseldorf (2009)
27. Winkler, T., Wegener, D., Eckstein, L.: Method development to analyse the vertical and lateral dynamic road-vehicle interaction of heavy-duty vehicles. Autom. Engine Technol. **3**, 129–139 (2018)
28. Winkler, T.: Generierung quasi-rollender Prüfstandsanregungssignale für ein gekoppeltes Reifen-Fahrbahn-System. Ph.D. thesis, RWTH Aachen University (2018)
29. Zegelaar, P.: The dynamic response of tyres to brake torque variations and road unevenesses, Ph.D. thesis, Delft University of Technology (1998)
30. Zheng, B., Chen, J., Runmin, Z., Xiaoming, H.: Skid resistance demands of asphalt pavement during the braking process of autonomous vehicles. MATEC Web Conf. **275**, 04002 (2019)

Characterization and Evaluation of Different Asphalt Properties Using Microstructural Analysis

Pengfei Liu, Tim Teutsch, Jing Hu, Stefan Alber, Dawei Wang, Gustavo Canon Falla, Markus Oeser, and Wolfram Ressel

Abstract Microstructural analyses of asphalt mixtures are described in this chapter using the X-Ray computer tomography (X-Ray CT) and related digital image processing (DIP) approaches. Different parameters of single elements like aggregates or air voids as well as characteristics of the whole grain and void structure, which can be determined, are introduced. These features can be linked to different mechanical, structural and functional properties. Changes of certain parameters by load application (tensile or compressive stress) or by artificial soiling (with porous structures), e.g. in before-and-after studies, are observed and certain conclusions are drawn. Fatigue and deformation of asphalt pavements and related deterioration effects in the microstructure like cracking are presented as exemplary use cases in this chapter as well as drainage and sound absorption of porous asphalt. The virtual reconstruction of asphalt structures based on three-dimensional (3D) images of real asphalt samples is also of great relevance for computational studies, e.g. for finite element modeling, and is shown exemplarily in this chapter. Simplification approaches of the microstructure are discussed in that context briefly as well.

Keywords Microstructure characterization · Air void analysis · Aggregate analysis · CT technology

Funded by the German Research Foundation (DFG) under grant OE 514/1, RE 1620/4, WE 1642/11 and LE 3649/2.

P. Liu · D. Wang · M. Oeser
Institute of Highway Engineering, RWTH Aachen University, Aachen, Germany

T. Teutsch · S. Alber (✉) · W. Ressel
Institute for Road and Transportation Science, University of Stuttgart, Stuttgart, Germany
e-mail: stefan.alber@isv.uni-stuttgart.de

J. Hu
School of Transportation, Southeast University, Nanjing, People's Republic of China

D. Wang
School of Transportation Science and Engineering, Harbin Institute of Technology, Harbin, People's Republic of China

G. Canon Falla
Institute of Urban and Pavement Engineering, Technische Universität Dresden, Dresden, Germany

M. Kaliske et al. (eds.), *Coupled System Pavement—Tire—Vehicle*,
Lecture Notes in Applied and Computational Mechanics 96,
https://doi.org/10.1007/978-3-030-75486-0_6

1 Introduction

The functional and mechanical properties of asphalt mixtures are closely related to its internal structure. The aggregates are the most important components for stress relief, stiffness and durability of pavements. They also provide the foundation for skid resistance on the road surface and are therefore exposed to various influences. The mastic (filler and bitumen) and their interaction with the aggregates add another very important factor to structural stability. Not less important are the air voids within the structure, as their distribution and shape affect the mechanical properties of the asphalt structure, especially when looking at porous asphalt (PA), where they are a major component.

Conventional approaches of testing and modeling consider the asphalt as uniform material with homogeneous properties, which indirectly reflect the behavior of the individual components. Since the interactions between the components are far more complex and the conventional approaches are no longer adequate to characterize asphalt mixtures, especially for research purposes more detailed and in-depth approaches have been developed in recent years. It is at this point, where the research described below sets in.

The aim of this chapter is to present different methods used in research to characterize the microstructure of asphalts. Thereby, the focus lies on the aggregates and the air voids. Due to its complexity and the very fine structure, which makes its characterization very difficult, the asphalt mastic is only considered indirectly in this research, although it has an essential influence on the asphalt properties.

1.1 X-Ray Computer Tomography

All investigations use the X-Ray computer tomography (X-Ray CT) as shared technique for characterizing the different parameters. The system described in [15] is used to acquire the actual structure of asphalt mixtures. Detailed descriptions of the following shortly explained procedures and techniques can be found within the chapter "Computational Methods for Analyses of Different Functional Properties of Pavements" and research by [3, 19, 21, 25, 26].

In the research described here, images of 1024 $\times$ 1024 pixels were acquired at a scan interval of 0.1 mm. The pixel size varies between 10 μm and 80 μm and is specified later. The images contain different gray values between 0 and 255, where the asphalt components are represented by different gray value ranges due to their material density. While aggregates having the highest density also have the highest gray values, air voids with lowest density are represented by the lowest gray values. An example for such an image is shown in Fig. 1a.

The most important and critical step in the microstructural characterization process follows. For individual characterization, the aggregates or air voids must be reliably segmented and extracted separately from the images. During this process,

Fig. 1 Extraction process for components from an X-Ray CT image: **a** original image, **b** binary image of aggregates, **c** binary image of air voids [15]

the pixels containing the gray values representing a specific component need to be detected and converted into binary images that only contain aggregates (see Fig. 1b), air void areas (see Fig. 1c) or mastic.

Since the Otsu threshold method [24] segments the components most effectively but operates with a global threshold, an improved approach to this method was developed within the context of this research in order to segment the images. These were analyzed afterwards using a commercial software.

1.2 Materials and Load Tests

The research in this study is mainly focused on the change of the asphalt structure due to traffic loads. Therefore, the materials respectively specimens described are analyzed before and after exposure to certain stress states. For the research on PA described in Sect. 4, the material is not further described here and can be found in the mentioned references.

Specimens for fatigue damage analysis For the analysis on the influence of air voids and cracks on the fatigue behavior described in Sect. 3.1, drill core specimens were taken from a test track, called the *ISAC test track*, that has been built at the Institute of Highway Engineering at the RWTH Aachen University to conduct various tests. The asphalt mixture is a Stone Mastic Asphalt (SMA) and has a nominal maximum aggregate size of 11 mm (SMA 11 DS). It consists of diabase aggregates and a PG 50/70 bitumen binder. Its aggregate gradation curve is shown in Fig. 2a. It was paved with a temperature of 170 °C. A detailed description of the test track and the material is shown in [15].

For the fatigue damage analysis, the drill cores of the full pavement were cut into several specimens, which contain various air void contents, according to the former depth inside the pavement.

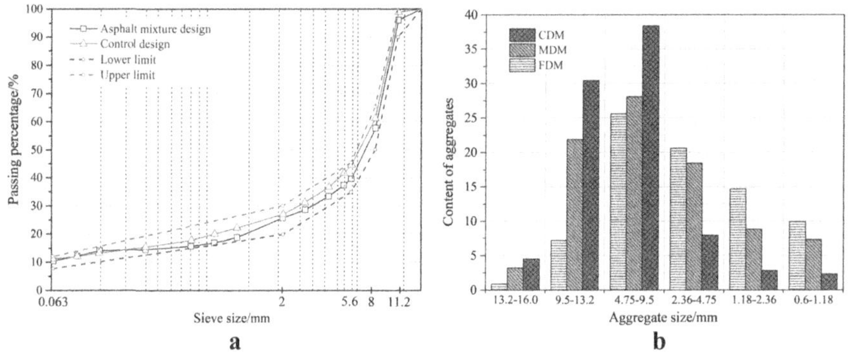

Fig. 2 **a** Aggregate gradation of the SMA 11 DS [15], **b** aggregates constitution of different asphalt mixtures [14]

Because only damages, that occur until micro cracks initiate, are considered in this study, indirect tensile tests were performed on the specimens by applying sinusoidal loads between 0.035 MPa and 0.5 MPa with a frequency of 0.1 Hz for two hours. Commonly fatigue tests are performed at 20 °C, but for micro crack analysis it is more advantageous, if they are conducted in three low temperature stages (−10 °C, 0 °C and 10 °C).

Specimens for interface stripping analysis The following information about the specimen is described more detailed in [14]. To examine the behavior of the interface between asphalt mastic and the aggregates described in Sect. 3.2, three asphalt mixtures with different aggregate gradation were tested. The basalt aggregates differ in their percentages of different grain size classes and are further referred to as coarse-dense mixture (CDM), mean-dense mixture (MDM) and fine-dense mixture (FDM). The percentages of the distribution in the particle size classes of the mixtures are shown in Fig. 2b.

The three mixtures were manufactured into specimens by compacting it with a gyratory compactor at a temperature of 170 °C with 100 cycles at a compacting angle of 1.16° ± 0.02°, a compacting pressure of about 600 kPa ± 18 kPa and a rotating rate of 30 r/min ± 5 r/min. Afterwards, drill cores with a diameter of 75 mm and a height of 50 mm were extracted. Then, three specimens of each mixture were deformed with a displacement controlled uniaxial compressive test, at a speed of 0.02 mm/s at 60 °C.

2 Microstructural Parameters of Asphalt

Asphalt mixtures are typical heterogeneous composite materials consisting of aggregates with an irregular shape and random distribution, asphalt binder, and air voids. There are numerous approaches and parameters for characterizing the internal structure and its components. Although the mastic plays an important role, it is only considered indirectly in this study, as it is a complex subject to determine a geometric characteristic. Therefore, this research focuses on the characterization of aggregates and the air voids on the influence of the structural performance using the following parameters.

2.1 *Aggregate Characteristics*

Orientation and positioning The major components of asphalt mixtures are aggregates, therefore, they have a great importance on its mechanical properties [19]. The shape of an aggregate can be described, for example, by a simplifying mathematical shape such as an ellipse (2D) (see for example [23]) or an ellipsoid (3D). An approach for three-dimensional shape characterization with the help of ellipsoids is discussed in chapter "Computational Methods for Analyses of Different Functional Properties of Pavements". Therefore, each ellipsoid representing an aggregate has a specific center that defines its position within the asphalt structure and three axes that define orientation and shape.

Studies of the spatial distribution by [9, 11, 28], have shown that the aggregate orientation affects the shear resistance and stiffness of the asphalt structures. Figure 3a shows, how the major axis of an aggregate is defined, whereby for the orientation angle equals 0° the aggregate is considered "standing" in vertical direction, while at an angle equals 90°, the aggregate is considered "lying" in horizontal direction.

Contact points As shown by [8], the contact points between the aggregates have an important influence on the stability and determine the dissipation of the forces within the asphalt structure. The contact behavior between the aggregates is characterized by calculating the pixel distance (L) between two particles with an optimized boundary search method, as shown in Fig. 3b. If L gets shorter than a defined surface distance threshold, the two examined aggregates are considered sharing a contact point.

Fractal dimension The common parameters used to characterize the shape of aggregates, such as roundness, sphericity, angularity or flat and elongated particles percentage (see also chapter "Computational Methods for Analyses of Different Functional Properties of Pavements") are usually based on simplifications and two-dimensional methods. Nevertheless, these cannot specify the complex three-dimensional geometry of aggregates. Besides the method used, other three-dimensional approaches, such as described by [17, 18], have been developed in the last few years.

Within this study, the fractal dimension is used, as it is an effective approach to analyze, for example, the irregular shape of aggregates or cracks in asphalt structures. Specifically, the 3D box-counting method is used to divide a three-dimensional model of an aggregate into spatial grids of equal length, to determine the fractal dimension from the variation of these grids, which represents the complexity of the model [14]. The methodology is illustrated in Fig. 4. The method is applicable to all three-

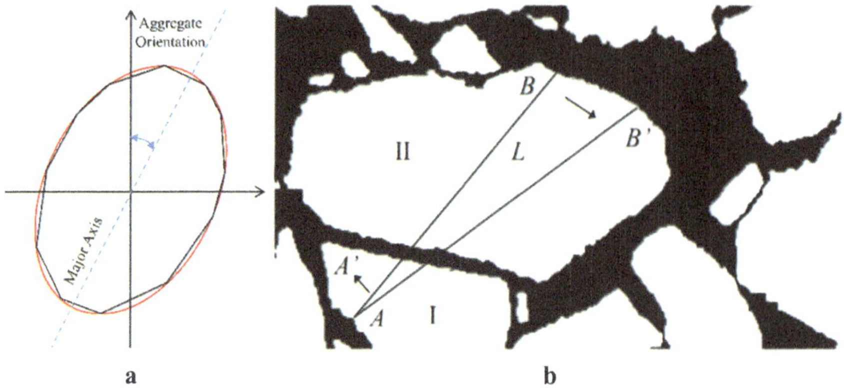

Fig. 3 **a** Orientation of the major axis, **b** calculation of contact relations between two aggregates [19]

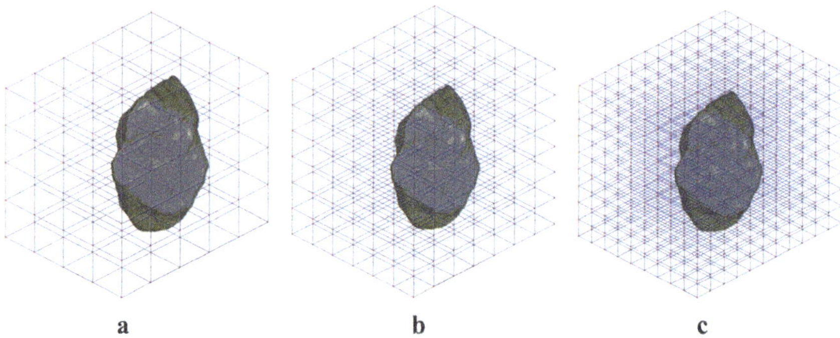

Fig. 4 Spatial morphology analysis of 3D model using 3D box-counting, **a** side length r = 3.25 mm, **b** side length r = 2.43 mm, **c** side length r = 1.62 mm, after [14]

dimensional objects that are self-contained and have a more or less spherical shape. So besides aggregates, it is also used to characterize air voids located within the structure.

Complex structures with multiple end points, such as the air void structure of PA (see [26]) or the asphalt mastic, cannot be characterized using this method.

2.2 *Air Void Characteristics*

The air void characteristics have an important influence on damage initiation and propagation due to loading as well as on drainage and sound absorption. Besides the fractal dimension mentioned above, the following described parameters are important for air void characterization within this study.

Shape index The shape index (SI) describes the morphology of the air voids in the structure and is calculated using the equation, as mentioned in [19]

$$SI = \frac{l_{mmd}}{l_{mnd}} \ . \tag{1}$$

It is calculated as shown in Eq. (1) by dividing the maximum length (l_{mmd}) by the minimum length or width (l_{mnd}) of an air void. The parameter is used to determine, whether the air void has a more elongated or more rounded shape.

Tortuosity As the SI can only provide information about the ratio between the length and width of an air void, the spatial characteristics are not really taken into account. Therefore, additionally the tortuosity (τ) is calculated, by dividing the total length (l) of the connective air void's centroids by the shortest distance between the beginning and the end of the air void (l_{min}), as defined by the equation

$$\tau = \frac{l}{l_{min}} \ . \tag{2}$$

If the tortuosity equals 1, the air void is considered straight, while the higher the ratio gets, the air void becomes more twisted.

Damage ratio Besides other influences, that effect structural damage, in this study the influence of the air void area and cracks on fatigue damage is specified by the damage ratio (DR) defined by the equation

$$DR = \frac{S_{ca} + S_{eaa}}{S_{iaa}} \ . \tag{3}$$

The parameters S_{ca} and S_{eaa} represent the area of cracks respectively air voids after fatigue damage, while S_{iaa} is defined as the area of the initial air voids.

3 Influence of Aggregate and Air Void Characteristics on Damage Mechanisms

3.1 Fatigue Damage of Asphalt Mixtures with Different Air Voids

The following explanations about fatigue damage of asphalt mixtures are primarily described in more detail in [15]. The fatigue damage of pavements, as a result of repeated traffic loads, significantly reduces their performance and stability. In order to gain more detailed knowledge of this damaging mechanism, the inner structure of asphalt mixtures has been analyzed using X-Ray CT technology and Digital Image Processing (DIP). The pixel size of the images is set to 80 μm.

It turns out that the air void distribution and characteristics have a considerable influence on the fatigue resistance. Of course, temperature plays a major role, but

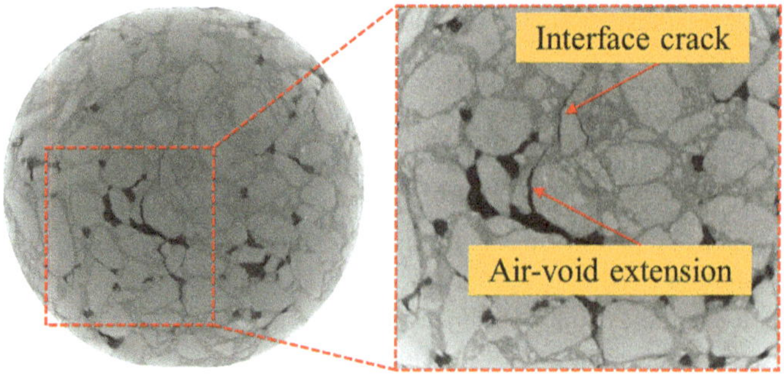

Fig. 5 Two types of damage to the asphalt structure [15]

these air void related aspects are predominately related to the compaction ratio. This section describes a method that provides an effective evaluation of fatigue damage.

Microstructural air void characteristics For the transformation analysis of air voids and the appearance of new cracks in and between the asphalt mastic and aggregates, assumed being caused by fatigue damage, three specimens from a depth of 90 mm were scanned before and after loading with the CT and processed with the DIP. The information obtained from the images, as shown in Fig. 5, is divided into two types: new interface cracks and the extension of the air voids. Since the pixel size is 80 μm as mentioned above, cracks wider than that can be detected. Using images with a higher resolution even finer cracks could be detected. After the extraction of the air voids respectively cracks, the SI is calculated for every one individually using Eq. (1). Air voids with an SI greater than 10 and an area smaller than 5 mm^2 are considered as cracks. The results of the three specimens at three temperature stages, in terms of air void area as a function of SI, are compared in Fig. 6. Figure 6a shows the results at -10 °C, Fig. 6b at 0 °C and Fig. 6c at 10 °C.

The diagrams in Fig. 6a–c show that the fatigue damage of the material significantly increases the cracks inside the structure. In addition, the average SI and the average area of the air voids increase, which leads to the conclusion that fatigue damage transforms the morphology and the number of air voids simultaneously. It can be expected that air voids represent a kind of predetermined fracture point within the material structure, because they disrupt the path of force transfer.

A comparison of different void contents shows that a higher void content leads to a lower fatigue resistance. This is clearly shown in Fig. 6d, since with a higher percentage of initial air voids, the cracks that occur under load also increase. Furthermore, it should be mentioned that this behavior is temperature depended, because more cracks occur at 10 °C than at 0 °C or, above all, -10°C.

In addition, it can be seen that the increase in cracks decreases as the initial number of voids increases. This is explained by the fact that the possible contact points between the aggregates and the asphalt mastic are decreased by increasing air void content, which decreases the possibility of crack initiation and propagation.

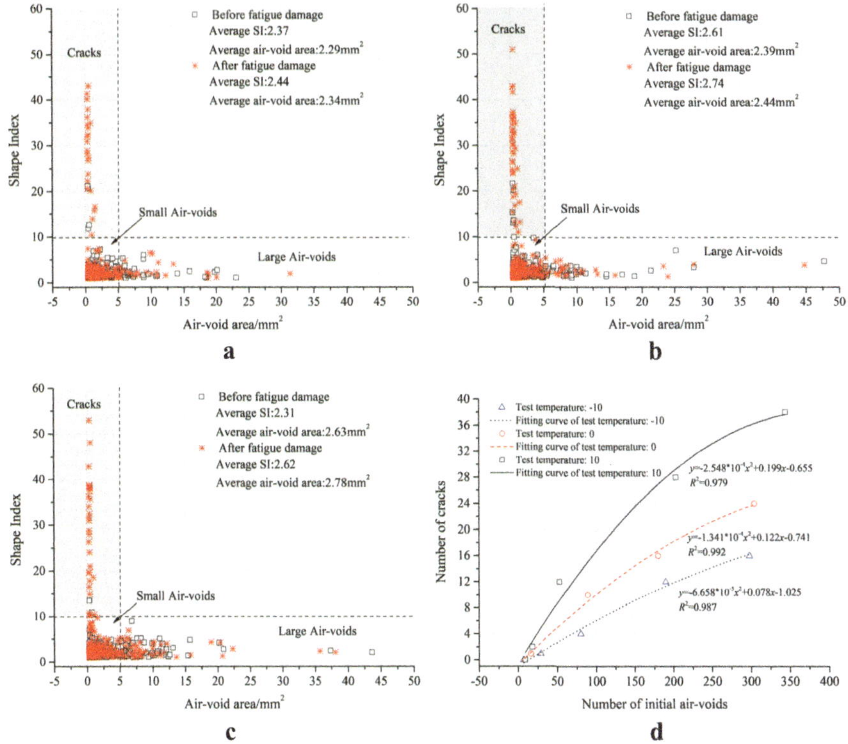

Fig. 6 Comparison of air void area and SI before and after fatigue damage, **a** -10 °C test temperature, **b** 0 °C test temperature, **c** 10 °C test temperature, **d** compared to initial air voids [15]

Impact of air void morphology For characterizing the influence of the air void morphology on the fatigue damage behavior, the X-Ray CT scans of four specimen were analyzed by calculating the average fractal dimension (see Sect. 2.1) and the DR as described in Sect. 2.2 by using Eq. (3).

Figure 7 shows the damage ratio as a function of the fractal dimension depending on the test temperature. It appears that there is a linear correlation between the two parameters, where with increasing average fractal dimension the DR also increases. This indicates that the morphology of the air voids in fact plays a considerable part regarding the fatigue damage. This characteristic can be explained by the stress concentrations caused by the complex shape of the air voids, resulting in increased cracking at the interface between aggregates and asphalt mastic.

An additional study [13] shows the influence of aggregate shape and size on the air void distribution and morphology. It reveals that in SMA, a higher percentage of coarse aggregates leads to a lower number of air voids, but these are significantly more complex in their characteristics. This leads to the conclusion that there is a sophisticated relationship between aggregate shape, aggregate size distribution and the durability of an asphalt structure. This is supported by the research of [19] on the

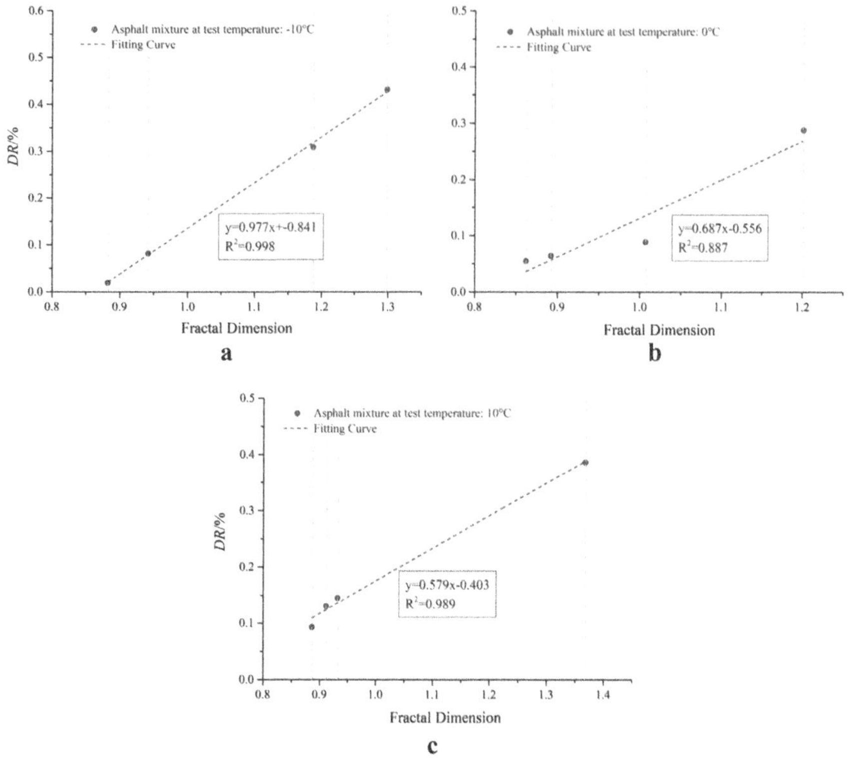

Fig. 7 Damage ratio as a function of the fractal dimension, **a** -10 °C test temperature, **b** 0 °C test temperature, **c** 10 °C test temperature [15]

influence of compaction on the stability characteristics of pavement. It shows that the degree of compaction has an important influence on the void distribution as well as on shape and morphology.

As the research of [20] shows, it should be mentioned that there is a non-linear relationship between tensile strength and fracture energy at the same temperature. This phenomenon is called over-compaction, which is likely to cause an increase in the complexity of the air void morphology, so that its negative effect on stress resistance exceeds the positive effect of the reduced number of voids.

3.2 Interface Stripping Damage of Asphalt Mixtures at High Temperatures

The research on interface stripping damage outlined here is described in more detail in [14]. The dynamic loading and related fatigue damage of an asphalt structure are

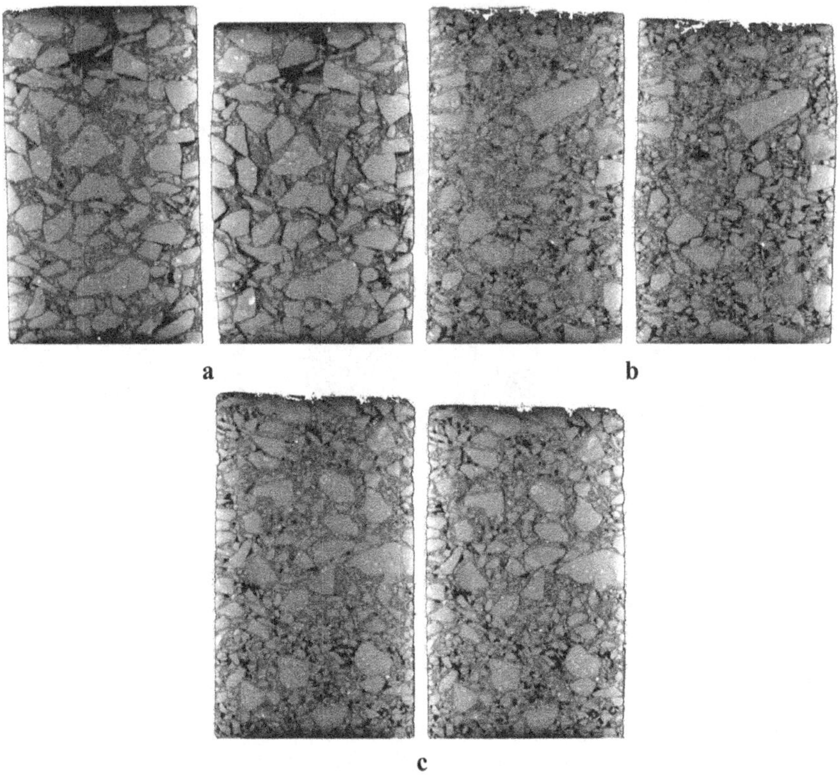

Fig. 8 X-Ray CT images of crosssections before and after the uniaxial loading test of: **a** CDM, **b** MDM and **c** FDM [14]

accompanied by an adhesive failure, called interface stripping. During this process, the cracks in the structure increase, between the aggregates and the mastic, leading to a decreasing durability.

Microstructure reconstruction To characterize the interface stripping during the change of the asphalt structure, the specimens are scanned before and after the uniaxial compression test. To acquire the small displacements and cracks within the structure, the images pixel size is set to 10 μm. Figure 8 shows cross-section images as results of the test process described in Sect. 1.2, by an example for each mixture, before and after the loading test. Finally, the aggregates and air voids are extracted from the images using the DIP methods in Sect. 1.1. Aggregates smaller than a nominal size of 1.18 mm, have no significant impact on the mechanical behavior and are considered as asphalt mastic. This specific definition of the asphalt mastic is only used in this chapter.

Modeling of the microstructure After the extraction of the aggregates, air voids and thereby the mastic from the X-Ray CT images, they are used to create a Finite Element Model (FEM) of the asphalt structure. Figure 9 shows the steps in the devel-

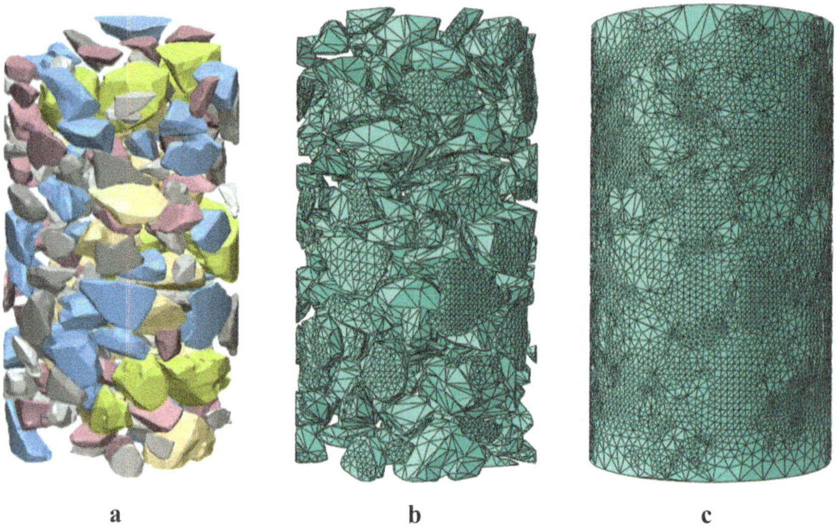

Fig. 9 FEM development process of the CDM specimen, **a** shell model, **b** finite element model of aggregates, **c** finite element model of mastic [14]

opment of the FEM. First, a shell model of the aggregates (Fig. 9a) is computed, which is then used to set up the FEM of the aggregates (Fig. 9b). The volume between the individual aggregates is used to create the FEM of the mastic (Fig. 9c). Due to the complex shape of the air voids, their modeling requires disproportionate effort, so they are not modeled individually within this simulation. Details about the FE modeling are presented in [14].

Then, the contact specifications between the individual components are defined. Hard contact behavior is defined among the aggregates, while a failure potential and friction are defined between the aggregates and the mastic. In order to model the adhesive fracture behavior on the contact area of mastic and aggregates, a surface-based cohesive behavior with linear softening feature is implemented and "the traction-separation law was used to define the constitutive response of the interface" [14].

The aggregates are modeled as linear elastic material with fixed parameters (Young's modulus: 55000 MPa, Poisson's ratio: 0.25) and the mastic as linear viscoelastic material at high temperature, since the tests were performed at 60 °C. The mastic performance is simulated with a generalized Maxwell model, where the parameters are determined based on Prony series. The generalized Maxwell model is commonly used to simulate the linear viscoelastic behaviour, where the corresponding parameters can be measured by dynamic modulus test. A detailed description of the parameter determination is described in [14, 16].

Interface damage analysis The FE model described above was used to characterize the interface stripping. Figure 10a shows the simulation results. In general, it appears that the number of micro cracks increases with increasing deformation. In

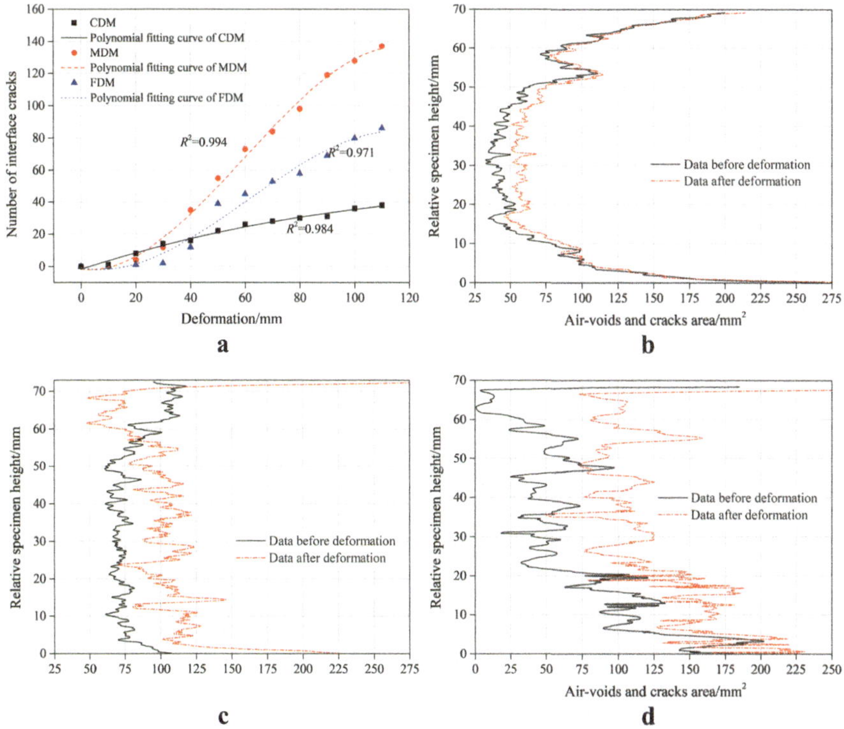

Fig. 10 **a** Number of cracks as a function of the deformation. Air void changes in different asphalt mixtures **b** FDM, **c** MDM and **d** CDM [14]

more detail, the process can be divided into three stages. Between 0 and 20 mm, a few cracks appear, because the material deforms mainly in the elastic-plastic range. This is followed by the deformation range from 20 to 90 mm, where the greatest percentage of cracks occur because the load resistance is mainly provided by the adhesive forces between the mastic and the aggregates. In the final deformation stage, between 90 and 110 mm, fewer cracks appear in relation to the deformation, since the load resistance is caused by the interaction between the individual aggregates, which have meanwhile interlocked, so the structure only deforms slightly.

Due to the lower percentage of fine aggregates, which means that the aggregates have to shift more than, for example, in the FDM structure in order to resist the load, the number of cracks is highest within the CDM structure. This method can quantify the cracks but cannot simulate the change in internal structural damage directly.

To characterize further the internal structural damage, additionally the air void distributions are analyzed, which were extracted using DIP and calculating their area before and after loading. Figure 10b–d show the distribution of internal damage. While it may not actually describe the interface stripping, it can indicate the influence of the aggregates on the change of the inner structure. Especially, the size of the aggregates is determined as a major factor. In the structure of the FDM, the dis-

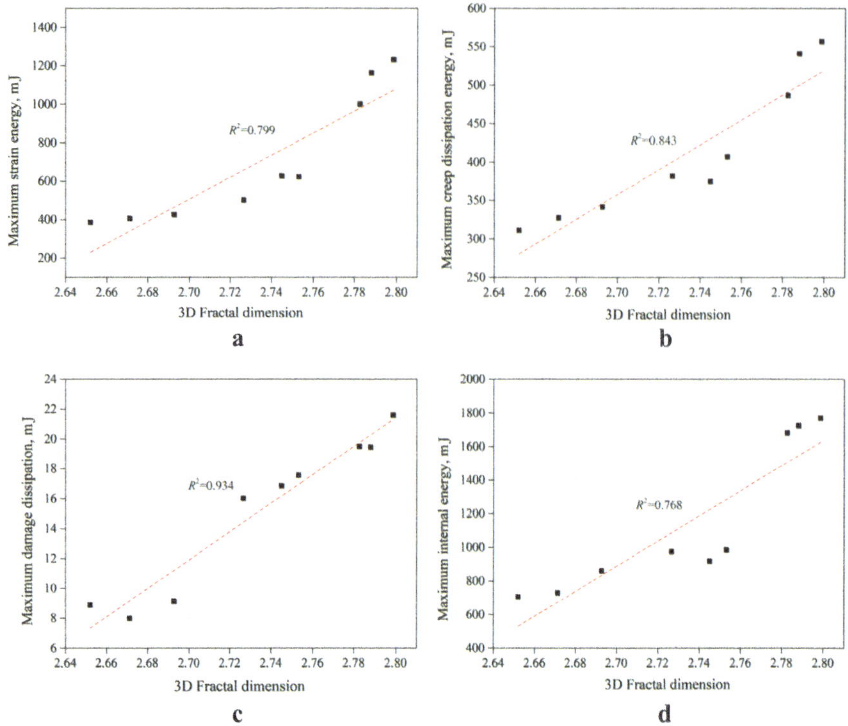

Fig. 11 Relations between fractal dimension and **a** maximum strain energy, **b** maximum creep dissipation energy, **c** maximum damage dissipation energy, **d** maximum internal energy [14]

tribution of the air voids is more uniform than in the CDM structure, which leads to the conclusion that a higher damage caused by deformation correlates with a higher percentage of coarse aggregates.

Impact of aggregate morphology Besides the morphology of the original air voids and the cracks that develop due to load, the morphology of the aggregates also plays a role in the transformation of the asphalt structure. As in the case of air voids, the fractal dimension is used to determine the shape of aggregates. Therefore, the aggregate shapes were determined for three samples of each mixture, i.e. a total amount of 9 samples. It shows that the averages of the fractal dimension increase with the increasing percentage of coarse aggregates.

Figure 11 shows the connection between the fractal dimension and the different stress parameters that resulted at maximum deformations from the FE simulations. The elastic deformation (Fig. 11a), the creep deformation of the asphalt mastic (Fig. 11b) and the interface damage cracks (Fig. 11c) are related in a linear correlation. Furthermore, this also applies between the maximum internal energy and the fractal dimension (Fig. 11d).

Since increasing fractal dimension represents more complex aggregates, which indicates an increase of the percentage of coarser aggregates, it can be assumed that aggregate shape has a significant influence on interface stripping distress, which means that more complex shapes support this damage mechanism.

The research described in this section is concluded by [22]. It includes a comprehensive analysis of a large number of specimens using all the factors and influences discussed in here. All results point to the fact that the degree of compaction has the greatest influence on the mechanical properties of asphalt.

4 Influence of Microstructural Parameters on Functional Properties of Pavements

Beneath effects on mechanical and structural properties, microstructural characteristics of asphalt also have an influence on certain functional properties of pavements like drainage and sound absorption. Especially for the case of porous pavements, e.g. PA, the void characteristics, which are also based on aggregate structures and shapes, play an important role for drainage and sound absorption as one of the decisive noise reducing effects of PA.

Typical void parameters, which can be determined by microstructural analysis are (interconnected) void content, number and shape of air voids, air void topology aspects and constrictions as described by [26] as well as pore diameters. Their statistical distributions are shown in Fig. 12 and described by [3, 7], just to give some examples and supplement the parameters described in the sections above.

In an acoustical sense, further special properties can be described regarding the porous void structure as a whole like pore size distributions, the structure factor or the tortuosity. Besides the void content, they have an influence on sound absorption behavior regarding frequency-dependent degree of sound absorption [2, 3].

The interconnected void content as well as the morphological pore structure have an influence on drainage parameters like discharge rates or retention volumes. Coarser PA structures—with bigger pore diameters due to higher nominal maximum aggre-

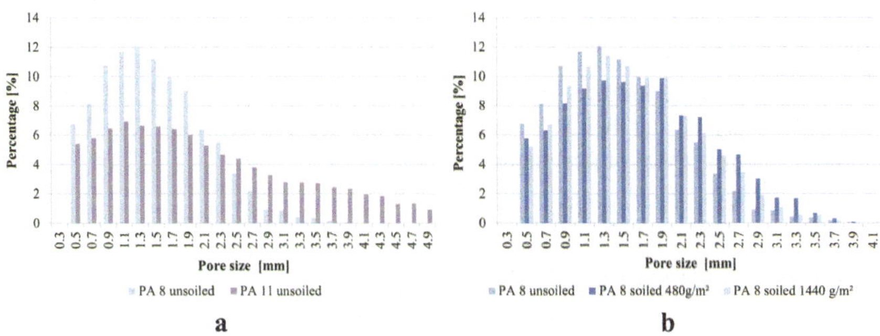

Fig. 12 Examples of pore size distributions of **a** PA 11 and PA 8 and **b** exemplary changes due to soiling [3]

gate size—show higher discharge rates and lower retention capabilities than finer ones in runoff experiments [6]. The hydraulic conductivity—as important parameter for drainage modeling (compare chapter "Computational Methods for Analyses of Different Functional Properties of Pavements") as well—can be analyzed by micro-structural considerations, even the significant difference of horizontal and vertical conductivity [26]. The infiltration rate, which is a quite common parameter to check drainage capabilities at the PA surface, usually depends on pore structures as well (e.g. [27]).

Because of rain water ingress into the porous structure of PA the problem of soiling and, thus, loss of drainage and acoustical properties is quite well-known in pavement engineering. Clogging effects of PA can be analyzed and made visible with micro-structural analysis as well (e.g. [5]). An effect of soiling on 3D fractal dimension was analyzed in [3], however, no dependency could be stated. With high soiling states, pores get clogged and water runoff is decreased in general. In case of intermediate and rather low soiling states, different effects can be observed. Several authors stated a decrease regarding infiltration rates for high soiling states as well (e.g. [1, 10]). But the influence of the drainage behavior as a whole including the runoff through the porous structure might not be affected too strong [6] as it was observed in artificial soiling tests (e.g. described by [4]) on different samples of PA with a size of 2.5 m^2.

Acoustical parameters also suffer from soiling, which is well-known to decrease noise reducing effects of PA significantly during their life-time. On the one hand, changes in sound absorption degree can be observed, while on the other hand, frequency shifts of sound absorption maxima depend on changes in the microstructure of the pores, which can be described and modeled in an acoustical way with structure factor and tortuosity [5]. Structure factor as well as tortuosity are increasing due to soiling which can be explained by the effects of different soiling mechanisms [5]. Pore size distributions also have an influence on sound absorption, which can be used in models like the one described by [12]. A possible change of pore size distribution because of soiling is shown in Fig. 12b. In this case the amount of smaller pores decrease, which might be explained by the fact, that small pores are filled with dirt first and the bigger ones remain. Other contradictory effects can appear as well like the increase of small pores, which might result from constrictions due to soiling (e.g. [7]).

Summed up, microstructural analysis techniques help to understand and observe different functional properties, here especially drainage and sound absorption of PA are addressed. Furthermore, also changes in asphalt structures, here by clogging of PA, can be analysed and therefore improve the understanding of long-term behavior and life-cycle processes of pavements.

5 Conclusions and Outlook

The microstructural characteristics of asphalt are determined by mix design and compaction issues. Different sizes and shapes of aggregates and their percentage in the total mix as well as bitumen content and compaction type and energy significantly influence the aggregate skeleton and the resulting air void structures. These properties can be described by different parameters for single elements (aggregates or voids) or the structure as a whole.

The determination and description of microstructural aspects are based upon imaging methods with X-Ray CT systems and subsequent DIP.

The microstructure of asphalt affects many different properties of an asphalt pavement. In this chapter, the following aspects are considered and described.

Fatigue damage cracking is discussed with air void analyses before and after indirect tensile testing. The shape and morphology of air voids are the decisive parameters in this case. The aim of these studies is a better understanding of the impact of initial void content and compaction degree on fatigue cracking.

Crack formation is also regarded in case of uniaxial deformation tests to analyze adhesion failure processes. The compression processes can be analysed based on changes of singular aggregates in the grain structure and, therefore, lead to a deeper understanding of this deterioration effect of asphalt pavements.

Moreover, microstructural analyses help to identify and quantify influencing parameters of the pore structure of (porous) asphalt on functional properties as drainage or sound absorption.

Beyond that microstructural analyses, using DIP methods is an important input for finite element or discrete element modeling approaches (compare Sect. 3.2). For certain questions, the exact microstructure must be reproduced. In other cases, simplified structures can help to improve and fasten computational calculations with an adequate precision. Furthermore, simplified structures can make it easier to relocate certain elements in before-and-after studies.

Last but not least, the description of asphalt structures in a simplified way helps to build up artificial structures for computational use, e.g. in terms of aggregate packing. Such coincident artificial structures make it possible to conduct and compare more analyses in a computational way than in a traditional experimental way. Particle flow modeling as shown in chapter "Numerical Simulation of Asphalt Compaction and Asphalt Performance", during construction and compaction processes of asphalt layers can improve the virtual build-up of asphalt structures and, moreover, help to understand compaction and construction in a microstructural and basic way.

As shown in this contribution, microstructural analyzing methods have an increasing importance in pavement engineering in order to gain basic understanding of material performance and will gain even more weight with an increasing availability and use of imaging methods in technical and engineering contexts.

References

1. Afonso, M.L., Fael, C.S., Dinis-Almeida, M.: Influence of clogging on the hydrologic performance of a double layer porous asphalt. Int. J. Pave. Eng. **21**, 736–745 (2020)

2. Alber, S.: Veränderung des Schallabsorptionsverhaltens von offenporigen Asphalten durch Verschmutzung. Ph.D. thesis, University of Stuttgart (2013)
3. Alber, S., Ressel, W., Liu, P., Hu, J., Wang, D., Oeser, M., Uribe, D., Steeb, H.: Investigation of microstructure characteristics of porous asphalt with relevance to acoustic pavement performance. Int. J. Transport. Sci. Technol. **7**, 199–207 (2018)
4. Alber, S., Ressel, W., Liu, P., Lu, G., Wang, D., Oeser, M.: Analyzing the effects of clogging of PA internal structure with artificial soiling experiments. Int. J. Transport. Sci. Technol. **8**, 383–393 (2019)
5. Alber, S., Ressel, W., Liu, P., Wang, D., Oeser, M.: Influence of soiling phenomena on air-void microstructure and acoustic performance of porous asphalt pavement. Construct. Build. Mater. **158**, 938–948 (2018)
6. Alber, S., Ressel, W., Schuck, B.: Explaining drainage of porous asphalt with hydrological modelling. Int. J. Pave. Eng. **2020**, 1–11 (2020)
7. Arbter, B.: Numerische Bestimmung der akustischen Eigenschaften offenporiger Fahrbahnbeläge auf Basis ihrer rekonstruierten Geometrie. Ph.D. thesis, University of Stuttgart (2014)
8. Beainy, F., Singh, D., Commuri, S., Zaman, M.: Laboratory and field study on compaction quality of an asphalt pavement. Int. J. Pave. Res. Technol. **7**, 317–323 (2014)
9. Ding, X., Ma, T., Huang, X.: Discrete-element contour-filling modeling method for micromechanical and macromechanical analysis of aggregate skeleton of asphalt mixture. J. Transport. Eng., Part B: Pave. **145**, 04018056 (2019)
10. Fwa, T., Lim, E., Tan, K.: Comparison of permeability and clogging characteristics of porous asphalt and pervious concrete pavement materials. Transport. Res. Record **2511**, 72–80 (2015)
11. Hassan, N.A., Airey, G.D., Khan, R., Collop, A.C.: Nondestructive characterisation of the effect of asphalt mixture compaction on aggregate orientation and segregation using X-ray computed tomography. Int. J. Pave. Res. Technol. **5**, 84–92 (2012)
12. Horoshenkov, K.V., Swift, M.: The acoustic properties of granular materials with pore size distribution close to log-normal. J. Acoust. Soc. Am. **110**, 2371–2378 (2001)
13. Hu, J., Liu, P., Wang, D., Oeser, M.: Influence of aggregates' spatial characteristics on air-voids in asphalt mixture. Road Mater. Pave. Design **19**, 837–855 (2018)
14. Hu, J., Liu, P., Wang, D., Oeser, M., Falla, G.C.: Investigation on interface stripping damage at high-temperature using microstructural analysis. Int. J. Pave. Eng. **20**, 544–556 (2019)
15. Hu, J., Liu, P., Wang, D., Oeser, M., Tan, Y.: Investigation on fatigue damage of asphalt mixture with different air-voids using microstructural analysis. Construct. Build. Mater. **125**, 936–945 (2016)
16. Hu, J., Qian, Z., Wang, D., Oeser, M.: Influence of aggregate particles on mastic and air-voids in asphalt concrete. Construct. Build. Mater. **93**, 1–9 (2015)
17. Jin, C., Yang, X., You, Z., Liu, K.: Aggregate shape characterization using virtual measurement of three-dimensional solid models constructed from X-ray CT images of aggregates. J. Mater. Civ. Eng. **30**, 04018026 (2018)
18. Kutay, M.E., Arambula, E., Gibson, N., Youtcheff, J.: Three-dimensional image processing methods to identify and characterise aggregates in compacted asphalt mixtures. Int. J. Pave. Eng. **11**, 511–528 (2010)
19. Li, T., Liu, P., Du, C., Schnittcher, M., Hu, J., Wang, D., Oeser, M.: Microstructural analysis of the effects of compaction on fatigue properties of asphalt mixtures. Int. J. Pave. Eng. **2020**, 1–12 (2020)
20. Liu, P., Hu, J., Falla, G.C., Wang, D., Leischner, S., Oeser, M.: Primary investigation on the relationship between microstructural characteristics and the mechanical performance of asphalt mixtures with different compaction degrees. Construct. Build. Mater. **223**, 784–793 (2019)
21. Liu, P., Hu, J., Wang, D., Oeser, M., Alber, S., Ressel, W., Falla, G.C.: Modelling and evaluation of aggregate morphology on asphalt compression behavior. Construct. Build. Mater. **133**, 196–208 (2017)
22. Liu, Q., Hu, J., Liu, P., Wu, J., Leischner, S., Oeser, M.: Uncertainty analysis of in-situ pavement compaction considering microstructural characteristics of asphalt mixtures. Construct. Build. Mater. **279**, 122514 (2021)

23. Masad, E., Fletcher, T.: Aggregate Imaging System (AIMS): Basics and Applications. Texas A & M University, College Station, Technical report (Texas Transportation Institute), Texas Transportation Institute (2005)
24. Otsu, N.: A threshold selection method from gray-level histograms. IEEE Trans. Syst. Man Cybernet. **9**, 62–66 (1979)
25. Ruf, M., Steeb, H.: An open, modular, and flexible micro x-ray computed tomography system for research. Rev. Scient. Ins. **91**, 113102 (2020)
26. Schuck, B., Teutsch, T., Alber, S., Ressel, W., Steeb, H., Ruf, M.: Study of air void topology of asphalt with focus on air void constrictions-a review and research approach. Road Mater. Pave. Design **22**, 425–443 (2021)
27. Winston, R.J., Al-Rubaei, A.M., Blecken, G.T., Hunt, W.F.: A simple infiltration test for determination of permeable pavement maintenance needs. J. Environ. Eng. **142**, 06016005 (2016)
28. Zhang, Y., Luo, X., Onifade, I., Huang, X., Lytton, R.L., Birgisson, B.: Mechanical evaluation of aggregate gradation to characterize load carrying capacity and rutting resistance of asphalt mixtures. Construct. Build. Mater. **205**, 499–510 (2019)

Numerical Friction Models Compared to Experiments on Real and Artificial Surfaces

Jan Friederichs, Lutz Eckstein, Felix Hartung, Michael Kaliske, Stefan Alber, Tobias Götz, and Wolfram Ressel

Abstract Friction between tire and pavement surface—also referred to as skid resistance in pavement engineering—is a complex phenomenon depending on many influencing parameters like speed, load or wetness of the surface as well as different effects like hysteresis and adhesion. Two different friction model approaches are used in this chapter, a microscale analytical model with special focus on microtexture influence and a multi-scale FE model considering both micro- and macrotexture wavelengths. Both approaches employ a generalized Maxwell model as material formulation for the tire rubber. Real and virtual textures of asphalt surfaces are replicated by 3D SLM printing on stainless steel plates. The virtual texture samples—which are still based on real asphalt surfaces—comprise pure microtextures (without macrotexture elements after filtering) and artificial combinations of sinusoidal waves with two different wavelengths. The printed surfaces are investigated by texture measurements for printing discrepancies with respect to the templates. Friction is measured with a linear friction test rig on these printed samples as well as on a real asphalt surface in dry and wet conditions. The measurements are used for calibration and validation issues by comparing them to the model calculations in wet and dry surface conditions.

Keywords Friction · Adhesion · Simulation · Experiments · Tire · Pavement · Interaction

---

Funded by the German Research Foundation (DFG) under grants KA 1163/30, RE 1620/4 and EC 412/1.

J. Friederichs (✉) · L. Eckstein
Institute for Automotive Engineering, RWTH Aachen University, Aachen, Germany
e-mail: jan.friederichs@ika.rwth-aachen.de

F. Hartung · M. Kaliske
Institute for Structural Analysis, Technische Universität Dresden, Dresden, Germany

S. Alber · T. Götz · W. Ressel
Institute for Road and Transport Science, University of Stuttgart, Stuttgart, Germany

M. Kaliske et al. (eds.), *Coupled System Pavement—Tire—Vehicle*,
Lecture Notes in Applied and Computational Mechanics 96,
https://doi.org/10.1007/978-3-030-75486-0_7

1 Introduction

Tire friction behavior depends on two interacting and complex materials, the road surface and the tread rubber. The friction coefficients, described by the ratio between horizontal and vertical forces of the rubber specimen in sliding events, vary versus the sliding velocity, the load pressure and the temperature, which makes the modeling of tire-road-friction very challenging, especially at humid surface conditions. Numerical and analytical studies [2, 8, 11] work out friction theories explaining the tire-road-friction based on a defined frame of experiments. However, it is difficult to elaborate a direct link between the multi-scale roughness of the real road and the resulting rubber friction due to changing weather and surface conditions. Therefore, the investigation of friction theories needs to be performed on individual surface topographies that wear only little and are easy to reproduce down to the microscale. 3D printing is an innovative procedure, which allows the buildup of any desired surface topography. First studies of [6] have shown that the replication of asphalt specimen using polymer material is only sufficient up to a wavelength of 0.5 mm. The friction coefficient for the investigated load and sliding velocity had an offset of around $\mu = 0.2$ between asphalt and replica. The following investigation presents a new 3D printing approach on the basis of stainless steel. The Selective Laser Melting (SLM) of single-component metallic material offers the advantage of microscale additive manufacturing together with a minimization of the wear parameter. Friction measurements are conducted on specifically built up surfaces to work out a less complex link between theoretical friction laws and friction experiments. In a further step, the measurements shall investigate the applicability of printed substrates in rubber friction studies. The comparison of asphalt and replica has already been investigated in chapter "Experimental and Simulative Methods for the Analysis of Vehicle-Tire-Pavement Interaction".

2 Experimental Setup

The general workflow of the experimental friction study is provided in Fig. 1. In a first step, the desired surface needs to be defined. For the investigation of multi-scale friction laws, sinusoidal geometries are generated. For the replication of asphalt, a typical stone mastic road segment was cut out of a road section, scanned and post-processed (see Sect. 2.1). Using CAD-software, the surfaces are extruded to a volume model for the feasibility of SLM manufacturing. The geometries are additively manufactured directly on mounting plates consisting of the same material. Then, the friction coefficients are measured according to chapter "Experimental and Simulative Methods for the Analysis of Vehicle-Tire-Pavement Interaction" at different sliding and moisture conditions.

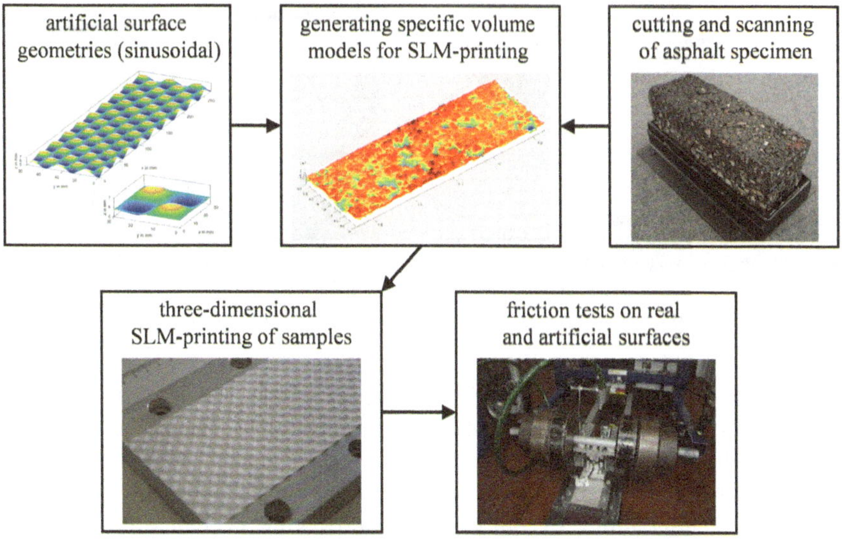

Fig. 1 Scheme of experimental setup

2.1 *Surface Preparation and Printing*

In this study, five different surface topographies have been built up with stainless steel (1.4404). To fit the functionality of the friction test rig described in chapter "Experimental and Simulative Methods for the Analysis of Vehicle-Tire-Pavement Interaction", horizontal dimensions of 270 mm × 80 mm are required. For the investigation of multi-scale friction dependencies (see Sect. 4), two sinusoidal surfaces (A and B) were generated in a two-dimensional grid with the local heights

$$z_A(x, y) = 0.5\,\text{mm} \cdot \sin\left(\frac{2\pi\, x}{10\,\text{mm}}\right) \cdot \sin\left(\frac{2\pi\, y}{10\,\text{mm}}\right) \tag{1}$$

and

$$z_B(x, y) = z_A(x, y) + \frac{0.1\,\text{mm}}{2} \cdot \left[\sin\left(\frac{2\pi\, x}{1\,\text{mm}}\right) + \sin\left(\frac{2\pi\, y}{1\,\text{mm}}\right)\right]. \tag{2}$$

According to the mathematical definition, Printing A represents a macroscopic sine surface and Printing B has an additionally overlaid microscopic sine surface.

For the microscale model approach for friction (Sect. 3), the replicas of microtextures of a typical rural road or motorway were regarded in the two Surfaces C and D. As high accuracy measurements of the microstructure are required, a micro mirror projector based on the principle of fringe projection has been used, whose simplified function is shown in Fig. 2a. Each mirror of the projector corresponds to one pixel of the camera. The micro mirror arrays used on both measuring fields have 1024 cells of

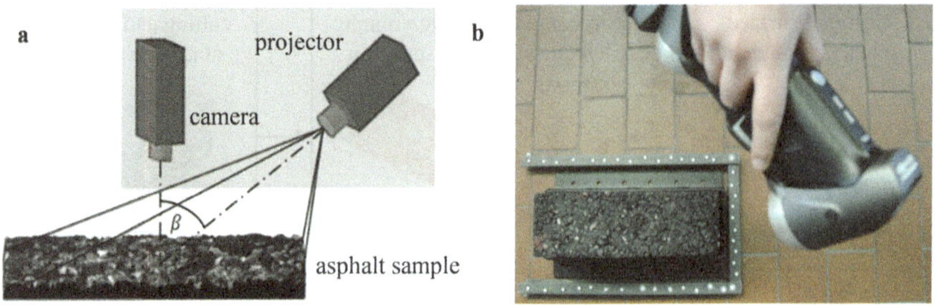

Fig. 2 **a** Simplified fringe projection method [12] and **b** full asphalt sample scanning

16 μm width each at a distance of 1 μm, so that the geometrical optically-generated cos2-shaped interferometric stripe patterns can be measured. The used system has a measurement area of 20 mm^2, a height resolution of 4 μm and a lateral resolution of 19 μm. For the isolation of the printable microtexture, the macrostructure is filtered out with a high pass filter as well as wavelengths below 0.1 mm with a low pass filter to avoid possible printing errors, so that the resulting texture is of a roughness between the wavelengths 0.1 and 0.5 mm. The small measurement areas are combined with other measurements to obtain the necessary testing dimension. Using digital combination methods, periodic structures as well as errors at the connecting areas are avoided. As a reference to a real pavement, an asphalt specimen and its replica are taken into consideration (Surface E). As the full asphalt specimen shall be replicated, a portable 3D laser scanner (HandySCAN 700) is used for measurement with a height resolution of up to 30 μm and a lateral resolution of up to 50 μm (see Fig. 2b). The post-processing, such as tilting or cutting, is performed using the software Geometric Wrap.

After the transition of the surface models into extruded volume models using CAD methods, the Fraunhofer Institute for Laser Technology (ILT) performed additive manufacturing using SLM of single-component metallic material according to the method of Gebhard [4]. Like all rapid prototyping methods, the CAD model is sliced into two-dimensional layers according to the printer's layer thickness of the metallic powder (30 μm). The printing method is sketched in Fig. 3 according to the works of Meiners [9]. In an iterative process, each layer geometry is fused lane by lane within the equally distributed powder material with a solid-state neodymium-doped yttrium aluminum garnet (ND:YAG) laser. The inert gas argon flows through the chamber to avoid unwanted melt reactions. After each layer, the construction cylinder is lowered and the leveling slider prepares the next powder layer. Research findings regarding process control and product quality can be found in [9].

2.2 *Topography Measurement of SLM Printings*

As the printed surfaces qualitatively appeared rougher than expected, possible printing asperities have been analyzed quantitatively using a texture scan on the basis

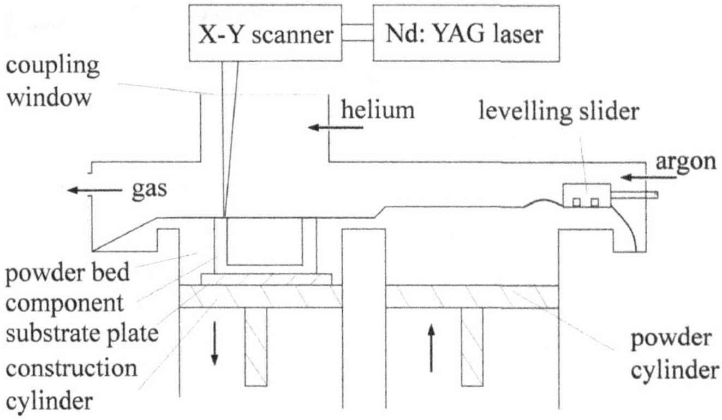

Fig. 3 Sketch of the used SLM method [9]

of fringe projection as described above. Due to the smooth and mathematically described surface geometries, Surfaces A and B are most suitable to determine printing asperities (see Fig. 4). According to Fig. 4a, small printing irregularities with a wavelength of ca. 0.2 mm are distributed randomly over the surface. These irregularities appear less significant at the overlaid sinusoidal Surface B. However, they occur to the same degree, which can be seen in the radially averaged power spectral density diagram in Fig. 5. The curves show peaks at the expected wavelengths of $10\sqrt{2}$ mm or 1 mm. However, starting with wavelengths of around 1 mm, the PSD values of the printed surfaces are larger than the PSD of the target surface. As this behavior is visible at all surfaces, an overlaid micro roughness due to process-related asperities needs to be considered in the following steps.

On the macrostructure level, several values exist to characterize the surface roughness. The following three one-dimensional roughness characteristics are considered in both horizontal dimensions in Table 1. RMS is the averaged root mean square of all lines of the height values, R_A is the segmentally averaged mean value, and R_Z is the segmentally averaged peak-to-peak value. The measurement of the full geometry was performed using a line scanner (resolution: $x = 0.1$ mm, $y = 0.4$ mm) mounted on a transverse track evenly moved in the longitudinal direction by an electric motor. The sinusoidal Surfaces A and B show plausible roughness values in longitudinal and lateral direction with a small offset due to the directional resolution of the scan data and resulting segmentation lengths. Generally, the roughness values of Surfaces C and D are smaller. Both surfaces show a characteristic height difference of 0.5 mm. However, the structural texture size of Surface C is 6 mm, which represents a smooth texture in comparison to the rough texture of Surface D with a structural texture size of 2.5 mm. As a result, the roughness values at least differ by the factor two. The roughness values of the asphalt replica are slightly smaller than those of the original asphalt sample, which may be a result of the printing depth limitation of 8 mm or the optical measurement in combination with the metallic material. After the friction measurement series, the scans based on fringe projection and line scanning

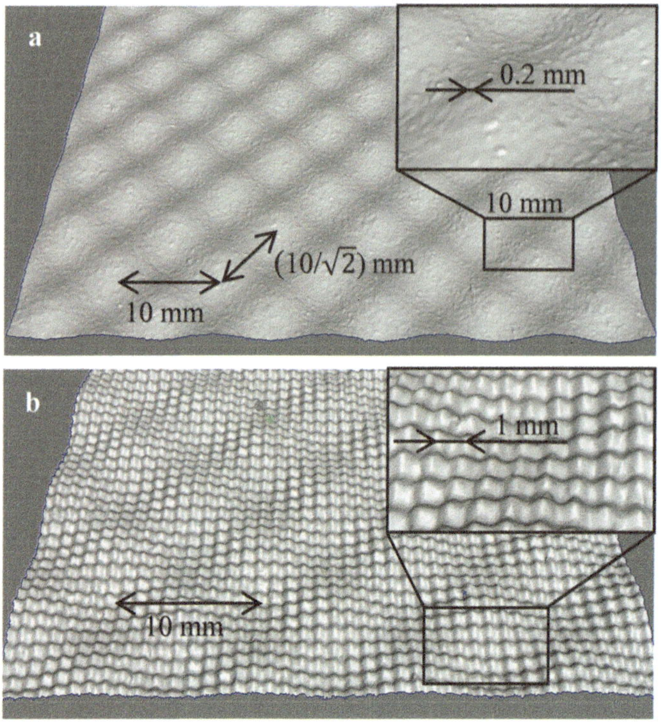

Fig. 4 3D scans of the sinusoidal SLM-printed **a** Surface A and **b** Surface B [3]

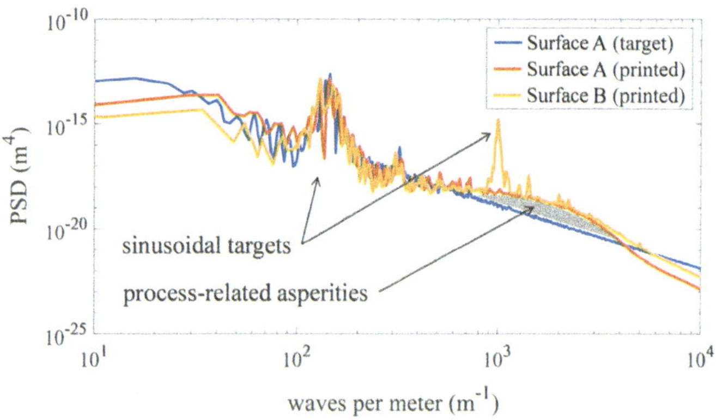

Fig. 5 Radially averaged power spectral density (PSD) of Surfaces A and B [3]

Table 1 Characteristic roughness values of considered surfaces [3]

Surface ID	$RMS\,(x)$	$RMS\,(y)$	$R_A\,(x)$	$R_A\,(y)$	$R_Z\,(x)$	$R_Z\,(y)$
A	0.2250	0.2250	0.1844	0.1592	0.1065	0.1056
B	0.2311	0.2311	0.1879	0.1672	0.1196	0.1259
C	0.1890	0.1879	0.1277	0.0980	0.1513	0.1465
D	0.0826	0.0804	0.0506	0.0351	0.0515	0.0362
E (original)	0.9497	0.9245	0.5457	0.5356	0.4805	0.5790
E (replica)	0.9066	0.8800	0.5069	0.5392	0.4074	0.5014

were repeated. The detected changes lie within the accuracy of the scanner systems, for which equal surface conditions for the performed measurement series can be assumed.

2.3 Friction Analysis on Artificial SLM-Surfaces

The friction measurements have been performed using the linear friction test rig introduced in chapter "Experimental and Simulative Methods for the Analysis of Vehicle-Tire-Pavement Interaction" at dry and wet conditions with two different tread rubber samples from a "standard" tire (S) and an "e-mobility" tire (E). In this case, the standard tire, which is widely spread among compact cars, has a ShoreA hardness of 68.8, while the softer e-mobility tire, which has a typical large diameter and a narrow tread pattern, has a ShoreA hardness of 61.0. Each rubber-surface pairing has a unique friction coefficient map depending on the sliding velocity and the local pressure. Figure 6 shows a comparison of the sliding friction maps (μ_{sliding}) for the standard tire tread for all investigated surfaces at dry conditions. The friction coefficients of the sinusoidal Surfaces A and B are rather independent of the load and dependent on the sliding velocity. The friction coefficients on the macro sine surface (A) are small at low velocities and reach peak values at around 50 mm/s. With the overlaid micro sine surface, the friction coefficients are more homogenous within the full friction map. Especially, the values at low sliding velocity increase significantly. Further simulative investigations on Surfaces A and B can be found in Sect. 4. The two surfaces for the investigation on the microscale model approach for friction show a deviating characteristic. The friction coefficients of the "smooth" Surface C with an average structure wavelength of 6 mm are more load dependent than the values of the "rough" Surface D with an average structure wavelength of 2.5 mm. Furthermore, the peak values of the rough Surface D are smaller and an area of local minimum values was measured for a sliding speed of 100 mm/s. Further simulative investigations on Surfaces C and D can be found in Sect. 3.

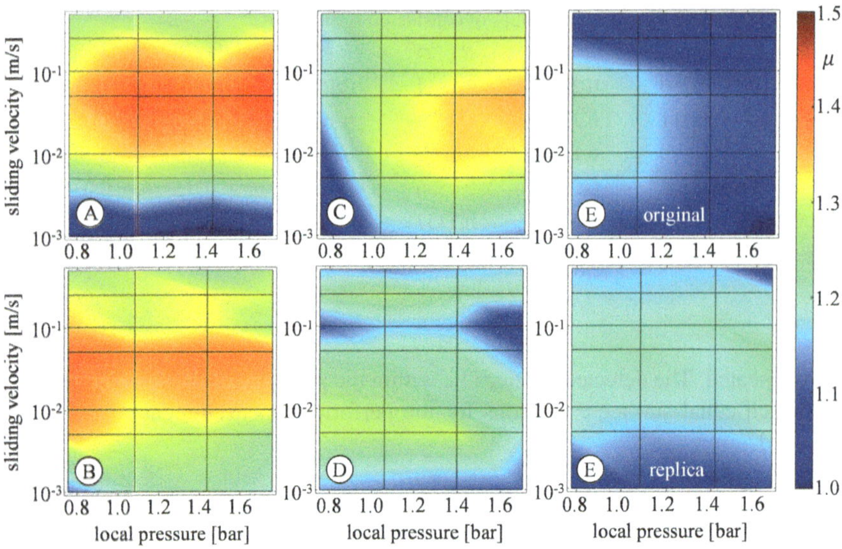

Fig. 6 Sliding friction maps at dry conditions with the standard tread sample S

For completeness, Fig. 6 also compares asphalt and its replica. A detailed analysis on the friction characteristics at dry and wet conditions can be found in the chapter "Experimental and Simulative Methods for the Analysis of Vehicle-Tire-Pavement Interaction". The friction maps of the static and the sliding friction coefficients (μ_{static} and μ_{sliding}) for both tire tread samples at dry and wet conditions can be taken from [1]. Figure 7 summarizes all measurement series for the rubber of the standard tire (S) and the e-mobility tire (E) with calculated mean values for each friction map. The full colored bars in the diagram represent the maximum value μ_{static}, while the more transparent bars represent the stationary value μ_{sliding}. An additional measurement series was performed on corundum P120, which serves as a reference for an intense microscale surface.

Friction mainly consists of adhesion and hysteresis. At wet conditions, the adhesion and with it the friction coefficients are assumed to be significantly reduced. Due to the hydrophobic material and the overlaid microstructure in the form of printing asperities, the hysteresis friction mainly contributes to the force transmission on the SLM-surfaces. As the remaining adhesion proportion is small, the friction reduction from dry to wet conditions is also rather small. This hypothesis can be additionally confirmed by the comparison of the original asphalt and the SLM-replica. Due to the smooth surface of the grains of the original asphalt, the proportion of adhesion contributing to the friction is higher than the replica. The overlaid printing asperities of the replica lead to an increased hysteresis friction. As a result, the dry friction coefficients are rather similar, while the wet friction coefficients vary significantly. The highest friction values were measured at the Surfaces B, D and P120. The surfaces are characterized by short-wave and intense asperities. Surfaces C and D have filtered

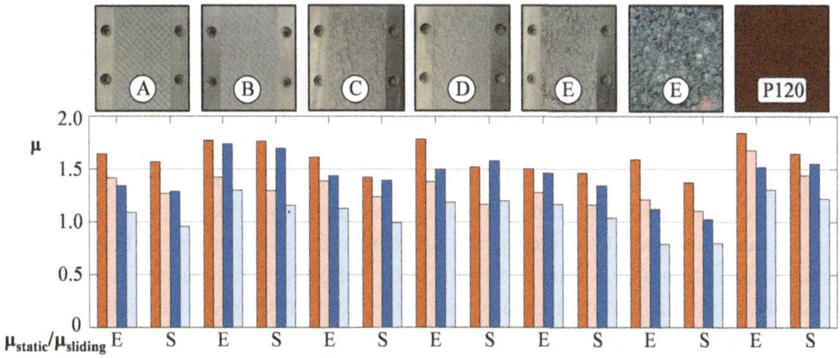

E	dry	1.645	1.418	1.771	1.423	1.614	1.391	1.784	1.386	1.504	1.277	1.593	1.211	1.843	1.680
	wet	1.345	1.087	1.741	1.302	1.438	1.125	1.715	1.289	1.464	1.167	1.120	0.790	1.521	1.305
S	dry	1.570	1.271	1.767	1.298	1.424	1.238	1.501	1.187	1.461	1.145	1.374	1.104	1.648	1.439
	wet	1.291	0.957	1.698	1.155	1.398	0.995	1.601	1.170	1.344	1.040	1.025	0.798	1.554	1.220

Fig. 7 Averaged friction coefficients for all measurement series

out macrotextures, overlaid short-wave printing irregularities, and correlate to the values of P120, especially at wet conditions. In combination with the fact that P120 consists of only the microstructure defined by its grain size, the microtexture is one of the main contributors to high friction values in both wet and dry conditions. As the thickness and the effective contact area of both rubber samples are similar, the main difference is the ShoreA hardness. As expected, the sliding friction coefficients of the softer e-mobility tread sample (E) are higher than the ones of the harder standard tire throughout all measurement series. The increased indentation depth of the soft rubber further contributes to the hysteresis friction. According to Fig. 7, the static friction coefficients may be less affected by the different ShoreA hardnesses. Further investigations on the frictional behavior of tread rubber on 3D-printed surfaces on the basis of stainless steel with regard to relevant conditions in driving dynamics can be found in [3]. The experimental friction coefficients contribute to the calibration and validation of the different friction simulation models, which are introduced in the next sections.

3 Microscale Approach for Friction

3.1 Microtexture Dataset

Skid resistance and friction between tire and road surface depend on both microtexture and macrotexture effects (see also chapter "Computational Methods for Analyses of Different Functional Properties of Pavements", Sect. 4.1). Testing friction on real asphalt pavement surfaces (as shown in Sect. 2) always includes all scales of texture. It is therefore not possible to gain quantitative information about the influence of microtexture experimentally, as the effects cannot be separated. The presented microscale approach focuses on the relevance of microtexture for the entire friction behavior of pavement surfaces neglecting macrotexture effects. 3D-printing offers the possibility to reproduce surfaces consisting only of certain (microscale) wavelengths of the texture spectrum of a real asphalt surface and perform corresponding friction measurements. Considering microtexture effects separately can help to obtain better information about their contribution to overall skid resistance. Two surfaces are analyzed in this part of the study: Surface C and Surface D (see Sect. 2 and [3]) showing different microtexture characteristics.

The surfaces were derived from a real stone mastic asphalt (SMA) specimen representing a typical rural road or motorway surface. A small section of this surface was analyzed by measuring (micro-)texture with a fringe projection system. An area of 20 mm^2, which corresponds to about the size of a single grain surface on the pavement surface, was measured with a high resolution of 4 μm in vertical direction and 19 μm in lateral direction. Since the measurement also includes macrotexture elements, it was necessary to filter the dataset to exclude wavelengths greater than 0.5 mm. The applied filtering method is basically described in [7]. A further restriction of the texture dataset is the possible resolution of the 3D-printing. Thus, texture wavelengths below 0.1 mm were also eliminated by filtering the data, as they could cause printing inaccuracies (compare also [3]). The measured texture was analyzed by the microscale approach (see Sect. 3.2) and the multi-scale method (see Sect. 4) in a computational way and printed on steel plates by SLM 3D-printing technique in order to have comparative measurements (see Sect. 2). As the measured texture section would be far too small for an experiment with the linear friction test rig, it was necessary to create a virtual texture with an adequate size (here: 270 mm $\times$ 80 mm). Therefore, the measured section was computationally combined several times, which should avoid a periodic shape and discontinuities at the edges of the combined elements (see Fig. 8a). Although some repetitive characteristics can be seen due to the method (see Fig. 8b), the resulting virtual texture is homogeneous, random and isotropic. The printed texture was measured and analyzed again with a texture scanner in order to compare it to the template. As some printing errors occurred for small wave-lengths, it was necessary to add these artefacts to the virtual texture before performing computational simulations.

3.2 Model Approach

The simulation of friction effects based on the surfaces described in Sect. 3.1 was carried out using a (microtexture) hysteresis friction model, which is described in more detail in chapter "Computational Methods for Analyses of Different Functional Properties of Pavements", Sect. 4.2. First developed in [12], the model has been enhanced in several further steps (see also [3, 5]), see also chapter "Computational

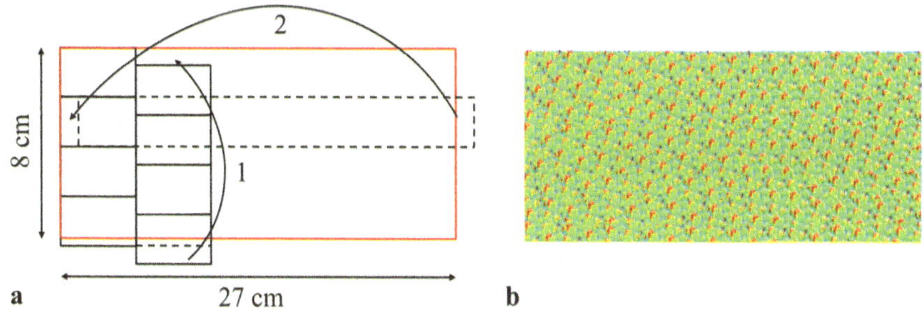

Fig. 8 Method of creating a virtual texture sample by **a** combining single elements and **b** the resulting texture

Methods for Analyses of Different Functional Properties of Pavements", Sect. 4.3. In the present study, a generalized Maxwell approach, which is also used in the multi-scale approach (compare Sect. 4), was implemented as basic rubber model [3]. It consists of a spring and a series of 15 Maxwell elements in parallel. Each Maxwell element is described by its shear modulus G (for the spring of the Maxwell element) and its dynamic viscosity ν (for the damper). The single spring is represented by a shear modulus parameter only. The parameters for these rheological elements are determined by fitting them to the mechanical behavior of the tires used in this study (see Sect. 2 and also [3]). Using the same tire rubber model improves the comparability of the different approaches—multi-scale and microscale. A basic comparison of both friction models—with a different consideration of rubber properties in the microscale case—and a comparison to experiments has also been presented in [5].

The hysteresis friction model used for the microscale approach first had to be calibrated to the experimental data for dry conditions (see also [3]). This was done by adjusting different parameters of the model. First, the tire parameters and load assumptions were modified, as the behavior of the tire rubber, especially the contact radii in the microscale, and the simulated load significantly affects the penetration depth and subsequently calculated friction results. Second, an adhesion approach had to be supplemented because the model only simulates hysteresis effects. At each contact point of rubber and surface, where hysteresis effects are calculated in the model, an adhesion friction term was added by a calibration term. Although calibration was necessary in order to reach a certain level of friction (which cannot be reached without considering adhesion), it can be stated that the model calculations fitted to the location of the friction peak at a certain velocity without calibration needs (see also [3]).

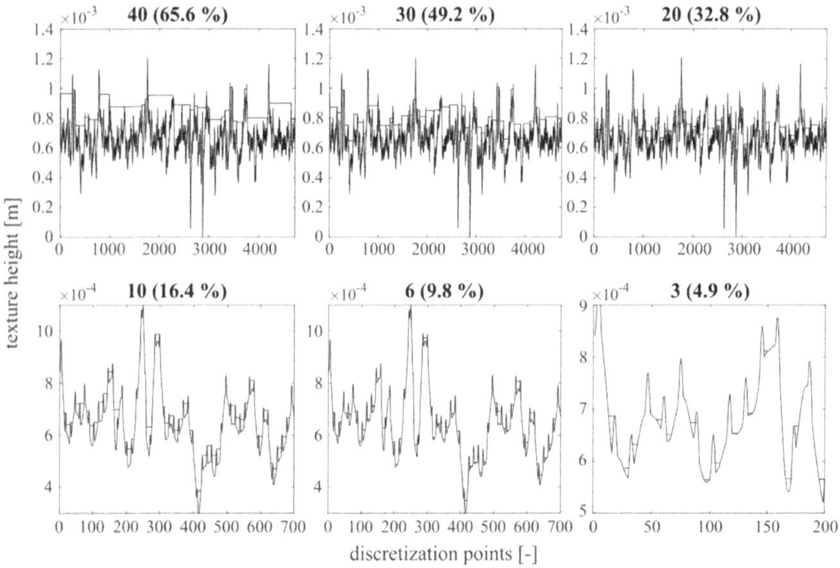

Fig. 9 Water filling states of microtexture profiles according to the algorithm

3.3 Wet Friction Model Approach

In order to consider friction at wet conditions, an algorithm is developed in order to simulate different water levels within the microtexture profiles. The algorithm assumes that the microtexture is filled step by step. Every filling step is in the range of about the vertical resolution of the microtexture and the amount of water is described as percentage of filling ratio of the entire microtexture. This approach is also described basically and briefly in [3]. Examples of different filling states are shown in Fig. 9. In fact, the microtexture characteristics change with the water level. Certain heights of asperities and depth of valleys of the profile are decreased by the water surface in the (micro-) texture valleys. Thus, contact points and subsequently hysteresis and adhesion effects are reduced by the water film in model simulations. Furthermore, the water leads to reduced penetration and deformation of the tire rubber which also reduces hysteretic forces. This way, calculated friction values decrease at wet conditions with increasing water level compared to dry conditions (original texture). An example of the difference between dry and wet conditions (30 % water level) simulations on the same profile showing these effects can be seen in Fig. 10. Viscous effects of the water film and transport of water by the tire, which also influences wet friction, are neglected and not considered in this approach.

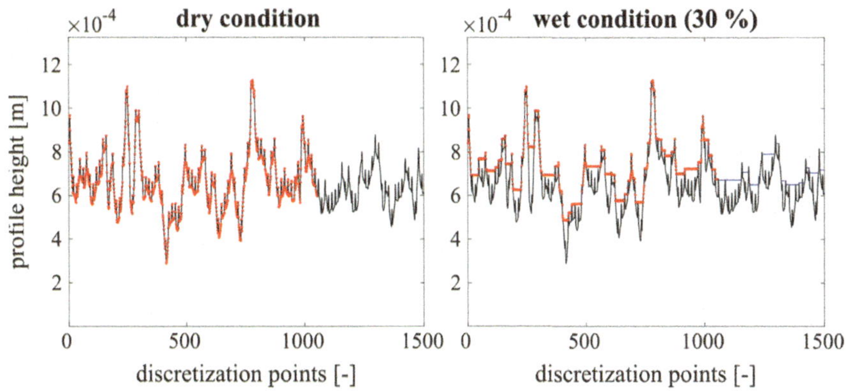

Fig. 10 Inhibited deformation and reduced contact area by trapped water

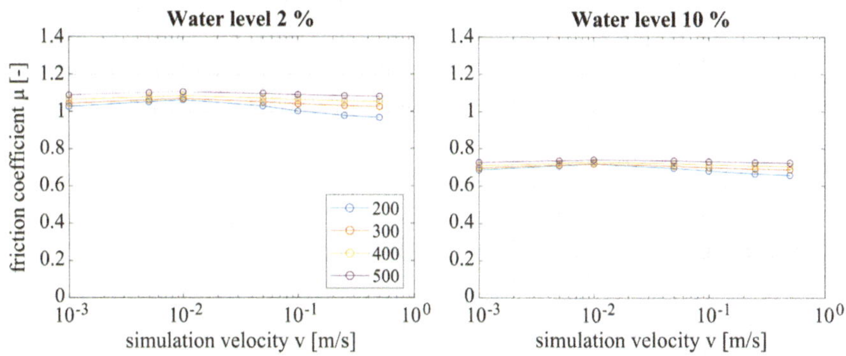

Fig. 11 Influence of simulated water level and decrease of friction with higher water level

3.4 Comparison to Experiments

In [3], the computational calculations were compared to the measured friction values of the printed Surfaces C and D at wet and dry conditions. Calculations with the microscale approach model (see Sect. 3.2) and the water filling algorithm (see Sect. 3.3) show the theoretical decrease of friction at wet conditions (see Fig. 11) at different water levels in the microtexture. Especially for wet conditions, it should be possible to calculate the water level (water level according to Sect. 3.3) by the measured decrease of friction and, thus, gain some information about the dependency of water films and friction levels on the microtexture scale. With the help of back-calculation approaches, it would be possible in a further calculation step to determine the water volume remaining in the microtexture (and, thus, a water film depth) in a three-dimensional sense.

In fact, the measured decrease of friction on wet surfaces was not as significant (see Sect. 2) for the microtexture Surfaces C and D (see Sect. 3.1) as expected. Possible reasons could be the hydrophobic character of the steel surface, which might diminish the ingress of water into the microtexture valleys. Thus, water on the surface would not have much influence on friction because it is removed by the tire rubber easily from the top of the microtexture. The hypothesis of hydrophobic material influences might be supported by comparing the measurement on the real asphalt surface and the printed (steel) replica (see Sect. 2). Real asphalt shows a higher decrease of (measured) friction than the steel replica at wet conditions. As both samples have the same texture—in micro- and macroscale—it seems to be possible that the material accounts for the observed difference.

Hence, a back-calculation of water filling states of the microtexture has not been possible with the experimental data. Although the (microscale) model approach shows plausible results (see Fig. 11), it could not yet be validated with the experimental data presented in Sect. 2. However, considering hydrophobic material effects, there might be a possibility to calibrate the model in an adequate way in order to fit experimental and computational data (see also [3]). Maybe the choice of another material for the 3D-prints or a procedure to make the steel samples more hydrophilic (and more realistic compared to real asphalt in a used condition) without changing the microtexture properties can improve the comparison of real asphalt and SLM samples.

4 Multi-scale Approach for Friction

With the application of the multi-scale approach for friction introduced in chapter "Multi-physical and Multi-scale Theoretical-Numerical Modeling of Tire-Pavement Interaction", Sect. 3, it is possible to obtain deeper insights into the different friction phenomena adhesion and hysteresis friction. Surfaces A and B (see Fig. 4) are printed via SLM in order to investigate the friction contributions with the multi-scale approach using simple two-dimensional trigonometric functions as defined in Eqs. (1) and (2) to describe the rough counter surface of the sliding tread rubber block. The linear friction tests on the Surfaces A and B (see Fig. 6) are used to compare the output of the simulations to the experiments. Unfortunately, the SLM created an unintended microscale roughness during the printing process. For example, the 3D scan of Surface A in Fig. 4 shows asperities in the range of 0.2 mm. Note that the microscale wavelength of Surface B is only 1 mm.

As material formulation for the tread rubber of the "standard" tire (S) (see Sect. 2.3), the generalized Maxwell model, which consists of one single spring and multiple Maxwell elements (spring and dashpot in series also called Prony series) in parallel, is used to capture the viscoelastic behavior of rubber [3]. The material parameters are provided by the tire manufacturer. Consequently, 15 Maxwell elements are applied to represent the time-dependent material response of rubber in an adequate frequency range. The mechanical behavior of the 16 springs in the generalized Maxwell model is described by Mooney-Rivlin hyperelasticity [10]. The finite element (FE) model of the tread block which is shown in Fig. 12 consists in total

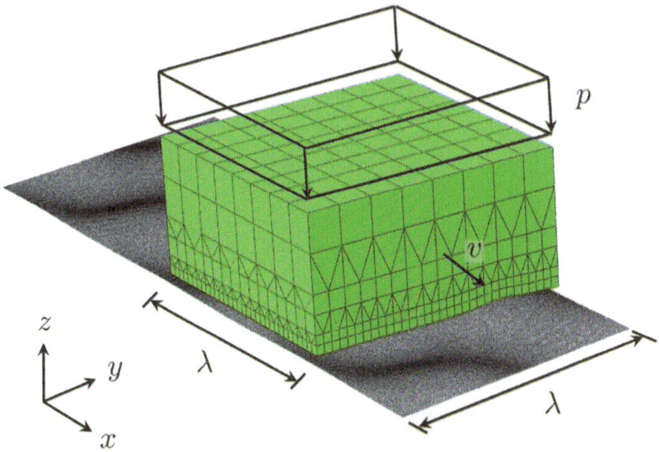

Fig. 12 Finite element model is virtually sliding over the two-dimensional sinusoidal rigid surface

of 3968 elements and is finer discretized in the contact area (32 × 32 elements) to avoid stress singularities. In the FE simulation, the counter surface (Surface A and B from Sect. 2) is assumed to be rigid. The analytical description of the corresponding rigid surface is given by Eqs. (1) and (2).

In Sect. 4.1, the adhesion parameters of the adhesion model described in chapter "Multi-physical and Multi-scale Theoretical-Numerical Modeling of Tire-Pavement Interaction", Sect. 3.1 are identified by using the friction tests of Surface A at dry and wet conditions. Followed by Sect. 4.2, in which the microscale friction of Surface B (last term in Eq. (2)) is homogenized to form the friction law for the macroscale. Finally, the linear friction tests at wet conditions are compared to the macroscale friction results from the multi-scale approach for different load levels and sliding velocities.

4.1 Adhesion Friction

For identifying the adhesion model parameters K (initial adhesional stiffness) and G_{tot} (total fracture energy), the linear friction test results at dry conditions are reduced by the corresponding frictional output at wet conditions

$$\mu_{\text{exp,adh}} \approx \mu_{\text{exp,dry}} - \mu_{\text{exp,wet}} \,. \tag{3}$$

It is assumed that adhesion is negligible (mainly hysteresis friction) during the friction tests at wet conditions. During the simulation, both hysteresis as well as adhesion friction occurred, so that the adhesion part is computed as the difference between the

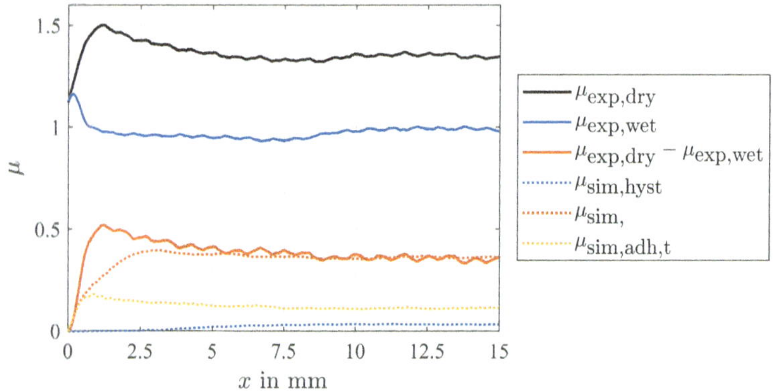

Fig. 13 Friction features during experiment and simulation using load of 50 kg and sliding velocity of 10 mm/s

total and the hysteresis friction

$$\mu_{\text{sim,adh}} = \mu_{\text{sim}} - \mu_{\text{sim,hyst}} \,. \tag{4}$$

The friction contribution due to internal dissipation of the viscoelastic material ($\mu_{\text{sim,hyst}}$) is computed in another simulation in which the adhesion model is deactivated. Figure 13 shows friction features during the linear friction test and the simulation as a function of the elapsed sliding distance using a load of 50 kg (0.17365 $^{\text{N}}/_{\text{mm}^2}$) and a sliding velocity of 10 mm/s. In Fig. 13 both, tangential adhesion (yellow dotted line) as well as normal adhesion contributes to the resulting adhesional friction coefficient during the simulation (red dotted line). Additionally, the unintended microscale roughness of the SLM-printed Surface A leads to the significant difference of both blue lines (hysteresis friction).

4.2 *Multi-scale Hysteresis Friction*

Since Surface B is described by Eq. (2), the scale identification results from two different sinusoidal waves. The microscale corresponds to the smaller sine wave with its wavelength of $\lambda_{\text{micro}} = 1$ mm and the macroscale has a wavelength of $\lambda_{\text{macro}} = 10$ mm. At the beginning, multiple FE simulations, where a block is sliding over the microscale wave with different loads (0.1 to 4 $^{\text{N}}/_{\text{mm}^2}$) and sliding velocities (1 to 500 mm/s), are performed. After the friction features of each simulation is homogenized over sliding time like described in chapter "Multi-physical and Multi-scale Theoretical-Numerical Modeling of Tire-Pavement Interaction", Sect. 3.2, the friction law for the macroscale is generated by using piecewise cubic spline interpolation. The resulting friction map is shown in Fig. 14, which is a function of contact pressure

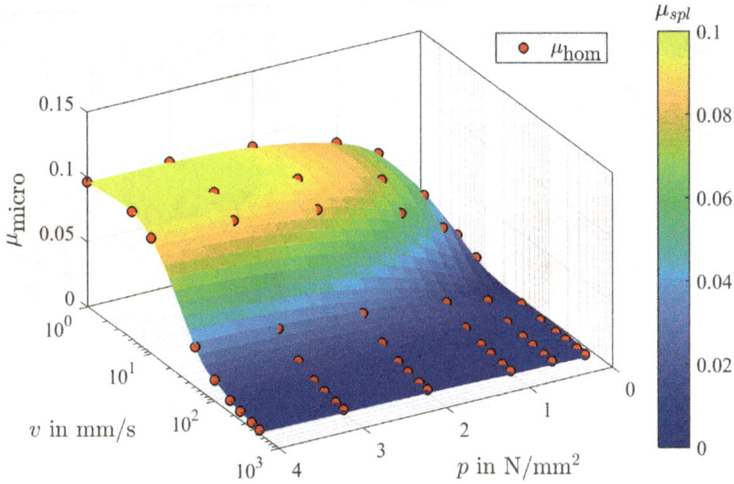

Fig. 14 Friction map of microscale

and sliding velocity. Each red dot in Fig. 14 represents a homogenized friction coefficient from the related microscale block simulation. Finally, the macroscopic friction coefficient results from the time homogenization after the macroscopic block (see Fig. 12 with block dimensions λ_{macro}) has been moved over macroscale sine wave described in Eq. (1) using the friction law illustrated in Fig. 14. The pressure as well as the velocity ranges of the microscale friction map are defined iteratively with respect to the contact pressure and sliding velocities that are required in the macroscale simulation. The friction test results at wet conditions (sliding friction) of Surface B are shown in Fig. 15a and the corresponding simulation output are presented in Fig. 15b. For both, the experiment and the simulation friction coefficient, a peak occurred for small sliding velocities (between 10 and 50 mm/s). In contrast to the laboratory output, the friction coefficients of the simulation are significantly smaller, which is mainly due to the influence of the (unwanted) microscopic roughness of the SLM-printed surfaces. The observed offset ($\Delta\mu \approx 1.1$) fits to the simulated water level of 2 % in Fig. 11a, where only the microscale of a different SLM-printed surfaces is analyzed. A significant influence of the contact pressure in the given range is not observed in either the experiments or the simulations, because the bottom face of the tread block has not been completely in contact with the SLM-printed Surface B at the highest load.

5 Conclusions and Outlook

With the additive manufacturing of stainless steel (1.4404) on the basis of SLM, it is possible to build up any desired low-wear surface for rubber friction experi-

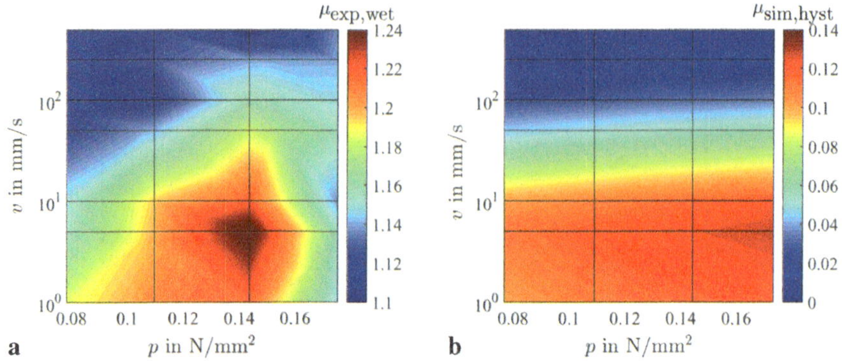

Fig. 15 Test results of linear friction tests **a** at wet conditions and **b** simulation results of multi-scale approach using SLM-printed Surface B

ments. While the macrostructure could be replicated with a high level of detail, the microstructure is increased due to the process-related printing asperities. This results in an increased proportion of hysteresis friction and a reduced proportion of adhesion friction. In combination with the more hydrophobic behavior of the material, the friction reduction at wet conditions is marginal. However, a good correlation of the friction coefficients with varying sliding velocities and loads at dry conditions could be achieved for an asphalt specimen and its replica. The friction coefficient maps have shown characteristic load and velocity dependencies for each rubber-surface pairing. A simulation model to analyze the influence of microtexture on the friction coefficients has been introduced. Due to the more hydrophobic behavior of SLM 3D-prints, the back-calculation of water filling states within the microtexture and its effects on the friction coefficients according to the model could not be validated by the experimental data. Therefore, less hydrophilic surfaces shall be considered in a next step. Using artificial surfaces with simple sinusoidal geometries, the parameters of a multi-scale friction model can be identified. In this way, adhesion and hysteresis friction can be analyzed separately. The superimposition of additional multiple sine waves can validate the multi-scale homogenization approach for rubber friction on rough texture without a scale identification algorithm. The hydrophobic properties of stainless steel and the process-related printing asperities challenged the experimental validation of the introduced friction models with regard to high wet friction coefficients. Therefore, first investigations using an alternative thermoplastic material and varying lubricants on a multiple sine wave surface have been performed (see Fig. 16). In this case, the process-related printing asperities could be reduced by a sandblasted post-treatment. Figure 16 shows, that the friction difference at dry and wet conditions is marginal also with a thermoplastic sample. It is assumed, that thermoplastic material is also less hydrophilic than asphalt mixes. With a rather academic intention, the proportion of adhesion friction can be reduced significantly by the usage of different lubricants (e.g. oil or liquid soap) instead of water, which correlates better to

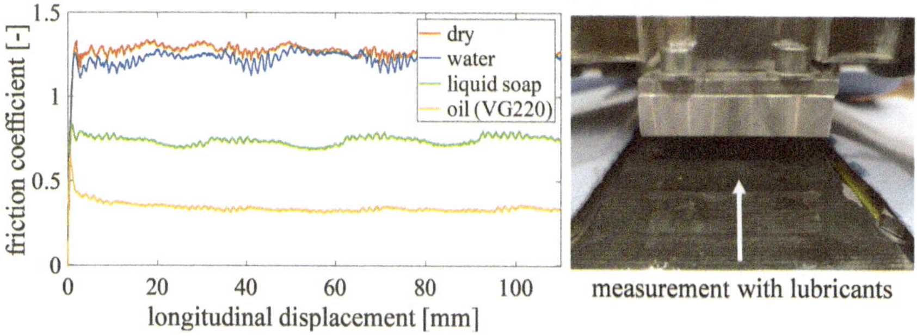

Fig. 16 Friction experiments on a 1D-sinusoidal surface on thermoplastic material at dry, wet and lubricated conditions with a load of 40 kg and a sliding velocity of 100 mm/s

the simulation data of the introduced friction models. This alternative experimental approach is a promising validation of rubber friction laws for future investigations.

References

1. https://tu-dresden.de/bu/bauingenieurwesen/sdt/forschung/for2089/download
2. Carbone, G., Putignano, C.: A novel methodology to predict sliding and rolling friction of viscoelastic materials: theory and experiments. J. Mech. Phys. Solids **61**, 1822–1834 (2013)
3. Friederichs, J., Wegener, D., Eckstein, L., Hartung, F., Kaliske, M., Götz, T., Ressel, W.: Using a new 3D-printing method to investigate rubber friction laws on different scales. Tire Sci. Technol. **48**, 250–286 (2020)
4. Gebhard, A.: Rapid prototyping: Werkzeuge für die schnelle Produktentwicklung, pp. 21–28. Hanser Fachbuchverlag, Munich pp (2000)
5. Hartung, F., Kienle, R., Götz, T., Winkler, T., Ressel, W., Eckstein, L., Kaliske, M.: Numerical determination of hysteresis friction on different length scales and comparison to experiments. Tribol. Int. **127**, 165–176 (2018)
6. Kanafi, M.M., Tuononen, A.J.: Application of three-dimensional printing to pavement texture effects on rubber friction. Road Mater. Pavement Des. **18**, 865–881 (2017)
7. Kienle, R., Ressel, W., Götz, T., Weise, M.: The influence of road surface texture on the skid resistance under wet conditions. Proc. Inst. Mech. Eng. Part J: J. Eng. Tribol. **234**, 313–319 (2020)
8. Lorenz, B., Oh, Y., Nam, S., Jeon, S., Persson, B.: Rubber friction on road surfaces: experiment and theory for low sliding speeds. J. Chem. Phys. **142**, 194701 (2015)
9. Meiners, W.: Direktes Selektives Laser Sintern einkomponentiger metallischer Werkstoffe. Ph.D. thesis, RWTH Aachen (1999)
10. Mooney, M.: A theory of large elastic deformation. J. Appl. Phys. **11**, 582–592 (1940)
11. Persson, B.N.: Theory of rubber friction and contact mechanics. J. Chem. Phys. **115**, 3840–3861 (2001)
12. Weise, M.: Einflüsse der mikroskaligen Oberflächengeometrie von Asphaltdeckschichten auf das Tribosystem Reifen-Fahrbahn. PhD thesis, Institute for Road and Transport Science, Universität Stuttgart (2015)

Multi-scale Computational Approaches for Asphalt Pavements Under Rolling Tire Load

Ines Wollny, Felix Hartung, Michael Kaliske, Pengfei Liu, Markus Oeser, Dawei Wang, Gustavo Canon Falla, Sabine Leischner, and Frohmut Wellner

Abstract An innovative consistent simulation chain is used in this chapter for the combination of the advantages of a microstructure finite element (FE) model of asphalt composites with a macrostructure FE model of pavement under tire rolling load. For this study, an existing microstructural FE model of a Stone Mastic Asphalt including coarse aggregates, asphalt mortar, and air voids was parameterized and validated beginning with experimental tests of asphalt mortar. In order to identify the macroscopic (homogenized) material properties of the asphalt mixture for use in the FE computations of two pavement structures under rolling tire load, this validated microstructural model is applied. These calculations are then evaluated using a new macro-micro-interface, which represents the rolling tire loading conditions for the microstructural model by generating time-dependent displacement boundary conditions. The results indicate that the introduced simulation chain allows for the investigation of the processes, stresses and strains inside the asphalt composite at realistic loading conditions. The experimental tests on the component level can be improved and a better comprehension of the interacting processes in asphalt mixtures under rolling tire load can be obtained by using the results.

Keywords Multi-scale computational approach · Asphalt pavements · Rolling tire load · Finite element method · Macro-micro-interface

Funded by the German Research Foundation (DFG) under grants KA 1163/30, OE 514/1, WE 1642/11 and LE 3649/2.

I. Wollny (✉) · F. Hartung · M. Kaliske
Institute for Structural Analysis, Technische Universität Dresden, Dresden, Germany
e-mail: ines.wollny@tu-dresden.de

P. Liu · M. Oeser · D. Wang
Institute of Highway Engineering, RWTH Aachen, Aachen, Germany

G. Canon Falla · S. Leischner · F. Wellner
Institute of Urban and Pavement Engineering, Technische Universität Dresden, Dresden, Germany

M. Kaliske et al. (eds.), *Coupled System Pavement—Tire—Vehicle*,
Lecture Notes in Applied and Computational Mechanics 96,
https://doi.org/10.1007/978-3-030-75486-0_8

1 Introduction

Asphalt mixtures, as a composite made of aggregates, asphalt binder, filler, additives, and air voids, are heterogeneous materials. The material properties of the components of the asphalt and their interaction lead to short-term material behavior under single tire loads and long-term material behavior after a multitude of tire loads at changing temperature states. Hence, for the investigation of short- and long-term material behaviors of asphalt, the asphalt material is tested and modeled under consideration of its different components. To numerically study the interaction between each component under different loading conditions [5, 15], detailed microstructural finite element (FE) models of asphalt mixtures are necessary. Therefore, the size of the microstructural FE models is usually consistent with the size of asphalt samples in laboratory tests. With consideration for the different components of asphalt, great efforts are needed to model a whole pavement structure under rolling tire load. As a result, the pavement structure under tire load is normally modeled as homogenous using macroscopic phenomenological material formulations (e.g. [10]). Laboratory tests of asphalt samples can determine the required macroscopic material parameters (see e.g. [1, 11]). In fact, the construction of the test machine and the available measuring system limit the capabilities of the laboratory tests. As shown in numerical computations, in the pavement structure, asphalt under rolling tire load is subjected to three-dimensional (3D) stress states with different stresses and frequencies and with changing directions of the principal stresses [9, 12]. However, asphalt tests under such complex and changing 3D stress states cannot be conducted in the laboratory tests yet to the authors knowledge.

This chapter is aimed at supplying a consistent simulation chain that closes the gap between the microstructural and macrostructural models (see Fig. 1).

The simulation chain proceeds from the component level (micro level, asphalt mixtures) to the setup of the structure level (macro-model, asphalt pavement) (Sect. 2). The material properties of the asphalt mortar for the microstructural model are identified by using laboratory tests at the component level which were conducted at the Institute of Urban and Pavement Engineering, TU Dresden (see chapter "Experimental Methods for the Mechanical Characterization of Asphalt Concrete at Different Length Scales: Bitumen, Mastic, Mortar and Asphalt Mixture"). The X-ray Computed Tomography (CT) scanning and image processing are all processed at the Institute of Highway Engineering, RWTH Aachen University (see chapter "Numerical Simulation of Asphalt Compaction and Asphalt Performance" and chapter "Characterization and Evaluation of Different Asphalt Properties Using Microstructural Analysis") for developing the microstructural FE model. Further, the macroscopic asphalt material properties, which are used in the macrostructural arbitrary Lagrangian–Eulerian (ALE) pavement model under rolling tire load, see chapter "Multi-physical and Multi-scale Theoretical-numerical Modeling of Tire-Pavement Interaction", are identified by using the microstructural FE model of asphalt (Sect. 3). A macro-micro-interface is developed to determine the results of the macrostructural ALE pavement model under rolling tire load, and the results are applied to

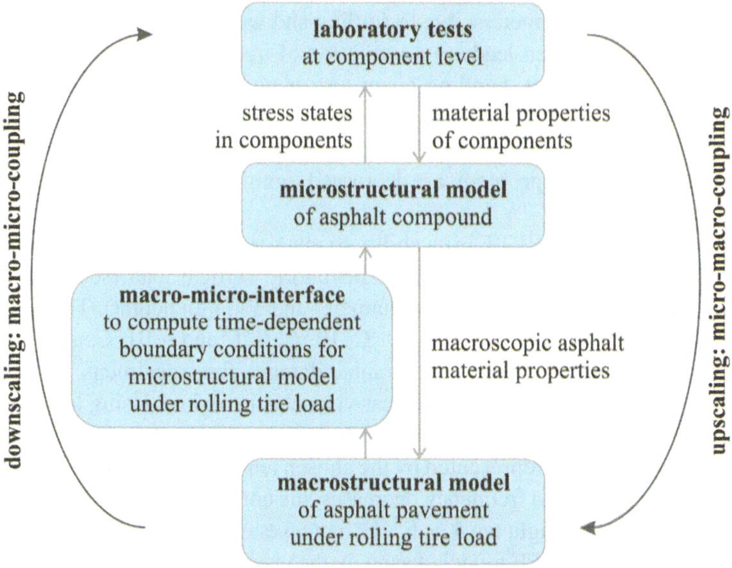

Fig. 1 Simulation chain to couple microstructural and macrostructural models [14]

the microlevel in the return route (Sect. 4). Time-dependent displacement boundary conditions for the microstructural model are computed, which represent the loading case of a tire overrun. With the help of the results of the microstructural FE asphalt model, the processes inside the asphalt structure at realistic loading can be investigated and the stress states inside the asphalt components to define proper laboratory test conditions can be identified. The simulation chain is installed and the macrostructure is modeled at the Institute for Structural Analysis, TU Dresden. Lastly, the results of the presented research are concluded in Sect. 5.

2 Upscaling Through Micro-Macro-Coupling

2.1 Determination of Material Parameters for Asphalt Mortar

During the modeling, asphalt at the microscale is considered as a mixture including coarse aggregates, asphalt mortar, and air voids. Asphalt mortar consists of asphalt binder (bitumen), filler, and aggregate particles with an average grain size of less than 2 mm. The coarse aggregates in this study are considered to be linear elastic material. The characterization of the viscoelastic behavior of the mortar in the laboratory needs to be investigated for the representation of the real asphalt mixture in the microstructural numerical model. Based on the previous experience of the authors, the mortar cannot be tested with the equipment used for bitumen testing, e.g. dynamic shear rheometer or viscometers, due to settling and workability problems. Also, the distortion of the specimen leads to a limitation of traditional tests of asphalt, such as the indirect tensile test, bending beam test or uniaxial test. Thus, the repeated load triaxial test (RLTT) is the only alternative solution regarding the limitations to be selected, although the preparation of the specimens is difficult. However, the distortion and settlement problems are decreased significantly by confining the mortar in all directions.

The advantage of the RLTT is its ability to measure the effects of a simultaneous constant hydrostatic pressure and cyclic deviatoric vertical load (see Fig. 2) on a cylindrical mortar specimen (150 mm diameter and 300 mm height). The frequency sweeps at four temperatures, namely 20 °C, 10 °C, 0 °C and −10 °C, are consistent in the testing protocol. In order to guarantee damage free specimens, the tests are conducted at strain levels within the linear viscoelastic regime. Using log mode, the tests are conducted with increasing frequency from 1 to 10 Hz. A wide range of material behavior can be well represented by the chosen temperatures and frequencies. As temperature increases and frequency decreases, the complex modulus $E^* = E' + iE''$ reduces and the phase angle $\tan\delta = E''/E'$ increases expectedly.

In the microstructural FE model of asphalt, the viscoelastic behavior of the asphalt mortar is described by a generalized Maxwell approach with storage modulus

$$E' = \gamma_\infty E_0 \sum_{i=1}^{n} \frac{(\omega\tau_i)^2 \gamma_i E_0}{1 + (\omega\tau_i)^2} \tag{1}$$

and loss modulus

a b

Fig. 2 Triaxial testing equipment: **a** triaxial cell, **b** mortar specimen [14]

Table 1 Prony-series parameters of the asphalt mortar [14]

$E_0 = 7{,}842.69$ N/mm^2		
$\gamma_1 = 0.111569$	$\gamma_2 = 0.322009$	$\gamma_3 = 0.566177$
$\tau_1 = 0.177154$ s	$\tau_2 = 0.015832$ s	$\tau_3 = 0.000553$ s

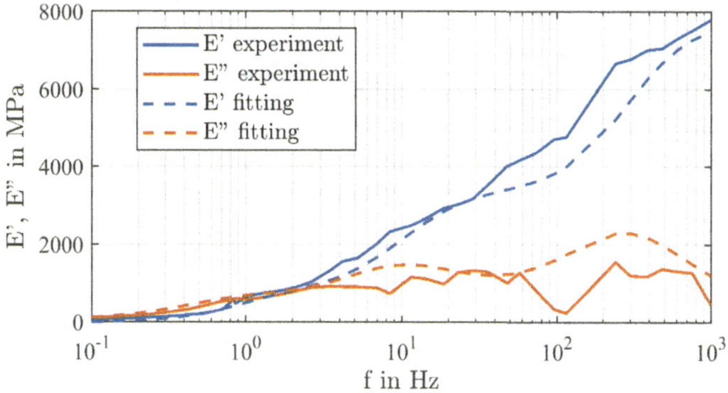

Fig. 3 Storage and loss moduli of asphalt mortar as output of triaxial tests and corresponding simulation results with identified material parameters at 10 °C [14]

$$E'' = \sum_{i=1}^{n} \frac{\omega \tau_i \gamma_i E_0}{1 + (\omega \tau_i)^2}. \tag{2}$$

Thereby, ω is the angular load frequency. The Prony-series parameters E_0 (instantaneous stiffness), γ_i (normalized stiffness ratio), γ_∞ (long-term stiffness ratio) and τ_i (relaxation time) of the mortar material are identified by a fitting routine. The parameter identification is conducted at a reference temperature of 10 °C. Thus, the Williams-Landel-Ferry transformation is applied to shift the test results at the other temperatures. The fitting tool minimizes the difference between the experimental moduli from the triaxial test results and Eqs. (1) and (2) with respect to the number of Maxwell branches n. Table 1 illustrates the optimal parameters of the three Maxwell branches for presenting the asphalt mortar. By applying the Prony-series parameters of Table 1 obtained from the optimization tool, Fig. 3 depicts the remaining deviation between the storage and loss moduli measured in the laboratory and Eqs. (1) and (2).

2.2 Microstructural Modeling of SMA 11 D S

In this study, samples of a Stone Mastic Asphalt (SMA) with a maximum grain size of 11 mm (SMA 11 D S) with diabase aggregates are used. The unmodified penetration graded bitumen of the type 50/70 is used as binder. The asphalt specimens are obtained from a test track with the geometry of 26 m × 1.2 m × 0.3 m (length

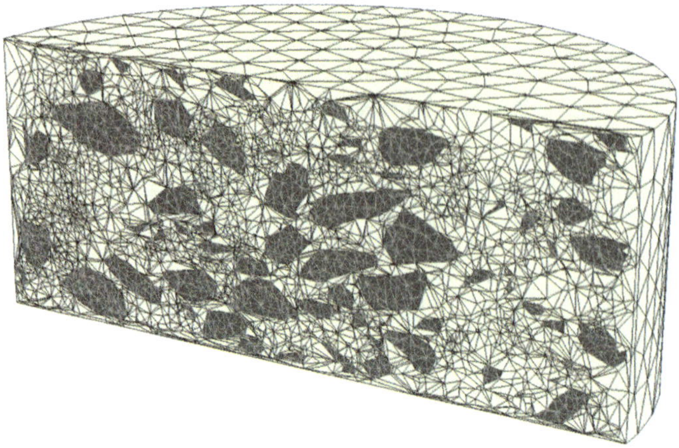

Fig. 4 Cut through assembled microstructural FE model [14]

$\times$ width $\times$ depth). The cylindrical cores with a height of 300 mm and a diameter of 150 mm are drilled from the test track. In [6], further information on the material design and the construction of the test track is described. For the investigation of the internal microstructure, the asphalt specimens are cut close to the surface of the test track, and X-ray CT scanning is used with 0.1 mm scanning intervals and a resolution of the gray images of 1,024 pixels $\times$ 1,024 pixels, whereby each pixel corresponds to 80 μm. The binary images, in which the aggregates and air voids are separated, are obtained by conversion from the gray images (see [2]).

The 3D FE model includes explicitly aggregate grains of size 2.36 mm and larger. Thus, mortar contains all aggregates with a size smaller than 2.36 mm. The two-dimensional (2D) binary images are stacked to reconstruct the microstructures of the aggregates and air voids. Boolean operations are processed for the creation of the asphalt mortar after the models of air voids and aggregate grains are imported into the input file. Figure 4 shows the construction of the microstructural model after assembling the aggregate grains and the asphalt mortar. Further information about the process can be found in [5, 6].

2.3 *Formulation of Macroscopic Material*

The inelastic behavior of asphalt at a specific temperature state is addressed by adopting the macroscopic material formulation, which is introduced in chapter "Multi-physical and Multi-scale Theoretical-numerical Modeling of Tire-Pavement Interaction". The formulation extracts the viscoelastic properties out of the asphalt model introduced in [16] by removing the plastic portion of the model and applying finite strain theory. The formulation is founded on the multiplicative split of the deformation gradient

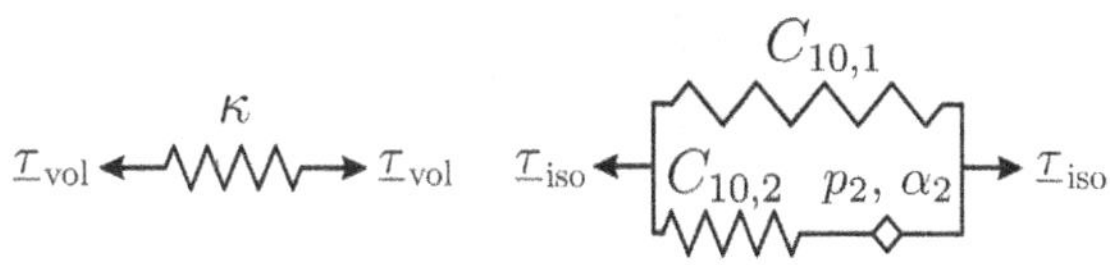

Fig. 5 Volumetric and isochoric part of the asphalt material model [14]

$$\underline{F} = \underline{F}_{vol} \cdot \underline{F}_{iso} = (J^{1/3}\underline{1}) \cdot (\underline{F}^e_{iso} \cdot \underline{F}^i_{iso}) \tag{3}$$

into a volumetric $\underline{F}_{vol}$ and an isochoric part $\underline{F}_{iso}$. The isochoric contribution of the deformation gradient again consists of an elastic part $\underline{F}^e_{iso}$ and an inelastic part $\underline{F}^i_{iso}$. The Jacobian J describes the volume change with respect to the reference configuration and $\underline{1}$ represents the second-order identity tensor. The Kirchhoff stress tensor

$$\underline{\tau} = J\underline{\sigma} = \underline{\tau}_{vol} + \underline{\tau}_{iso}, \tag{4}$$

where $\underline{\sigma}$ is the Cauchy stress tensor, consists of a volumetric $\underline{\tau}_{vol}$ and an isochoric part $\underline{\tau}_{iso}$ according to the rheology shown in Fig. 5.

The volumetric Kirchhoff stress tensor

$$\underline{\tau}_{\text{vol}} = J^{2/3}\kappa(J-1)\underline{1} \tag{5}$$

is characterized by the bulk modulus κ. The isochoric contribution of the asphalt model consists of two branches in parallel. Therefore, the Kirchhoff stress tensor

$$\underline{\tau}_{\text{iso}} = \underline{\tau}^e_{\text{iso},1}\left(C_{10,1}\right) + \underline{\tau}^i_{iso,2}(C_{10,2}, p_2, \alpha_2) \tag{6}$$

is composed of two different parts. The upper branch of the isochoric portion is described by the Neo-Hookean spring stiffness $C_{10,1}$. The bottom branch, which is a fractional Maxwell element, captures rate dependent viscoelastic effects. The bottom element consists of a Neo-Hookean spring with the stiffness $C_{10,2}$ and a fractional element, where p_2 is comparable to the viscosity and α_2 prescribes the order of the time derivative (see [8]).

2.4 Determination of Macroscopic Material Properties from the Microstructural Model

To ensure consistency of the simulation chain, the microstructural model determines the parameters of the macro-model. The application of the microstructural model is time-saving and cost-effective compared to that of laboratory tests. The microstructural model also allows for a better definition of various loading conditions as compared to experiments. Figure 6 shows the microstructural model under hydrostatic pressure ($\sigma_z(t) = \sigma_r(t) \leq 0$) with fixed nodes at the bottom.

The volumetric stress is computed by Eq. (5), which uses the resulting vertical and radial strains at the vertical center slice of the cylindrical microstructural model as the input. As the stress states of the microstructural model and macro-model become almost equal, the bulk modulus κ in Eq. (5) stops changing. The isochoric parameters are identified using several uniaxial cyclic microstructural simulations ($\sigma_r(t) = 0$). The frequencies of 1 and 10 Hz with absolute amplitudes of 1 and 2 MPa are investigated both for compression and pressure expansion during the simulations. In order to fit the vertical strains of each corresponding microstructural simulation to the vertical strains of the macro-model, the isochoric parameters of the macroscopic material formulation are identified via particle swarm optimization [4]. Table 2 lists the identified material parameters.

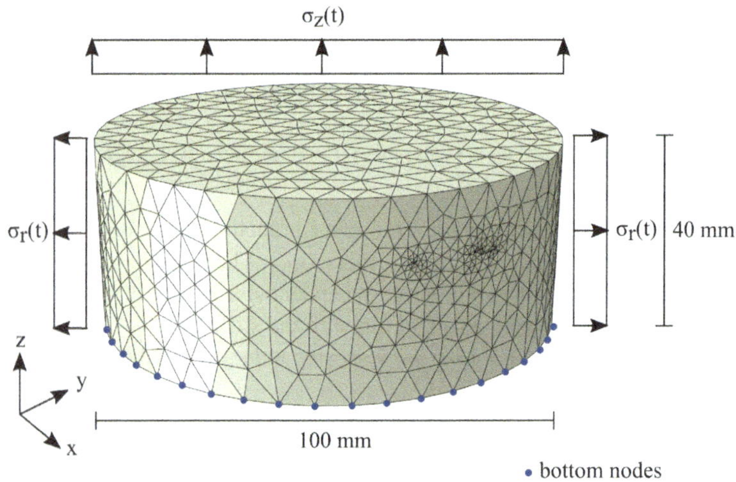

Fig. 6 Microstructural FE model of a triaxial test [14]

Table 2 Macroscopic material parameters of the SMA 11 D S [14]

κ [MPa]	$C_{10,1}$ [MPa]	$C_{10,2}$ [MPa]	p_2 [MPa]	α_2 [−]
7,413.866	194.688	2,100.723	684.694	0.58276

A validation with a cylindrical FE model depicted in Fig. 7 using the macroscopic material formulation is performed for $f = 10\,\text{Hz}$, $\sigma_{z,\min} = -1.5\,\text{MPa}$ (compression test). The bottom nodes are fixed. Then, the evaluations of the vertical and radial strains at the center slice nodes are processed. The averaged vertical displacements of the micro- and macro-model are compared and shown in Fig. 8. It should be noted that the macro-model parameters are not pre-adapted using the predefined load condition.

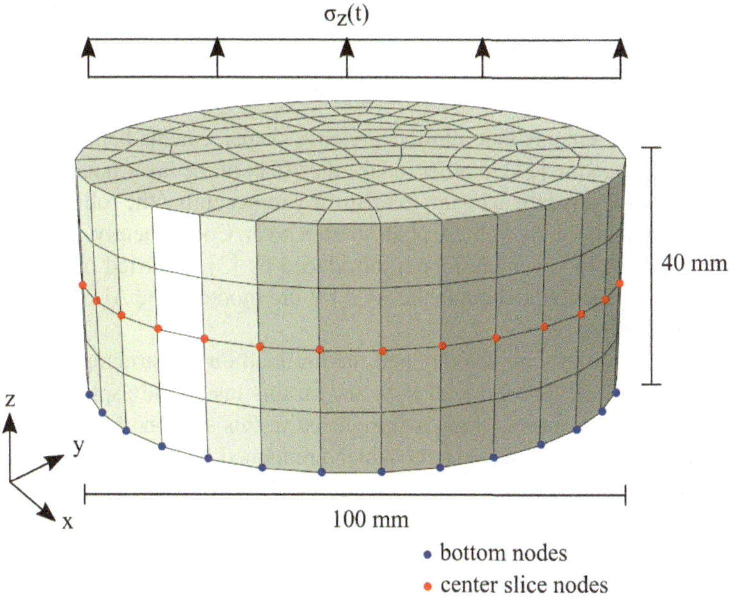

Fig. 7 FE model of the uniaxial test with a macroscopic material formulation [14]

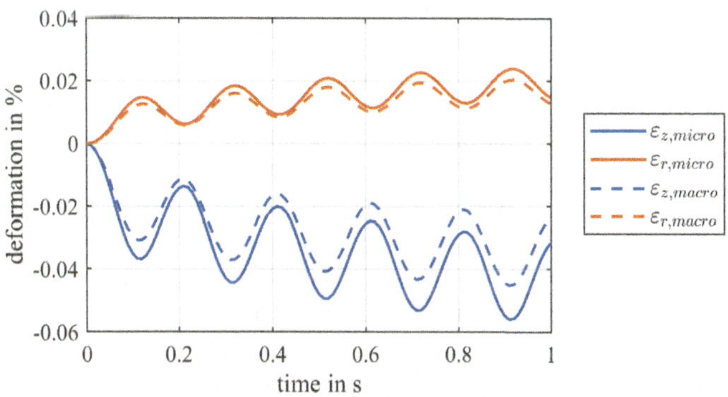

Fig. 8 Validation at $f = 5\,\text{Hz}$, $\sigma_{z,\max} = -1.5$ MPa and $\sigma_r = 0$ MPa [14]

3 Macrostructural Simulation of ALE Pavement Model

The rolling tire loaded macroscopic pavement computation is performed as an FE computation using an ALE formulation [7] for the coupled tire-pavement model. The computation of pavements under rolling tire load with the ALE formulation is much more efficient than the classical time-dependent Lagrangian FE computation, where the tire load is shifted in several time steps over the fixed FE pavement model. For details regarding the ALE kinematics and FE formulation, the reader is referred to the chapter "Multi-physical and Multi-scale Theoretical-numerical Modeling of Tire-Pavement Interaction". In the case of inelastic (history-dependent) materials such as asphalt, the material history has to be assembled along the material streamlines as described in [13].

The interface elements that allow for relative displacements between pavement layers are introduced in [11] in order to take the effect of the layer bond between the single pavement layers into account. Realistically representing the rolling tire load is especially vital in the study of the asphalt surface layer. Consequently, an application with the coupled tire-pavement model introduced by [3] is carried out. Thereby, a program interface is used to couple the ALE FE tire model to the ALE FE pavement model sequentially.

The effect of the driving velocity and the tire load on the structural behavior of the pavement as well as on the stresses and strains inside the asphalt material is investigated in an example, which is introduced in this section. Table 3 shows the structures of two different pavements, which are named Bk100 and Bk10 based on the German guideline RStO.

Table 3 Structure of investigated pavements [14]

Layer	Material	Bk100 [cm]	Bk10 [cm]
Surface layer	SMA 11 D S	4	4
Interface layer 3	bituminous emulsion		
Binder layer	SMA 11 D S	8	8
Interface layer 2	bituminous emulsion		
Base layer 2	AC 22 T S	11	7
Interface layer 1	bituminous emulsion		
Base layer 1	AC 22 T S	11	7
Interface	$G = 1$ MPa		
Frost protection layer	$E = 100$ MPa, $\upsilon = 0.35$	46	54
Soil	$E = 45$ MPa, $\upsilon = 0.35$	100	100

Figure 9 shows the ALE FE mesh of the macroscopic pavement model and the applied FE truck tire model. Within the FE pavement model, isoparametric 20-node brick elements are applied for the pavement layers while, correspondingly, isoparametric 16-node interface elements are used for the interface layers. The macroscopic material parameters identified in Sect. 2.4 based on the SMA 11 D S microstructural model are applied in the example for the asphalt surface and binder layer. For the asphalt base layer, the macroscopic asphalt material model and parameters for asphalt concrete (AC 22 T S) for 10 °C presented in [12] are used. This model includes additionally a plastic branch. The behavior of the layer bond of the bituminous emulsion is modelled in the example by the viscoelastic, temperature-dependent and normal pressure-dependent cohesive zone model described in [11] using the material parameters corresponding to 10 °C.

A discussion of the results of the macroscopic ALE FE pavement computation of the four investigated cases is conducted in this section. The vertical displacement of the pavement surface along the driving lane is shown in Fig. 10. The tire axis is located at $\chi_1 = 0\,\mathrm{m}$. As the driving velocities are reduced, the vertical displacements increase because of the viscoelastic asphalt material behavior. Furthermore, behind the tire axis overrun ($\chi_1 > 0\,\mathrm{m}$), the maximum displacement can be found, which is due to the viscoelastic properties not directly below the tire axis. With the increase of the tire load, the vertical displacements increase expectedly. The displacements reduce as the structural pavement stiffness (Bk100) increases. The vertical tire load affects the vertical stress in the asphalt surface layer more predominantly than the driving speed and the pavement construction (see Fig. 11).

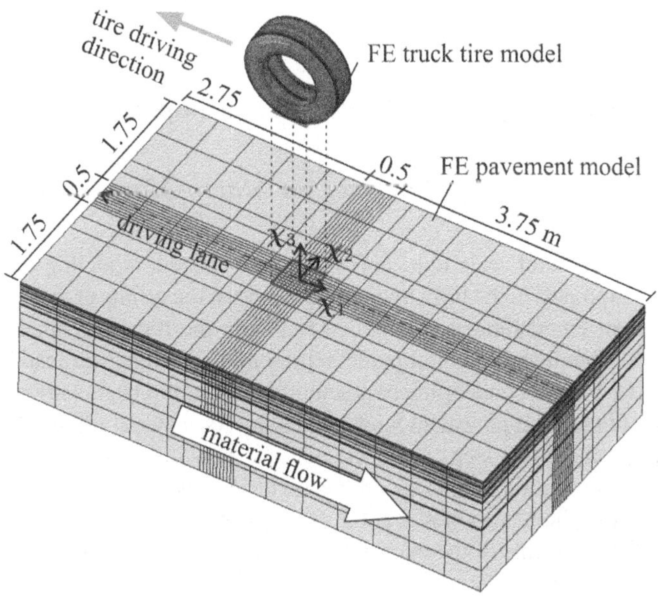

Fig. 9 Macrostructural FE mesh of pavement and truck tire [14]

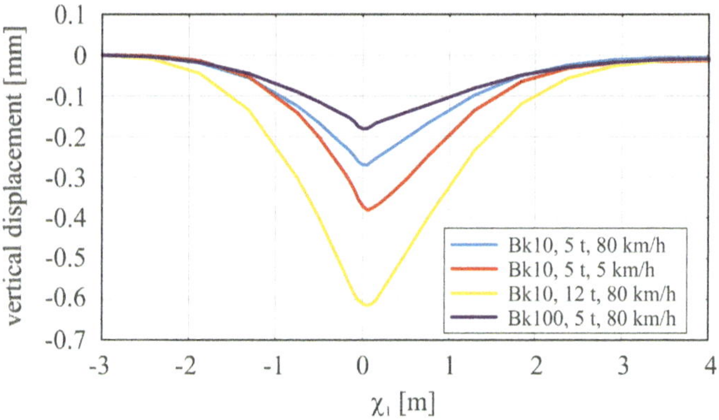

Fig. 10 Vertical displacement of the pavement surface along the driving lane [14]

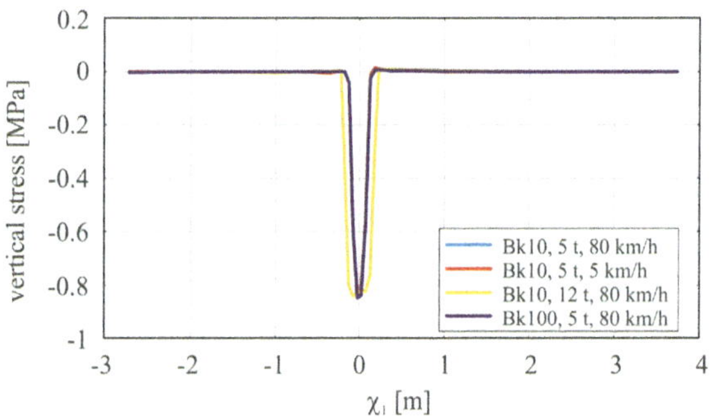

Fig. 11 Vertical Cauchy stress in the asphalt surface layer at $\chi_2 = -2.5$ cm and $\chi_3 = -1$ cm (according to Fig. 9) along the driving lane [14]

4 Downscaling Through Macro-Micro-Coupling

4.1 Development of the Interface for Macro-Micro-Coupling

As the macrostructural scale does not allow for a discrete modeling of the single pavement compounds in the form of grains and mortar, the results of the macroscopic ALE pavement computations are applied to obtain the time-dependent boundary conditions for the microstructural model representing a rolling tire load. As shown in Fig. 12, the time-dependent microstructural computation begins prior to loading and in front of the ALE pavement model boundary for material inflow at time t_0.

Next, prescribed time steps along the material flow direction through the ALE macro-model are applied to shift the microstructural model. As shown in Fig. 13, the displacements of the macro-model at the corresponding position are determined to obtain the displacements of the microstructural model at each time step. In order to

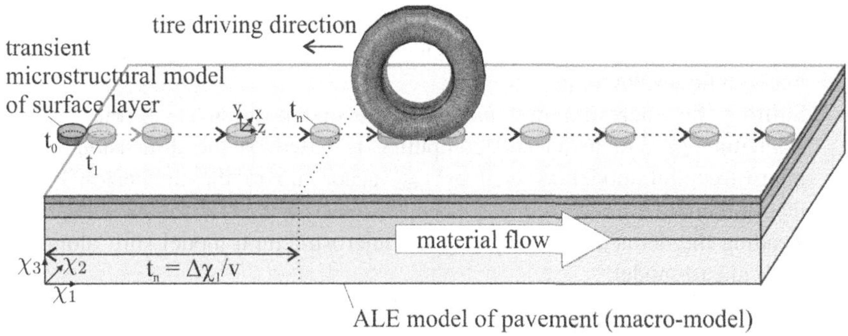

Fig. 12 Shift of the microstructural model through the ALE macro-model to obtain the time-dependent boundary conditions [14]

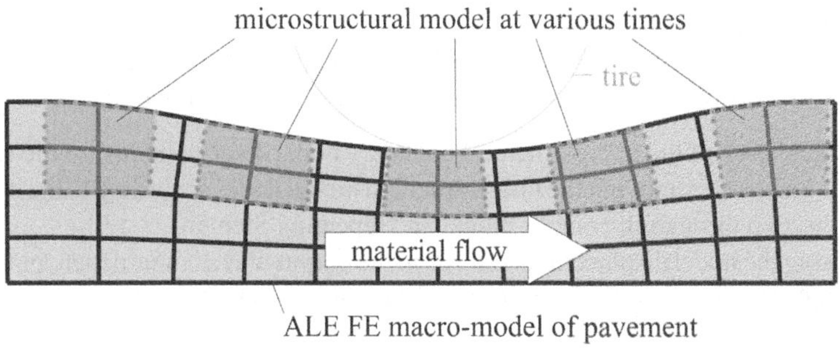

Fig. 13 Various microstructural model positions in a cut through macrostructural ALE FE pavement model [14]

allow free and independent displacements of the grains and mortar inside the microstuctural model, the corresponding displacements are only applied to the boundary nodes.

The following aspects need to be considered for the realization of this macro-micro-interface: (a) the coordinate systems of the microstructural model and of the macro-model may have different orientations (see Fig. 12), (b) different units might be used in the microstructural model and the macro-model and (c) the boundary nodes of the microstructural model rarely correspond directly to nodes of the macro-model, but more typically lie somewhere inside the finite elements of the macro-model.

Therefore, the implemented macro-micro-interface processes several steps to compute the time-dependent displacements of the boundary nodes of the microstructural model. The procedure is described in detail in [14] and summarized below:

1. Reading the nodal coordinates of the boundary nodes of the microstructural model and shifting them into the macro coordinate space and units.

2. Reading the data from ALE macro-model computation to get the geometry (nodal coordinates, element assignments), results (nodal displacements) and material flow velocity v.
3. Shifting the microstructural model to the first position at t_l (Fig. 12) by evaluating its dimension and computing t_l based on the dimension of the microstructural model as well as t_{max} according to the dimension of the macro-model.
4. Reading the defined time steps for the microstructural model shift along the ALE macro-model.
5. Calculation of the microstructural boundary nodal displacements for time t_l by:
   (a) Searching the corresponding position (element number and local coordinates) inside the macro-model.
   (b) Applying the isoparametric shape functions and the nodal displacements of the macro-model to compute the displacements according to the local coordinates.
   (c) Transformation of the nodal displacements into the coordinate system and units of the microstructural model and storage on an output file.
6. Calculation of the microstructural boundary nodal displacements for the further time steps t_n by moving the microstructural model nodes according to the defined time step through the macro-model and repeating Step 5.
7. Using the nodal displacements stored in the output files as time-dependent nodal displacement boundary conditions of the microstructural model.

4.2 Validation of the Interface for Macro-Micro-Coupling

The time-dependent displacements of the boundary nodes of the microstructural model can be computed by the presented interface. Next, these displacements are intended to be used as a time-dependent loading on the microstructural model. At that point, the resulting internal deformation and stress state of the microstructural model are assumed to correspond in average to the deformation and stress state inside the macro-model. This assumption is validated in the following by applying the macro-micro-coupling procedure to a cylindrical subdomain model called “macro part”. The macro part uses the same macroscopic asphalt material model and parameters as the macroscopic ALE pavement model. Within this validation example, altogether 200 time steps are prescribed to shift the macro part along the driving lane of the ALE macro-model. The different dimensions of both FE meshes and the vertical stress directly under the tire can be observed in Fig. 14. The tread ribs of the truck tire result in the nonuniformity of the contact stress distribution in the tire footprint, which can be seen in Fig. 14. This nonuniform stress distribution can be replicated although only the boundary displacements are applied to the macro part model.

Figure 15 shows the vertical stress distribution in a section cut of the macro part model, when it is directly loaded under the tire axle, as well as highlighting the

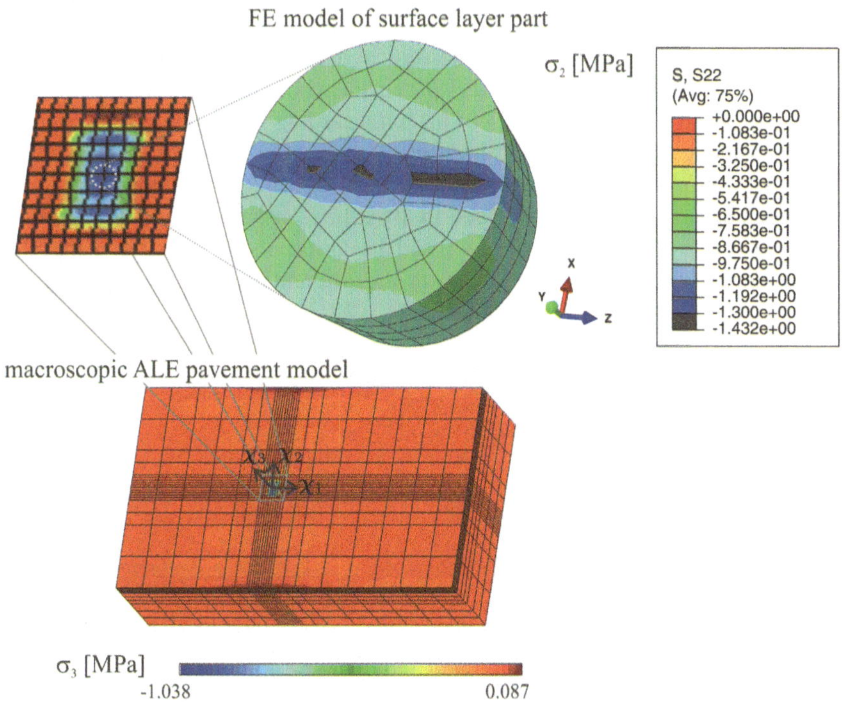

Fig. 14 Vertical Cauchy stress at the macroscopic ALE pavement model (Bk10, 5 t and 80 km/h) and the FE model of the surface layer part (macro part) [14]

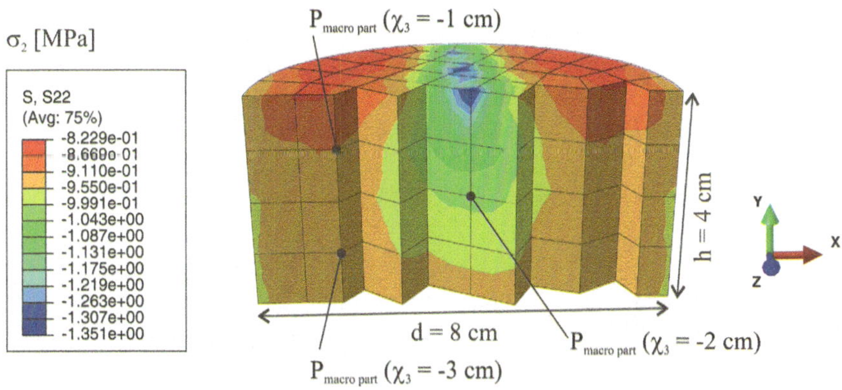

Fig. 15 Vertical Cauchy stress in a cut through the FE model of the macro part at a position directly under the tire axle [14]

points that are used to evaluate the results. Because of the tire tread, nonuniform stress distribution can be observed. The vertical displacements in the middle of the macro part and the macroscopic ALE pavement model are compared and shown in Fig. 16. Next, $t = (\chi_{1\text{macro}} + 3.04\,\text{m})/v$ (according to χ-coordinate system in Fig. 9) is adopted to calculate the corresponding time for the displacements of the ALE macro-model. The horizontal and vertical stresses in the longitudinal direction are compared and shown in Figs. 17 and 18. It can be seen that small differences occur between the two points of the macro part and the corresponding locations in the ALE macro-model, which presumably stem from the different meshing. Nevertheless, the

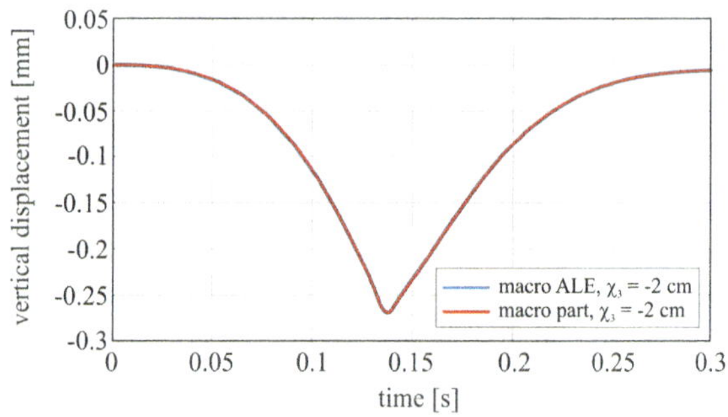

Fig. 16 Vertical displacement of the ALE macro-model and macro part at $\chi_3 = -2\text{cm}$ [14]

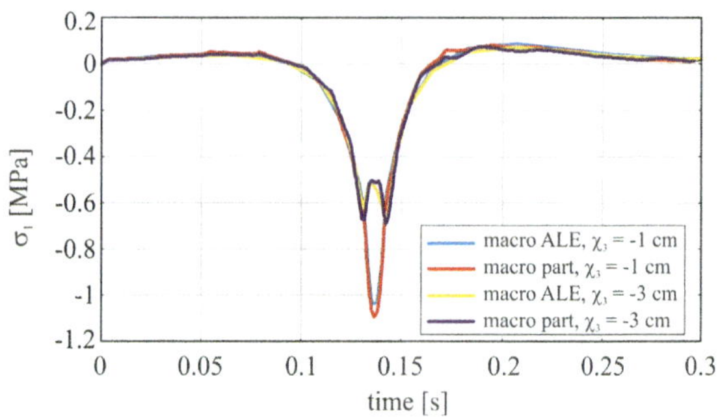

Fig. 17 Horizontal Cauchy stress σ_1 in the longitudinal direction of the ALE macro-model and macro part at $\chi_3 = -1$ cms and $\chi_3 = -3\,\text{cm}$ [14]

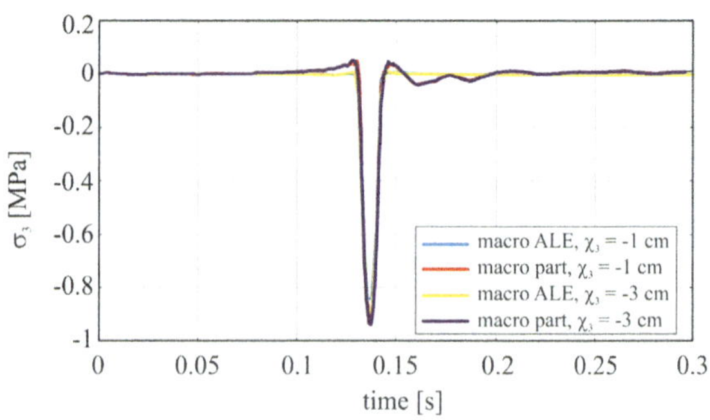

Fig. 18 Vertical Cauchy stress σ_3 in the longitudinal direction of the ALE macro-model and macro part at $\chi_3 = -1\,\text{cm}$ and $\chi_3 = -3\,\text{cm}$ [14]

macro part model reflects the general stress state in the surface layer under the rolling tire load well and validates the proposed macro-micro-coupling interface.

4.3 Microstructural Simulation of the Asphalt Under Rolling Tire Load

By applying the macro-micro-interface to the microstructural model of the SMA surface layer in Sect. 2.3 and all four cases of the macrostructural ALE pavement computations given in Sect. 3, the mechanical response of the microstructure of the asphalt mixtures is simulated. For shifting the microstructural model through the ALE macro-model, again, 200 time steps are chosen.

For example, as shown in Fig. 19, the resulting von Mises-stresses in the asphalt mortar are evaluated for all investigated road conditions and constructions for the selected time step, when the microstructural model is directly loaded underneath the tire axis. The normalized frequency that corresponds to a certain von Mises-stress value q is obtained by counting the number of asphalt mortar FE with a centroidal von Mises-stress value in the range of q $\pm$ 0.005 MPa and dividing them using the total number of asphalt mortar elements. The inelastic material of the macro-model has more relaxation time at the velocity of 5 km/h than at 80 km/h and thus results in lower stress values at 5 km/h. The higher von Mises-stresses are expectedly led from the highest tire load of 12 t (yellow line). The stresses increase slightly to higher values in case of a higher stiffness of the road construction Bk100.

In Fig. 20, using a macro-model of Bk10, a tire load of 5 t and a tire velocity of 80 km/h, both the hydrostatic pressure and von Mises-stresses of the microstructural model are plotted for three chosen time increments. The two microstructural model outputs at $\mathrm{t} = (2.733\,\mathrm{m} + 0.333\,\mathrm{m})/80\,\mathrm{km/h} = 0.138\,\mathrm{s}$ correspond to the moment in time when a maximum displacement is reached in the vertical direction along the upper surface of the macro-model. The findings on the left ($\mathrm{t} = 0.123\,\mathrm{s}$) and right ($\mathrm{t} = 0.153\,\mathrm{s}$) are close to the contact patch of the tire. The hydrostatic pressure state changes significantly within the three time steps. The detail at $\mathrm{t} = 0.138\,\mathrm{s}$ shows,

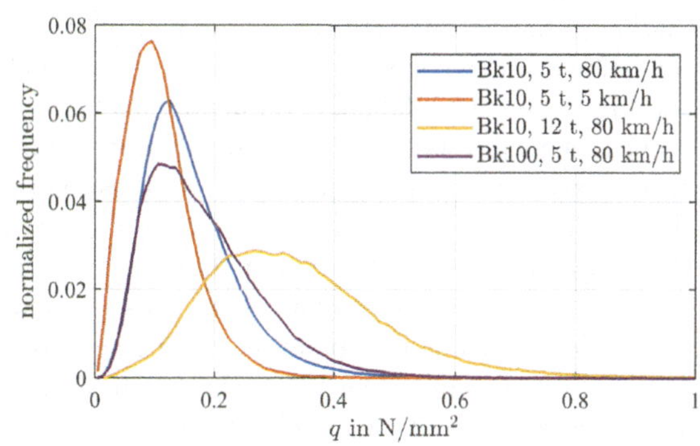

Fig. 19 Normalized frequency of mortar element's von Mises-stresses in the microstructural model [14]

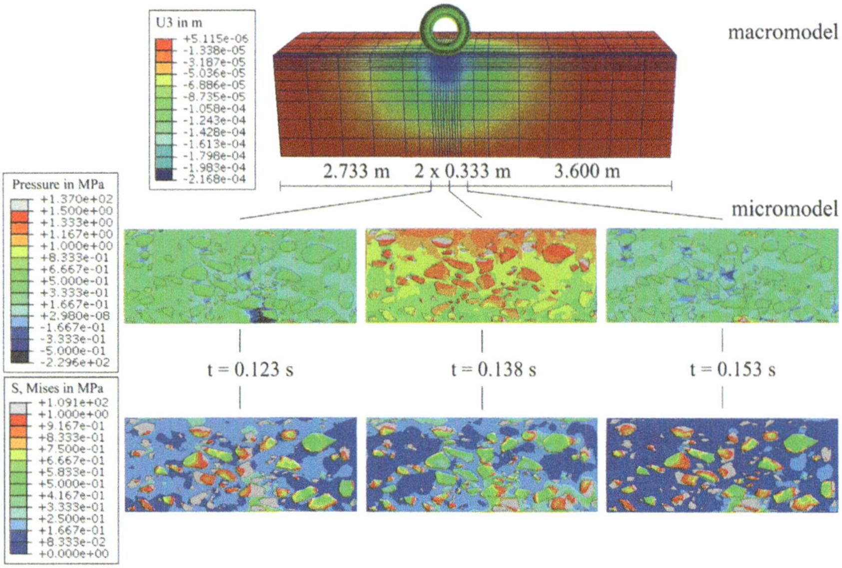

Fig. 20 Evaluation of hydrostatic pressure and von Mises-stresses of the microstructural model using Bk10, tire load of 5 t and velocity of 80 km/h [14]

further, the decreasing pressure from the top (tire footprint) to the bottom. Some different behaviors can be observed in the deviatoric portion, which is represented by the von Mises-stress. When comparing the von Mises stress at $t = 0.123$ s to $t = 0.153$ s, the stress values within the mortar below the tire footprint only increase slightly. A higher stiffness causes the aggregate to receive a higher hydrostatic pressure and deviatoric stresses.

5 Conclusions and Outlook

The investigation of asphalt material behavior under realistic rolling tire loading conditions can be innovatively approached by applying the presented consistent simulation chain to couple a microstructural FE model of asphalt and a macrostructural FE pavement model under rolling tire load. To ensure simulation chain consistency, numerical tests of the microstructural model are conducted to determine the macroscopic asphalt material properties. These numerical asphalt tests are more advantageous than laboratory tests, due to no limitations on certain load cases. Although numerical investigations of the asphalt material behavior using microstructural models do not replace laboratory tests of asphalt samples completely, they provide high potential to support the research on asphalt material behavior and the identification of macroscopic material properties. Beyond the numerical tests shown in the present research, many other load cases such as shear, torsion etc. could be considered. The effects of different tire rolling velocities, tire loads, and pavement constructions on the deformation and stress states of the pavement construction can be illustrated by applying the macroscopic asphalt material properties in the macrostructural pavement computations under rolling tire load.

The main goal and accomplishment of this study have been to successfully and accurately replicate the real loading conditions of asphalt layers under rolling tire load in a miscrostructural numerical model that can be used for further testing and research. Applying the introduced macro-micro-interface yields the required boundary conditions, which is validated by the testing results. Despite some small variations in the resulting internal stress test states, the macro-micro-interface does effectively enable the simulation of the rolling tire loading conditions. By applying these time-dependent boundary conditions to the microstructural model in Sect. 4.3, an idea of the stress states of the single components that occur inside in the asphalt mixture at realistic loading conditions is obtained. This will be crucial for the setup of future laboratory tests at the component level. Therefore, the microstructural and macrostructural models can be optimized iteratively by using the presented simulation chain. Deeper information on the processes that take place inside the asphalt composite material under rolling tire load and affect the pavement durability can be obtained by this kind of computations. Obviously, the quality of the microstructural and of the macrostructural FE models has an influence on the quality of the results.

Further investigation of the microstructural behavior of various microstructural asphalt models, such as other mixtures and other distributions of grains, mortar, and air voids of the same mixture, can be carried out by using the introduced simulation chain. In addition, the investigations of the influence on macrostructural behavior with a changing microstructure are included. It is also available for different macroscopic pavement structures and different tire loads. The load case investigated in this chapter for the microstructural model is the loading in the asphalt surface layer along the center of the driving lane. In further investigations, the loading conditions in other asphalt layers and other positions (e.g. lateral to the tire driving lane) shall be studied.

References

1. Darabi, M., Abu Al-Rub, R., Masad, E., Little, D.: A thermodynamic framework for constitutive modeling of time- and rate-dependent materials. Part II: numerical aspects and application to asphalt concrete. Int. J. Plast. **35**, 67–99 (2012)
2. Hu, J., Liu, P., Wang, D., Oeser, M.: Investigation on fatigue damage of asphalt mixture with different air-voids using microstructural analysis. Constr. Build. Mater. **125**, 936–945 (2016)
3. Kaliske, M., Wollny, I., Behnke, R., Zopf, C.: Holistic analysis of the coupled vehicle-tire-pavement system for the design of durable pavements. Tire Sci. Technol. **43**, 86–116 (2015)
4. Kennedy, J., Eberhart, R.: Particle swarm optimization. In: Proceedings of IEEE International Conference on Neural Networks, 4. IEEE Press, Piscataway, NJ (1995)
5. Liu, P., Hu, J., Wang, D., Oeser, M., Alber, S., Ressel, W., Canon Falla, G.: Modelling and evaluation of aggregate morphology on asphalt compression behaviour. Constr. Build. Mater. **133**, 196–208 (2017)

6. Liu, P., Hu, J., Wang, H., Canon Falla, G., Wang, D., Oeser, M.: Influence of temperature on mechanical response of asphalt mixtures using microstructural analysis and finite-element simulations. J. Mater. Civ. Eng. **30**, 04018327 (2018)
7. Nackenhorst, U.: The ALE-formulation of bodies in rolling contact—theoretical foundations and finite element approach. Comput. Methods Appl. Mech. Eng. **193**, 4299–4322 (2004)
8. Oldham, K.B., Spanier, J.: The Fractional Calculus: Theory and Applications of Differentiation and Integration to Arbitrary Order. Academic Press, San Diego, CA (1988)
9. Wang, G., Roque, R.: Impact of wide-based tires on the nearsurface pavement stress states based on three-dimensional tire-pavement interaction model. Road Mater. Pavement Des. **12**, 639–662 (2011)
10. Wang, H., Li, M.: Comparative study of asphalt pavement responses under FWD and moving vehicular loading. J. Transp. Eng. **142**, 04016069 (2016)
11. Wollny, I., Hartung, F., Kaliske, M.: Numerical modeling of inelastic structures at loading of steady state rolling: thermo-mechanical asphalt pavement computation. Comput. Mech. **57**, 867–886 (2016)
12. Wollny, I., Hartung, F., Kaliske, M., Canon Falla, G., Wellner, F.: Numerical investigation of inelastic and temperature dependent layered asphalt pavements at loading by rolling tyres. Int. J. Pavement Eng. **22**, 97–117 (2021)
13. Wollny, I., Kaliske, M.: Numerical simulation of pavement structures with inelastic material behavior under rolling tyres based on an Arbitrary Lagrangian Eulerian (ALE) formulation. Road Mater. Pavement Des. **14**, 71–89 (2013)
14. Wollny, I., Hartung, F., Kaliske, M., Liu, P., Oeser, M., Wang, D., Canon Falla, G., Leischner, S., Wellner, F.: Coupling of microstructural and macrostructural computational approaches for asphalt pavements under rolling tire load. Comput.-Aided Civ. Infrastruct. Eng. **35**, 1178–1193 (2020)
15. You, T., Al-Rub, R., Darabi, M., Masad, E., Little, D.: Three-dimensional microstructural modeling of asphalt concrete using a unified viscoelastic-viscoplastic-viscodamage model. Constr. Build. Mater. **28**, 531–548 (2012)
16. Zopf, C., Garcia, M.A., Kaliske, M.: A continuum mechanical approach to model asphalt. Int. J. Pavement Eng. **16**, 105–124 (2015)

Simulation Chain: From the Material Behavior to the Thermo-Mechanical Long-Term Response of Asphalt Pavements and the Alteration of Functional Properties (Surface Drainage)

Ronny Behnke, Michael Kaliske, Barbara Schuck, Stefan Alber, Wolfram Ressel, Frohmut Wellner, Sabine Leischner, Gustavo Canon Falla, and Lutz Eckstein

Abstract In this chapter, a simulation chain is described and applied to an asphalt test track and a highway pavement structure. The simulation chain consists of different modules reaching from the experimental identification of the asphalt material, its numerical modeling on the material scale via adequate models to finally the structural scale of the pavement and the vehicle-tire system, which is numerically assessed in the framework of a coupled vehicle-tire-pavement system. Relative dynamic effects of vehicle-tire-pavement interaction have been investigated based on a multibody analysis of the vehicle driving on a rough pavement surface (external stimulus). For the pavement simulation, equivalent tire loads are used in an arbitrary Lagrangian Eulerian framework for tire and pavement. Finally, the rut formation is computed by varying different influence factors (climate temperature, vertical tire force, type of asphalt material of the surface layers etc.). With the help of the simulated deformed pavement geometry (whole service life), surface drainage characteristics are finally analyzed and assessed via a surface drainage module, e.g. to compute and predict the alteration of the pavement runoff during the service life of the pavement.

Funded by the German Research Foundation (DFG) under grants KA 1163/30, RE 1620/4, WE 1642/11, LE 3649/2 and EC 412/1.

R. Behnke (✉) · M. Kaliske
Institute for Structural Analysis, Technische Universität Dresden, Dresden, Germany
e-mail: ronny.behnke@tu-dresden.de

B. Schuck · S. Alber · W. Ressel
Institute for Road and Transport Science, Chair for Road Design and Construction, University of Stuttgart, Stuttgart, Germany

F. Wellner · S. Leischner · G. Canon Falla
Institute of Pavement Engineering, Technische Universität Dresden, Dresden, Germany

L. Eckstein
Institute for Automotive Engineering, RWTH Aachen University, Aachen, Germany

M. Kaliske et al. (eds.), *Coupled System Pavement—Tire—Vehicle*,
Lecture Notes in Applied and Computational Mechanics 96,
https://doi.org/10.1007/978-3-030-75486-0_9

Keywords Vehicle-tire-pavement interaction · FEM simulation · Rutting · Surface drainage

1 Introduction

Usually, the tire-pavement interaction is analyzed using 'static' tire loads showing no transient relative variations. Advanced coupled tire-pavement models take into account the tire passing in terms of finite element (FE) discretized structures, but still time-independent loading is assumed (no relative dynamics stemming from the suspension system or external excitation signals). The effect of moving loads and the load-induced dynamic excitation was studied e.g. in [40]. Consequences for the pavement (including dynamic effects due to moving loads) were analyzed in [23, 42]. In [38], the influence of the tire type on the rutting of the pavement was investigated. Comparison of different FE and constitutive modeling techniques for rutting prediction highlights the influence of the loading and its correct representation within numerical pavement simulations [1, 37].

For the case of steady state motion, numerical attractive analysis schemes have been presented in the past, which focus on the ground state motion (translation of tire and pavement as well as translation and rotation of the tire). In this context, the use of arbitrary Lagrangian Eulerian (ALE) techniques for the analysis of steady state rolling tires is a standard approach, see chapter "Multi-physical and Multi-scale Theoretical-numerical Modeling of Tire-Pavement Interaction". As a rule, the steady state motion analysis already captures the basic frequencies of loading and unloading of a pavement by a rolling tire with time-independent vertical load. However, transient features (variations) in the vertical force are not captured in this context. Measurement data or simulated input data from a dynamic multibody system analysis [8], focusing on a typical truck-trailer combination, reveal additional frequencies, which are induced by the relative dynamic features of the vehicle-tire-pavement system. This leads to several loading frequencies and a variation in the load amplitude, which appear in a superposed manner to the pavement loading by the rolling tire.

For a more detailed but still practicable dimensioning of new pavement structures and the study of the impact of future changes in the boundary conditions (e.g. traffic intensity, traffic type, climate change), advanced numerical analyses, e.g. a holistic approach for the vehicle-tire-pavement system [21, 46], are required and have been proposed within the FOR 2089. In this scenario, adequate numerical and experimental techniques on different length as well as time scales are combined to gain insight into the short- and long-term behavior of the whole system. Potential vehicles are numerically represented by multibody models to compute vertical forces caused by the passing of the vehicle for different load states, given surface roughness and health state of the pavement. With the help of the load information, the tire-pavement interaction is studied in an FE discretized configuration of a tire and a pavement part. With the help of time homogenization, the long-term behavior of the pavement system is computed and objective quantities, such as the final rut depth, are predicted. The

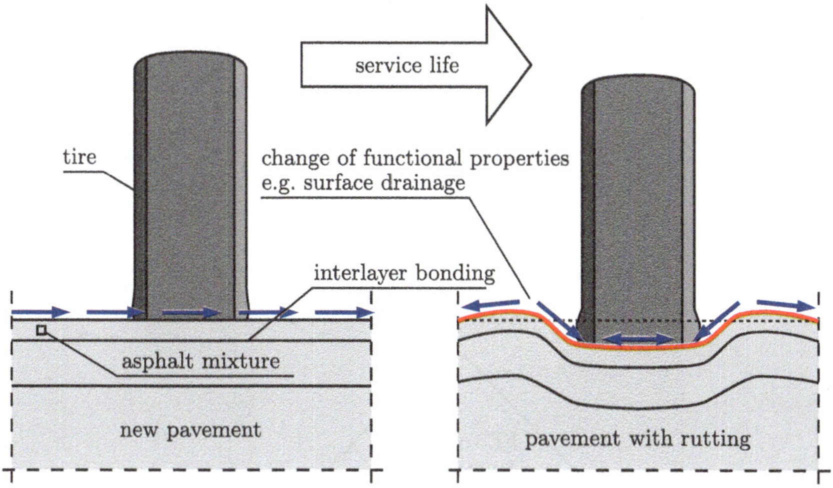

Fig. 1 Evolution of the cross-section (pavement structure) during its service life

numerical procedure is described in detail in [7]. By further analysis steps, functional properties (e.g. surface drainage) of the pavement can be assessed and predicted for the whole service life of the pavement [3].

Outline. In this contribution, a simulation chain for the modeling of the material behavior, the thermo-mechanical long-term response of asphalt pavements and the alteration of functional properties in terms of surface drainage is set up, see Fig. 1. The modules of the simulation chain are introduced in Sect. 2. In Sect. 3, the simulation chain in the form of underlying numerical models is used to carry out a sensitivity analysis of two different asphalt pavement structures (an asphalt pavement test track and a highway pavement of load class Bk100 according to RStO). Different influence factors (climate temperature, vertical tire force, type of asphalt material of the surface layers etc.) are considered and alteration of the pavement runoff during the service life of the pavement is analyzed. Results are presented in Sect. 4 together with a discussion.

2 Computation Procedure

In this section, the proposed computation procedure of the simulation chain (Fig. 2) is presented and basic quantities are introduced. The computation procedure consists of an FE based representation of tire and pavement as introduced in chapter "Multi-physical and Multi-scale Theoretical-numerical Modeling of Tire-Pavement Interaction". Input quantities of the tire-pavement model are vertical tire forces, a time function of the pavement surface temperature distribution, driving speed and

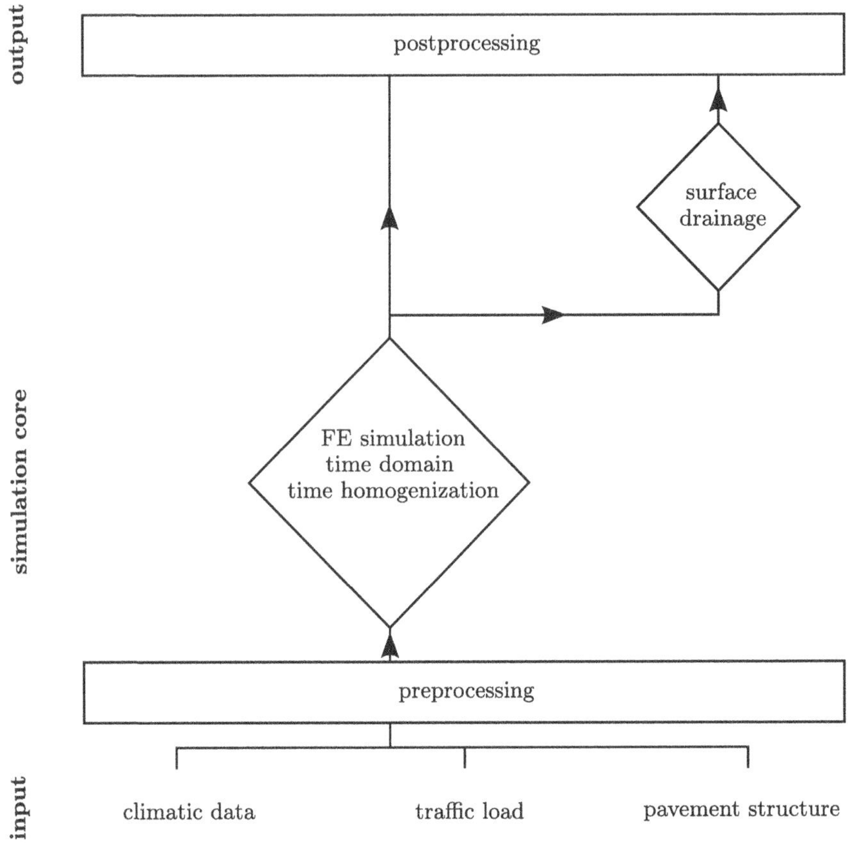

Fig. 2 Overview of the computation procedure (simulation chain)

sequence of tire loads (traffic data) and the type of the pavement structure (geometry and material information). Its short-term behavior is assessed e.g. in [48, 49]. For the computation of the long-term evolution, time homogenization is used in combination with a reference cross-section of the considered pavement [7]. The further investigation focuses on the long-term behavior and the sensitivity analysis of the pavement with respect to its long-term features in terms of rut formation caused by repetitive tire loadings on the pavement. Via an assessment of the computed deformed surface of the pavement over its service life, additional analyses, e.g. in terms of its surface drainage behavior, can be carried out [3]. Further details are provided in several literature references for each subpart of the procedure.

2.1 Climatic Data

For a given pavement cross-section of the road network, measured or standardized climatic data (temperature-time function) is used for the analysis, e.g. to realistically capture the temperature-dependent material behavior (asphalt) and the influence of the temperature on the final structural response of the pavement.

For the subsequent analysis, the climatic data (temperature-time function) has been generated from the relative frequency of the pavement surface temperatures provided in [14]. For the numerical analysis, the relative frequency has been transferred to a function of time with a periodicity of days and years, see Fig. 3. Note that an additional global temperature trend (e.g. global warming) can be superposed as well. For more details, the reader is referred to [7].

2.2 Traffic Load

In the following, relative dynamic effects are computed and assessed via a multibody model of a vehicle in order to subsequently study the effective load transmission to the pavement. Usually, dynamic multibody analyses are carried out on quarter-truck, half-truck or full-truck models of the real vehicle driving on a smooth or rough rigid pavement surface. In these classical models, the geometry and shape of the tire is often neglected by using single-contact-point tire models as approximation of the real tire.

Within the FOR 2089, dynamic tire forces have been computed for a truck-trailer combination from a dynamic multibody analysis of the full vehicle model depicted in Figs. 4 and 5. The full vehicle simulation includes spatially resolved physical sub-models for the representation of the spatially discretized tire. The multibody model of the vehicle including suspension system and tires has been built up from several

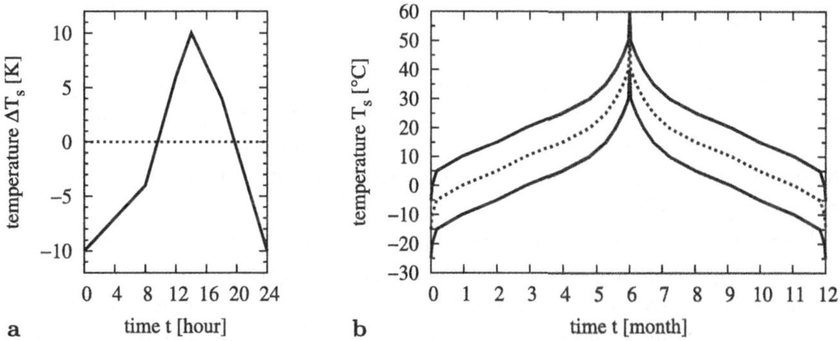

Fig. 3 Variation of pavement surface temperature (prescribed boundary condition) of climate Zone III according to the German analytical pavement design guideline RDO Asphalt [14]: **a** day, **b** year

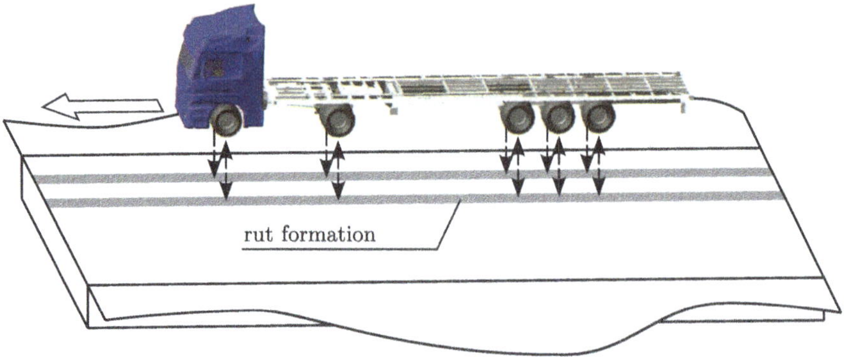

Fig. 4 Vehicle-tire-pavement system with multibody dynamics model of the vehicle

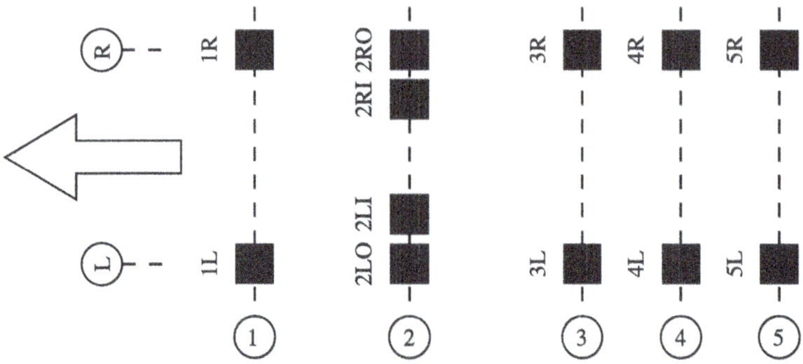

Fig. 5 Tire positions for truck-trailer combination (see Fig. 4): L—left, R—right, 1–5 axles of the truck/trailer

physical submodels (e.g. FTire, CDtire or RMOD-K for the discretized tire). Within the multibody analysis, the rigid and rough pavement surface shows a geometrical micro- and macro-texture. The correct description of the pavement surface is of high importance for the relative dynamic effects of the vehicle-tire-pavement system since the texture of the pavement acts as main external stimulus for vertical vehicle motion during driving [18]. The relative dynamic motion of the vehicle results in a transient load transmission of vehicle loads to the pavement. The load acts on the microscale (local deterioration of single components of the asphalt layer due to stress peaks in the tire-pavement contact area) and on the macroscale (global fatigue and rutting of the asphalt layers).

For the numerical simulation on the vehicle scale (spatial resolution of tire, vehicle, suspension system as discretized bodies), a virtual test rig for vehicle testing has been set-up and allows the study of the influence of high-resolution pavement texture (high level of detailed information) on the simulated vertical driving behavior of e.g. truck-trailer combinations. Further details are provided in [44].

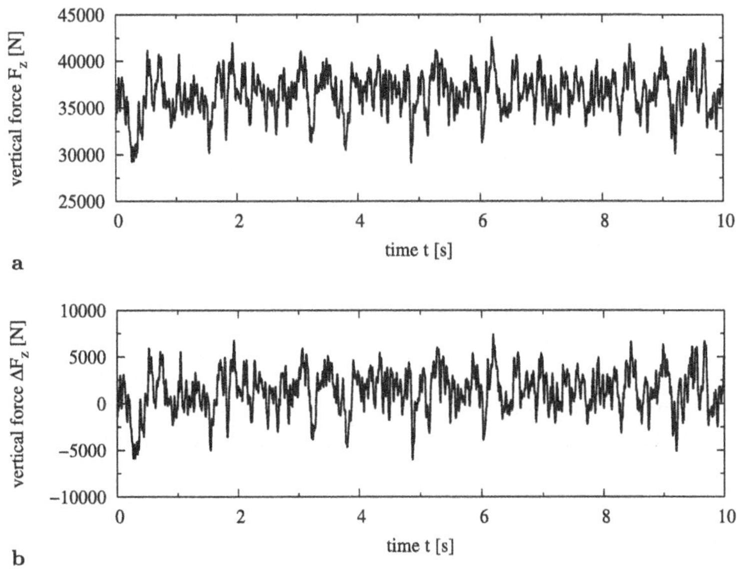

Fig. 6 Force signal of tire position 1L (see Fig. 5): **a** force $F_z(t)$, **b** relative dynamic part $\Delta F_z(t)$

For the multibody analysis described in the following, the pavement texture has been recorded by contactless 3D digital stripe projection with a vertical resolution up to 4 μm. Based on a MATLAB algorithm, the pavement surface is reconstructed and saved in the form of a supported data format for multibody analyses, see chapter "Experimental and Simulative Methods for the Analysis of Vehicle-Tire-Pavement Interaction". As data format, the open source format OpenCRG (curved regular grid) has been used. According to the CRG data, the pavement texture can be outputted as a function of the longitudinal coordinates of the pavement. For the present analysis, the micro- and macro-texture of the pavement represents the surface of a German motorway (Bundesautobahn—BAB) with micro-rough characteristics. In addition to these micro-rough characteristics, a topology of a real BAB is superposed on the macroscale. The topology data has been measured by a 3D laser measurement system and corresponds to the longitudinal direction of a real BAB (A61 Geilenkirchen, Germany) with wave lengths > 10 cm and height differences up to 8.9 cm (which explains the occurrence of relatively significant dynamic load coefficients).

The afore-described surface configuration in combination with the truck driving speed ($v = 80$ km/h) leads to a high frequency excitation of the tires and axles of the truck-trailer combination (fully loaded) during straight line driving. During the numerical multibody analysis, vertical tire forces are recorded for the tire positions given in Fig. 5 for a characteristic time period $T = 10$ s of the stationary/stabilized stochastic force signals with a characteristic amplitude-frequency spectrum. In Fig. 6, the absolute vertical force for the tire position 1L and its relative dynamic part are exemplarily plotted as a function of the characteristic time T.

The computation of the corresponding dynamic load coefficients (DLC) allows to assess the relative dynamic part of the tire forces. Values for the DLC can be obtained from on-site measurements or simulations [9, 22, 26]. Influencing quantities for the relative dynamic part are the behavior of the driver or autonomous pilot functions of the vehicle, traffic flow and speed, tire and pavement damage/usage, surface profile of the pavement, change of eigenfrequencies of the tire-suspension system due to alterations or active control of the tire-suspension system etc. Previous studies showed the sensitivity of the DLC with respect to static tire load [12], tire type [33, 43], tire inflation pressure, vehicle speed, type of tire suspension system, air flow around the tire (longitudinal), imperfections of the wheel/tire, pavement roughness and pavement damage conditions [41] to mention only a few of them. In general, each component of the vehicle-tire-pavement system contributes with its changing properties over time (alteration, ageing) to the final dynamic tire load and, hence, the current DLC. In this context, even the dynamic properties of the pavement (open half space) play a role and might become significant [27].

From the force-time signal for each tire position (Fig. 5), the associated DLC is computed by

$$\Delta^*_{\mathrm{DLC}} = \frac{\Delta F_z}{F_{z,\mathrm{stat}}} \tag{1}$$

with the maximum variation of the dynamic vertical force

$$\Delta F_z = \max \left| F_z(t) - F_{z,\mathrm{stat}} \right| , \quad t \in [t_0; t_{\mathrm{end}}] . \tag{2}$$

The computed values are provided in Table 2. In previous studies and construction standards [11], a range of $\Delta_{\mathrm{DLC}} \in [0.05; 0.30]$ has been revealed from experimental and numerical studies. In the ideal case of no external excitation and zero eigenoscillations, the DLC would be zero. For the example considered, the static forces are given in Table 1 and the DLC according to Eq. (1) are summarized in Table 2.

Besides the DLC, the sequence of different tires (axles) is important. For rutting predictions, not only the absolute value, but also the application duration (loading

Table 1 Static forces $F_{z,\mathrm{stat}}$ for each tire position (see Fig. 5), total static force $F_{z,\mathrm{stat,tot}} =$ 374 226 N

Axle	$F_{z,\mathrm{stat}}$ [N]			
	L			R
	LO	LI	RI	RO
1	0.3519E+05			0.3624E+05
2	0.2793E+05	0.2820E+05	0.2716E+05	0.2750E+05
3	0.3218E+05			0.3226E+05
4	0.3197E+05			0.3197E+05
5	0.3182E+05			0.3182E+05

Table 2 Dynamic load coefficient from total dynamic variation according to Eq. (1) for each tire position (see Fig. 5)

Axle	Δ^*_{DLC}			
	L			R
	LO	LI	RI	RO
1	0.210			0.192
2	0.185	0.206	0.246	0.228
3	0.165			0.195
4	0.185			0.179
5	0.176			0.176

rate) and the frequency spectrum of the loading are important. Furthermore, different load groups might occur. Here, relative dynamic variations per axle and relative dynamics per vehicle (i.e. the axle load of one load group is linked with each other) occur, which result in coupled and uncoupled load groups. To standardize this effect, equivalent standard axle loads (ESAL) are commonly used in order to transform the transient load signal to a periodic one.

With the help of ESAL, short-term processes can be linked to long-term processes of rutting and damage formation of the pavement. The real loading is characterized by e.g. vehicle type, number of vehicles, distance of vehicles, traffic hours etc. for which an equivalent or representative loading (load pattern) has to be derived. With the help of the DLC values, a static design with dynamic correction can be carried out. However, the derivation is not trivial and mostly depends on experience and empirically motivated equations.

2.3 Tire-Pavement Model

In the following, only the tire-pavement subsystem is considered and the dynamic design tire load (e.g. obtained from the multibody vehicle-tire-pavement model) is assumed as constant over time, see Fig. 7. The tire-pavement subsystem is analyzed using the framework described in chapter "Multi-physical and Multi-scale Theoretical-numerical Modeling of Tire-Pavement Interaction", see also [7].

For the steady state motion problem, the ALE framework is employed for both, tire and pavement, see Fig. 7. The FE model consists of an FE discretized truck trailer tire of type 385 65 R22.5 and the pavement structure of loading class Bk100 for high traffic, see Fig. 8. The tire is rolling on a representative pavement part (width 2 m for a driving lane of 4 m) at constant speed of 80 km/h. The tire pressure is 7 bar and as vertical load, 50 kN (mass 5 t) are considered for a tire. Corresponding to the loading class Bk100, $32 \cdot 10^6$ load cycles (32 Million 10-t-ESAL) are considered for the 30 years of the service life of the pavement. The computation procedure is described

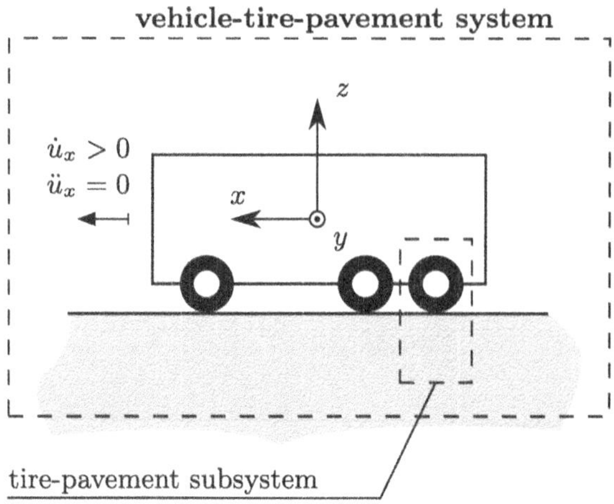

Fig. 7 Vehicle-tire-pavement system and tire-pavement system as subsystem

in chapter "Multi-physical and Multi-scale Theoretical-numerical Modeling of Tire-Pavement Interaction" and additional details on the theory can be found in [7]. The underlying computational models are provided in [7, 47]. The experimental characterization of the asphalt mixes is available in chapter "Experimental Methods for the Mechanical Characterization of Asphalt Concrete at Different Length Scales: Bitumen, Mastic, Mortar and Asphalt Mixture" as well as in [6].

2.4 Submodules for Functional Properties of the Pavement

Besides the mechanical analysis (see Sect. 2.3), further modules are available to assess e.g. the alteration of functional properties of the pavement during its service life. In this way, the surface drainage of the pavement has been captured as a function of time, see chapter "Computational Methods for Analyses of Different Functional Properties of Pavements" and [3]. A combination with other submodels or modules is possible.

3 Sensitivity Analysis

In Fig. 2, the overview of the sensitivity analysis is outlined. Based on the afore-described numerical FE discretized framework, several numerical simulations have been carried out by varying input factors and extracting the resulting sensitivity of the objective quantities which are rut depth (absolute value of difference between maximum and minimum of the deformed pavement surface – rut) and surface drainage characteristics.

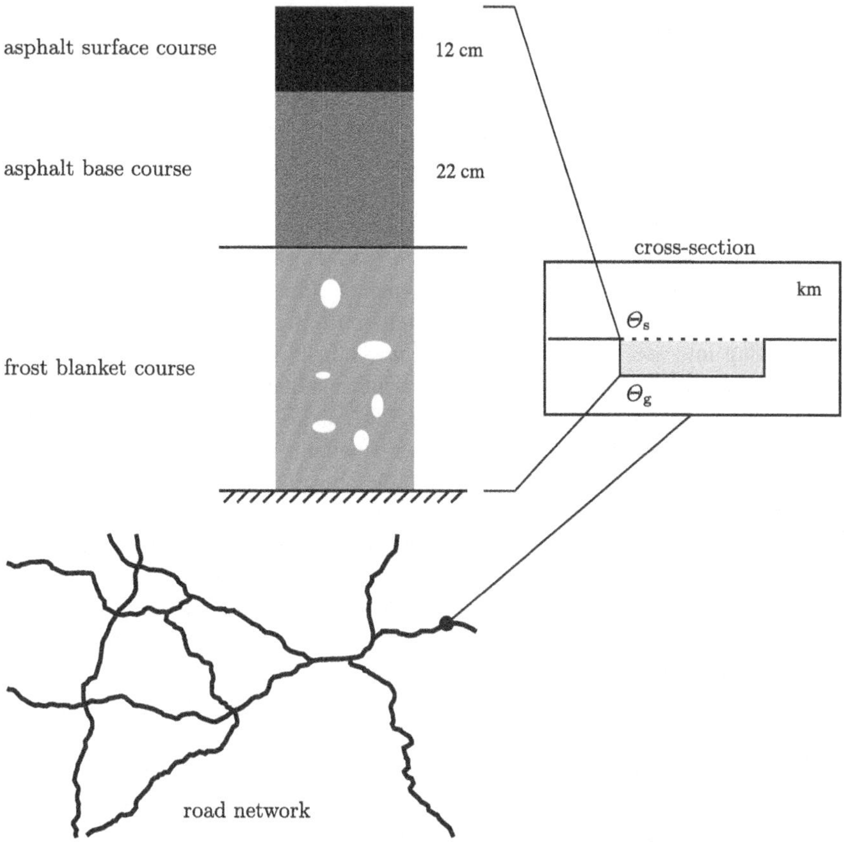

Fig. 8 Cross-section (pavement structure) of the road network (highway) to be analyzed (ground temperature Θ_g and pavement surface temperature Θ_s)

In the literature, several case studies and sensitivity analyses regarding long-term performance and structural behavior are available [5, 15, 17, 19, 24, 29, 31, 32, 34, 36]. In these studies, the sensitivity has been assessed based on different approaches. First, influence quantities have been obtained by exploring design software tools available for the structural design of pavement structures based on the current construction standards [15] or exploring the mechanistic-empirical design guides directly [25, 50, 51]. Other approaches use simplified or complex FE based models as underlying basic element for studying variations in the structural response by numerical simulations [30]. In this context, the influence of the tire-pavement contact stress [16], the material characteristics [10] and the structural design of the layers [39] play an important role.

3.1 Model Reduction

For the analysis carried out in the following, a model reduction from the 3D tire-pavement system to the reference cross-section of the pavement has been applied and an equivalent loading F_{eq} induced by the tire has been derived within 5 steps, see the procedure depicted in Fig. 9.

In case of repeated passings of tires on a pavement section with the same or a different tire type, additional variations in the load pattern (axles) have to be considered for the pavement (Step 1). Relative dynamic effects of the dynamic tire loads (F_{dyn}) can be represented via a constant average tire load F_{mean} and corresponding DLC taking into account the relative dynamic effects (Steps 2 and 3) in terms of the constant force F_{design}. The problem of various loading types and sequences (load groups) is usually overcome by introducing ESTL and EASL values, where the latter can also directly be taken from construction standards or measurement data (Step 4). For the definition of the equivalent load, the objective quantity has to be known, i.e. the equivalent load quantity might not be valid for other objective quantities or changes in the boundary conditions of the investigated system. Furthermore, the transformation to ESTL values per time period has to take into account the influence of load pauses (e.g. distance L of the load groups), i.e. the sensitivity of the pavement material with respect to loading and recovery times has to be incorporated in this approach (Step 5). Different approaches to compute ESTL values have been proposed and discussed in the literature, see e.g. [20] based on fatigue criteria and the number of tires per axle [4].

With the help of the ESTL, a design or validation of the fatigue life of the pavement can be normally conducted in terms of a standard validation protocol. This scenario represents a transformation of the load signal to a uniform, homogenized load signal (Step 5) of the pavement for which experimental, numerical or semi-empirical results of the objective quantity (e.g. rut depth, number of induced cracks) are known. For a more detailed study taking into account subpopulations and types as well as the real number of load cycles of each subgroup, a large database with information on tire population, tire loads and cargo loads of vehicles has to be available.

The reduced model obtained by these five steps has been validated by focusing on the difference between the outcome of the dynamic rolling simulation [7] and the outcome of the simulation using the reduced model.

3.2 Objective Quantities

As objective quantities, the rut depth (pavement surface) at the end of the service life (30 years) of the pavement structure is considered. Based on the deformed surface of the asphalt pavement, an analysis with respect to the alteration of the surface drainage has been carried out.

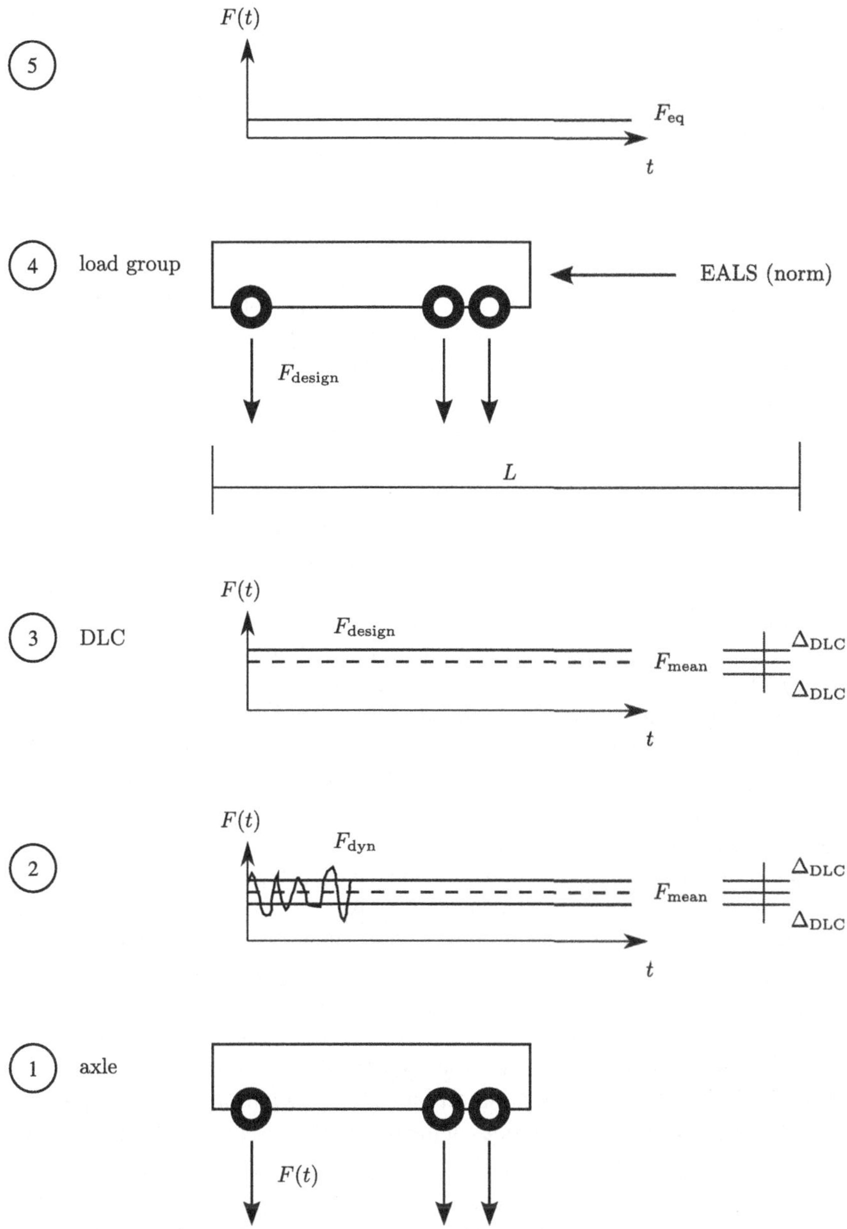

Fig. 9 Design value of traffic load

4 Results and Discussion

4.1 Surface Drainage Properties of an Asphalt Test Track

For this example, the calculated surface deformations (single rut) of a thin asphalt layer on a deformable subgrade layer (asphalt pavement test track, see chapter "Multi-physical and Multi-scale Theoretical-numerical Modeling of Tire-Pavement Interaction") have been used as input.

Subsequently, the pavement surface of a single lane with a width of 3.5 m and length of 10 m with varying cross and longitudinal slopes has been generated computationally. Two ruts with identical geometry were then positioned in a distance of 1.5 m on the pavement surface to represent the track width of a vehicle (see Fig. 10a). Ruts cause bulges above the normal level of the pavement surface, as can be seen in Fig. 1. These raised edges were set to linearly decrease to normal level in 30 cm. The bulges cause a further deformation of the pavement section next to the ruts and cause a local depression in the pavement middle. Along with the ruts, this local depression affects drainage and can lead to water accumulation in these surface areas (see also the vertically scaled pavement surface in Fig. 10b).

To model the impact of rut formation over the service life of the pavement, the generated pavement surfaces can be imported into a drainage model, the Pavement Surface Drainage Model (PSRM) [45] introduced in chapter "Computational Methods for Analyses of Different Functional Properties of Pavements". The PSRM, a finite-volume model for dense pavement surface runoff using the depth-averaged shallow water equations, can simulate the whole drainage process and resulting water depths over nearly arbitrary pavement surfaces. The resulting water depths represent the water level above the texture peaks.

Table 3 shows the parameters imported into the PSRM to simulate a representative pavement surface. The chosen rain intensity is taken from statistical rain analyses in Germany [28] and represents a (short) heavy rainfall with a duration of 15 min and a return period of 1 year. The pavement age and, thus, the years of rut formation under traffic and environmental load are simulated over the course of 30 years.

Fig. 11 shows the water depths along a cross-section of the pavement width simulated with the input data in Table 3 and a rut formation of 30 years (this also corresponds to Fig. 12d). The ruts cause very high water depths inside of the ruts, here up to 6.5 mm. The bulges on the sides of the rut drain the water into the local depres-

Fig. 10 Rut deformation of a pavement surface over the service life: **a** unscaled overview, **b** 10x vertically scaled

Table 3 PSRM input parameters for a representative pavement surface

Rain intensity	i	[mm/min]	0.75
Rain duration	D	[s]	900
Mean texture depth	MTD	[mm]	0.4
Cross slope	c	[%]	2.5
Longitudinal slope	s	[%]	1.0
Pavement age	n	[years]	1–30

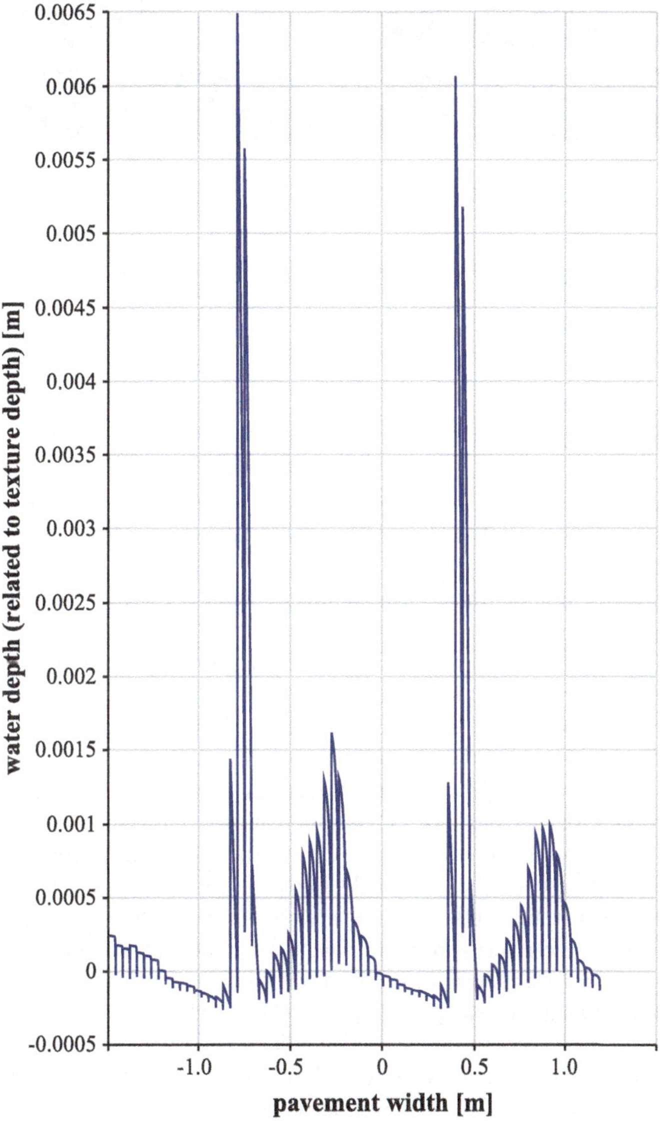

Fig. 11 Water depth along the pavement width with deep ruts for $n = 30$ years

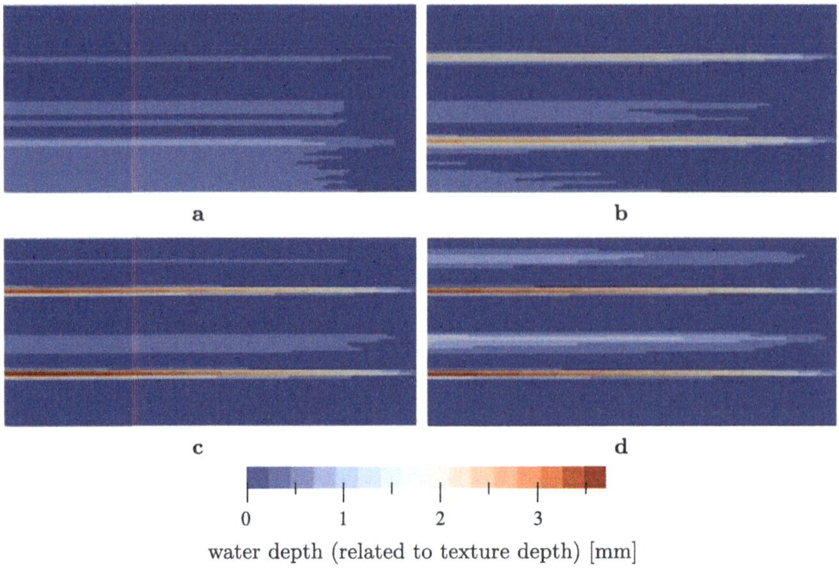

Fig. 12 Effects of rut formation over the simulated service life (1 − 30 years) on pavement surface drainage, see [3]: with rut formation of **a** 1 year, **b** 5 years, **c** 15 years, **d** 30 years

sions in the middle of the pavement and on the sides. Here, also high water depths of over 1.5 mm are reached that are close to the critical level of 2 mm still assumed for German highways [13]. The bulges are not only depressions that fill with water in drainage, but also retard the water flow along the cross-section and, thus, reduce the drainage capacity.

Rutting leads to impairment of the drainage function early in the service life of the pavement. As can be seen in Fig. 12, the water film depths inside the ruts are significant even after 5 years and do not significantly increase in the last 15 years of the pavement's service life (for input parameters see Table 3). The scale of Fig. 12 is adjusted to lower water depth levels, so that effects outside of the rut can be visualized, while Fig. 11 shows the full values achieved in the simulation.

Figure 13 compares four pavement surfaces after 30 years of rut formation and the same values for rain intensity, rain duration and mean textures depth as depicted in Table 3. The cross slope was chosen as 2.5% and 5.0% with a longitudinal slope of 1.0% and 4.0%. As can be seen easily, higher longitudinal slopes increase runoff and lead to lower water depths inside the two ruts. The influence of the depression in the middle of the road is more dominant with lower cross slopes, as here the raised edges of the ruts cause a local backwater effect. The influence of pavement parameters on drainage capacity is often interrelated and ambiguous. As an example: for cross slope $c = 2.5\%$, the area in the middle of the pavement shows lower water depth and a smaller affected area size with increasing longitudinal slope (Fig. 13a and b). A cross slope of 5.0%, however, leads to similar water depths and a smaller

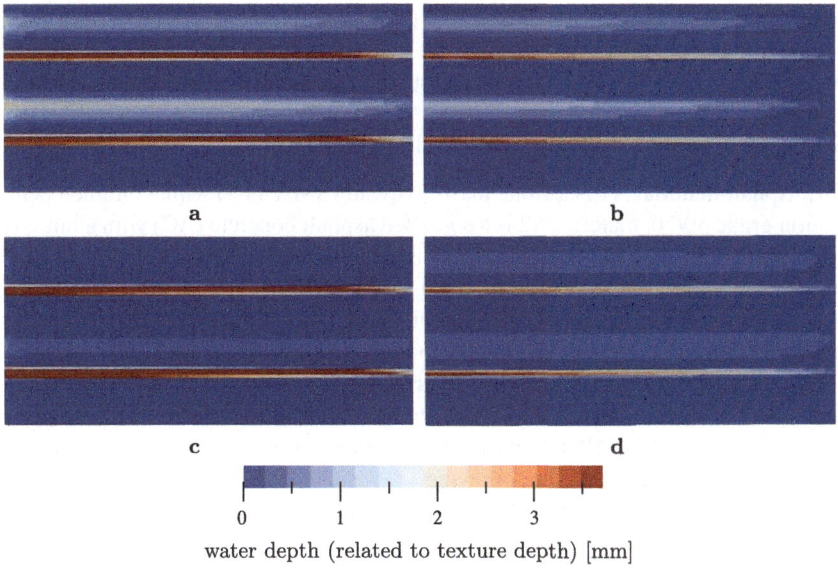

Fig. 13 Effects of cross and longitudinal slope on pavement surface drainage after 30 years of rutting, see [3]: with **a** $c = 2.5\% / s = 1.0\%$, **b** $c = 2.5\% / s = 4.0\%$, **c** $c = 5.0\% / s = 1.0\%$, **d** $c = 5.0\% / s = 4.0\%$

area size for the rise in longitudinal slope (Fig. 13c and d). Higher longitudinal slopes lead to an increase of flow path length. The influence of this length depends on the cross slope, thus, this could explain this effect.

A more detailed analysis can be found in [3]: This study takes the interaction of pavement parameters (rutting, cross and longitudinal slope, mean texture depth, pavement age) and rain parameters (intensity and duration) into account. Further information on how these interrelated parameters influence drainage can also be found in e.g. [2, 35].

4.2 Rut Depth Variations for Bk100

In the following, the Bk100 pavement structure (see Sect. 2.3) as a relevant pavement structure for German motorways is considered during a sensitivity analysis with respect to influence quantities regarding the rut depth at the end of the service life (30 years). As influence quantities, the following input geometry and model parameters are used:

- vertical tire design load F_{eq} including load pauses and number of 10-t-ESAL (10 t) for Bk100,

- surface temperature of the pavement in terms of different local temperature zones according to [14],
- interlayer bonding behavior between asphalt layers assumed as perfectly rigid (1) or without bonding (0),
- type of the asphalt surface layer (asphalt materials A1 to A3, see [6]).

The asphalt material A1 is a stone mastic asphalt (SMA 11 S) with a bitumen penetration grade 50/70, material A2 is a so-called asphalt concrete (AC) with a bitumen penetration grade 50/70 and material A3 is a stone mastic asphalt (SMA 11 S) with polymer-modified binder. These materials are typical asphalt mixtures commonly used in Germany for asphalt surface layers. More information on the asphalt mixtures and their numerical modeling (temperature-dependent continuum mechanical formulation with volume-preserving deformations) are provided in [6].

The frost blanket course is assumed to have a linear elastic material behavior (layer stiffness of 100 MPa). Inelastic deformations of the frost protection layer and the ground are neglected in the following.

Subsequently, the results of the analysis with respect to the influence quantities are provided in Tables 4, 5 and 6.

Table 4 Results: Study of the influence of different temperature zones [14] on rut depth after 30 years

Temperature zone	Material (layer bonding)	Rut depth [cm]	Variation regarding A1 (1) (Zone I) [%]
Zone I	A1 (1)	2.693	0
	A1 (0)	2.757	+2.38
	A2 (1)	5.879	+118.31
	A2 (0)	5.945	+120.77
	A3 (1)	3.964	+47.20
	A3 (0)	4.032	+49.72
Zone II	A1 (1)	2.687	−0.22
	A1 (0)	2.757	+2.38
	A2 (1)	5.868	+117.90
	A2 (0)	5.931	+120.24
	A3 (1)	3.956	+46.90
	A3 (0)	4.022	+49.35
Zone III	A1 (1)	2.672	−0.78
	A1 (0)	2.740	+1.75
	A2 (1)	5.838	+116.78
	A2 (0)	5.900	+119.09
	A3 (1)	3.935	+46.12
	A3 (0)	3.999	+48.50

Table 5 Results: Influence of global temperature trend on rut depth after 30 years

Temperature zone	Material (layer bonding)	Global temperature trend [K/year]	Rut depth [cm]	Variation regarding A1 (1) (Zone I), Table 4 [%]
Zone I	A1 (1)	+0.1	2.723	+1.11
	A1 (1)	−0.1	2.659	−1.26
	A1 (0)	+0.1	2.792	+3.68
	A1 (0)	−0.1	2.725	+1.19

Table 6 Results: Influence of vertical load on rut depth after 30 years

Temperature zone	Material (layer bonding)	Variation of vertical load [%]	Rut depth [cm]	Variation regarding A1 (1) (Zone I), Table 4 [%]
Zone I	A1 (1)	+10	2.982	+10.73
	A1 (1)	−10	2.462	−8.58
	A1 (0)	+10	3.054	+13.41
	A1 (0)	−10	2.527	−6.16

In Table 4, the influence of the temperature zones according to [14] (local climate and variation of the pavement surface temperature) is provided. The best resistance with respect to rutting is obtained for material A1 while material A2 shows an increase up to 120.8% in rut depth due to the lower material resistance on the material scale. In average, the loss of interlayer bonding results in an increase of about 2% in rut depth at the end of the service life of the pavement structure considered in this example. Note that in the present case, mix rutting predominates since the material of the frost protection layer is assumed as linear elastic leading to no additional inelastic deformation of the subgrade soil in the long term (e.g. as in the case of subgrade rutting).

In Table 5, the influence of a global, linear temperature trend on the rut depth is highlighted. Here, an increase or decrease by 0.1 K/year is considered, leading to an global increase or decrease of the annual average temperature by 3 K at the end of 30 years. For the investigated pavement structure (A1, Zone I, according to the German analytical pavement design guideline RDO Asphalt [14]), the variation is about 1%. Note that especially due to the highly nonlinear temperature dependency of the asphalt material, this relation will also be highly nonlinear, e.g. leading to larger rut depths for increasing global warming.

In Table 6, the influence of the vertical load F_{eq} is investigated. Here, a variation of 10% is assumed and the resulting change of the ruth depth is computed for the pavement structure (A1, Zone I). From the analysis, an increase by 10% and a decrease by 8% (in average) can be observed, respectively. Hence, the relation is slightly nonlinearly correlated and this might drastically change for higher temperatures (e.g. global warming), where the nonlinearity of the asphalt material becomes even more pronounced.

5 Conclusions

In this chapter, a simulation chain reaching from the material behavior to the thermo-mechanical long-term response of asphalt pavements and the alteration of functional properties (e.g. surface drainage) has been set up from the submodels and experimental data obtained out of the previous chapters. Finally, the rut formation has been computed by varying different influence factors (climate temperature, vertical tire force, type of asphalt material of the surface layers etc.). With the help of the simulated deformed pavement geometry (whole service life), its surface drainage characteristics have been computed based on the outcome of the tire-pavement analysis (deformed pavement surface).

Future aspects of the methodology presented might concentrate on the incorporation of uncertainty measures and their quantification within a polymorphic uncertainty analysis. Furthermore, alteration of the material (ageing and healing [52], crack formation or structural failure) have not yet been taken into account and might be incorporated into the present analysis by further submodels. A combination of the ansatz with structural health monitoring (SHM) for a permanent investigation of the pavement's service life would also be meaningful.

Acknowledgements The authors thank Kalidhasan Rajendran and Qian Xu for the data assessment and preparation of parts of the raw data.

References

1. Al-Rub, R., Darabi, M., Huang, C.W., Masad, E., Little, D.: Comparing finite element and constitutive modelling techniques for predicting rutting of asphalt pavements. Int. J. Pavement Eng. **13**, 322–338 (2012)
2. Alber, S., Schuck, B., Ressel, W.: Importance of pavement drainage and different approaches of modelling. In: Chen, X., Yang, J., Oeser, M., Wang, H. (eds.) Functional Pavements. pp. 403–406. CRC Press, Taylor & Francis Group, London (2021)
3. Alber, S., Schuck, B., Ressel, W., Behnke, R., Canon Falla, G., Kaliske, M., Leischner, S., Wellner, F.: Modeling of surface drainage during the service life of asphalt pavements showing long-term rutting: a modular hydro-mechanical approach. Adv. Mater. Sci. Eng. **2020**, 8793652 (2020)
4. Amorim, S., Pais, J., Vale, A., Minhoto, M.: A model for equivalent axle load factors. Int. J. Pavement Eng. **16**, 881–893 (2015)
5. Anderson, E., Daniel, J.: Long-term performance of pavement with high recycled asphalt content: case studies. Transp. Res. Rec. **2371**, 1–12 (2013)

6. Behnke, R., Canon Falla, G., Leischner, S., Händel, T., Wellner, F., Kaliske, M.: A continuum mechanical model for asphalt based on the particle size distribution: numerical formulation for large deformations and experimental validation. Mech. Mater. **153**, 103703 (2021)
7. Behnke, R., Wollny, I., Hartung, F., Kaliske, M.: Thermo-mechanical finite element prediction of the structural long-term response of asphalt pavements subjected to periodic traffic load: tire-pavement interaction and rutting. Comput. Struct. **218**, 9–31 (2019)
8. Blundell, M., Harty, D. (eds.): The Multibody Systems Approach to Vehicle Dynamics. Butterworth-Heinemann, Oxford (2015)
9. Buhari, R., Rohani, M., Abdullah, M.: Dynamic load coefficient of tyre forces from truck axles. Appl. Mech. Mater. **405–408**, 1900–1911 (2013)
10. Chen, E., Li, K., Wang, Y.: Influence of material characteristics of asphalt pavement to thermal stress. Appl. Mech. Mater. **256–259**, 1769–1775 (2013)
11. Das, A.: Structural design of asphalt pavements: principles and practices in various design guidelines. Transp. Dev. Econ. **1**, 25–32 (2015)
12. Dong, Z.: Study on the dynamic wheel load of multi-axle vehicle based on distribute loading weight. Adv. Mater. Res. **159**, 35–40 (2011)
13. Forschungsgesellschaft für Strassen- und Verkehrswesen (FGSV) (ed.): Richtlinien für die Anlage von Autobahnen (RAA). FGSV-Verlag, Köln (2008)
14. Forschungsgesellschaft für Straßen- und Verkehrswesen (FGSV), Arbeitsgruppe Infrastrukturmanagement (ed.): Richtlinien für die rechnerische Dimensionierung des Oberbaus von Verkehrsflächen mit Asphaltdeckschicht—RDO Asphalt 09. FGSV, Köln (2009)
15. Guclu, A., Ceylan, H., Gopalakrishnan, K., Kim, S.: Sensitivity analysis of rigid pavement systems using the mechanistic-empirical design guide software. J. Transp. Eng. **135**, 555–562 (2009)
16. Guo, M., Zhou, X.: Tire-pavement contact stress characteristics and critical slip ratio at multiple working conditions. Adv. Mater. Sci. Eng. **2019**, 5178516 (2019)
17. Guo, R., Nian, T., Zhou, F.: Analysis of factors that influence anti-rutting performance of asphalt pavement. Constr. Build. Mater. **254**, 119237 (2020)
18. Hu, L., Yun, D., Liu, Z., Du, S., Zhang, Z., Bao, Y.: Effect of three-dimensional macrotexture characteristics on dynamic frictional coefficient of asphalt pavement surface. Constr. Build. Mater. **126**, 720–729 (2016)
19. Ioannides, A., Tallapragada, P.: An overview and a case study of pavement performance prediction. Int. J. Pavement Eng. **14**, 629–644 (2013)
20. Judycki, J.: Determination of equivalent axle load factors on the basis of fatigue criteria for flexible and semi-rigid pavements. Road Mater. Pavement Des. **11**, 187–202 (2010)
21. Kaliske, M., Wollny, I., Behnke, R., Zopf, C.: Holistic analysis of the coupled vehicle-tire-pavement system for the design of durable pavements. Tire Sci. Technol. **43**, 86–116 (2015)
22. Kulakowski, B., Streit, D., Wollyung, R., Kenis, W.: A study of dynamic wheel loads conducted using a four-post road simulator. Road Transp. Technol. **4**, 301–307 (1995)
23. Lak, M.: Numerical prediction of ground vibrations generated by road traffic and pavement breaking. Ph.D. thesis, KU Leuven, Leuven (2013)
24. Lee, J.H., Baek, S.B., Lee, K.H., Kim, J.S., Jeong, J.H.: Long-term performance of fiber-grid-reinforced asphalt overlay pavements: a case study of Korean national highways. J. Traffic Transp. Eng. **6**, 366–382 (2019)
25. Li, X., Zhang, R., Zhao, X., Wang, H.: Sensitivity analysis of flexible pavement parameters by mechanistic-empirical design guide. Appl. Mech. Mater. **590**, 539–545 (2014)
26. Lu, Y., Yang, S., Li, S., Chen, L.: Numerical and experimental investigation on stochastic dynamic load of a heavy duty vehicle. Appl. Math. Model. **34**, 2698–2710 (2010)
27. Madsen, S.: Dynamic modeling of pavements with application to deflection measurements. Ph.D. thesis, Technical University of Denmark, Lyngby (2016)
28. Malitz, G., Ertel, H.: KOSTRA-DWD-2010 Starkniederschlagshöhen für Deutschland (Bezugszeitraum 1951 bis 2010). Deutscher Wetterdienst, Offenbach (2015)
29. McDonald, M., Madanat, S.: Life-cycle cost minimization and sensitivity analysis for mechanistic-empirical pavement design. J. Transp. Eng. **138**, 706–713 (2012)

30. Mwanza, A., Muya, M., Hao, P.: Towards modeling rutting for asphalt pavements in hot climates. J. Civ. Eng. Arch. **10**, 1075–1084 (2016)
31. Nazzal, M., Iqbal, M., Kim, S., Abbas, A., Akentuna, M., Quasem, T.: Evaluation of the long-term performance and life cycle costs of GTR asphalt pavements. Constr. Build. Mater. **114**, 261–268 (2016)
32. Norouzi, A., Kim, D., Kim, Y.: Numerical evaluation of pavement design parameters for the fatigue cracking and rutting performance of asphalt pavements. Mater. Struct. **49**, 3619–3634 (2016)
33. Park, D.W.: Prediction of pavement fatigue and rutting life using different tire types. KSCE J. Civ. Eng. **12**, 297–303 (2008)
34. Ranadive, M., Tapase, A.: Parameter sensitive analysis of flexible pavement. Int. J. Pavement Res. Technol. **9**, 466–472 (2016)
35. Ressel, W., Wolff, A., Alber, S., Rucker, I.: Modelling and simulation of pavement drainage. Int. J. Pavement Eng. **20**, 801–810 (2019)
36. Schwartz, C., Li, R., Ceylan, H., Kim, S., Gopalakrishnan, K.: Global sensitivity analysis of mechanistic-empirical performance predictions for flexible pavements. Transp. Res. Rec. **2368**, 12–23 (2013)
37. Shakiba, M., Gamez, A., Al-Qadi, I., Little, D.: Introducing realistic tire-pavement contact stresses into pavement analysis using nonlinear damage approach (PANDA). Int. J. Pavement Eng. **18**, 1027–1038 (2017)
38. Smith, H.: Truck tire characteristics and asphalt concrete pavement rutting. Transp. Res. Rercord **1307**, 1–7 (1991)
39. Titi, H., Dakwar, M., Sooman, M., Tabatabai, H.: Long term performance of gravel base course layers in asphalt pavements. Case Stud. Constr. Mater. **9**, e00208 (2018)
40. Valašková, V., Melcer, J., Lajčáková, G.: Moving load effect on pavements at random excitation. Procedia Eng. **111**, 815–820 (2015)
41. Wang, G., Roque, R., Morian, D.: Effects of surface rutting on near-surface pavement responses based on a two-dimensional axle-tire-pavement interaction finite-element model. J. Mater. Civ. Eng. **24**, 1388–1395 (2012)
42. Wang, H.: Analysis of tire-pavement interaction and pavement responses using a decoupled modeling approach. Ph.D. thesis, University of Illinois, Urbana-Champaign (2011)
43. Wang, H., Al-Qadi, I.: Impact quantification of wide-base tire loading on secondary road flexible pavements. J. Transp. Eng. **137**, 630–639 (2011)
44. Winkler, T.: Generierung quasi-rollender Prüfstandsanregungssignale für ein gekoppeltes Reifen-Fahrbahn-System. Ph.D. thesis, Institut für Kraftfahrzeuge, RWTH Aachen University (2018)
45. Wolff, A.: Simulation of pavement surface runoff using the depth-averaged shallow water equations. Ph.D. thesis, Universität Stuttgart (2013)
46. Wollny, I., Behnke, R., Villaret, K., Kaliske, M.: Numerical modelling of tire-pavement interaction phenomena: coupled structural investigations. Road Mater. Pavement Des. **17**, 563–578 (2016)
47. Wollny, I., Hartung, F., Kaliske, M.: Numerical modeling of inelastic structures at loading of steady state rolling—thermo-mechanical asphalt pavement computation. Comput. Mech. **57**, 867–886 (2016)
48. Wollny, I., Hartung, F., Kaliske, M., Canon Falla, G., Wellner, F.: Numerical investigation of inelastic and temperature dependent layered asphalt pavements at loading by rolling tyres. Int. J. Pavement Eng. **22**, 97–117 (2021)
49. Wollny, I., Hartung, F., Kaliske, M., Liu, P., Oeser, M., Wang, D., Canon Falla, G., Leischner, S., Wellner, F.: Coupling of microstructural and macrostructural computational approaches for asphalt pavements under rolling tire load. Comput.-Aided Civ. Infrastruct. Eng. **35**, 1178–1193 (2020)

50. Yang, X., You, Z., Hiller, J., Watkins, D.: Sensitivity of flexible pavement design to Michigan's climatic inputs using pavement ME design. Int. J. Pavement Eng. **18**, 622–632 (2017)
51. Zhang, C., Wang, H., You, Z., Ma, B.: Sensitivity analysis of longitudinal cracking on asphalt pavement using MEPDG in permafrost region. J. Traffic Transp. Eng. **2**, 40–47 (2015)
52. Zhang, D., Birgisson, B., Luo, X., Onifade, I.: A new long-term aging model for asphalt pavements using morphology-kinetics based approach. Constr. Build. Mater. **229**, 117032 (2019)

The manufacturer's authorised representative in the EU is Springer Nature Customer Service Centre GmbH, Europaplatz 3, 69115 Heidelberg, Germany. If you have any concerns regarding our products, please contact ProductSafety@springernature.com

Printed and bound by CPI Group (UK) Ltd, Croydon, CR0 4YY
28/09/2026
02224091-0012